Study Guide
to accompany

Biology

FOURTH EDITION

Solomon • Berg • Martin • Villee

Ronald S. Daniel
California State Polynechnic University, Pomona

Sharon Callaway Daniel
Orange Coast College, Costa Mesa

Ronald L. Taylor
Orange County Health Care Agency

Saunders College Publishing

Harcourt Brace College Publishers

Fort Worth Philadelphia San Diego New York Orlando Austin
San Antonio Toronto Montreal London Sydney Tokyo

Printed in the United States of America.

Daniel, Daniel & Taylor; Study Guide to accompany Biology, 4E.
Solomon, Berg, Martin & Villee.

ISBN 0-03-017304-3

 67 017 98765432

PREFACE

❑

This study guide provides a means to learn the content of biology, and to an extent, it also acquaints the beginning student with some of the experiential aspects of biology. It is designed for use with the fourth edition of the textbook *Biology* by Solomon, Berg, Martin, and Villee.

While it is of utmost importance that students learn the subject matter of biology, it is also important that they appreciate the ways that biologists think and perceive the world about them. Accordingly, we believe that students should be engaged in the processes of biology while learning its content. This study guide offers learning aids and study strategies that we have found effective in achieving these goals. The approach we have taken has also been influenced by reviewers of the two previous editions of this book, and by feedback from our students about pedagogical methods they find valuable.

The variety of exercises we have included accommodates the different cognitive learning styles among students. Consequently, study guide chapters contain the following sections: *Chapter Overview, Chapter Outline and Concept Review, Building Words, Matching, Table, Thought Questions, Multiple Choice,* and *Visual Foundations.* Each section has a specific purpose, and the whole is designed to provide a well-rounded learning experience.

Chapter Overview provides the student with a brief synthesis of the chapter *in toto* and demonstrates how major themes are related and how chapter topics are organized. It helps the student to keep "the big picture" in view.

The *Chapter Outline and Concept Review* section provides an outline of the textbook chapter, so students can quickly see the principal topics and concepts presented. Major concepts from each chapter are presented in the form of fill-in questions.

Building Words provides some of the prefixes and suffixes (word roots) that are used in the textbook. Word roots are used to construct terms that match the definitions provided.

The *Matching* section tests the student's knowledge of some of the key terms used in the chapter.

The *Table* section tests the student's knowledge of key themes found in each chapter.

The *Thought Questions* section tests the student's ability to understand, analyze, organize, and apply the material, both within the context of the information in any given chapter and, in some cases, within the context of information in other chapters as well.

Multiple Choice questions help the student develop study skills, and they provide a means to re-examine, integrate, and analyze textbook material. This section also helps the student evaluate his or her mastery of the subject matter, as well as identify material that needs additional study. The questions provide a fairly comprehensive sampling of the material presented in each chapter.

The *Visual Foundations* exercises are designed to involve the student directly in the learning process, specifically as pertains to visual and mechanical learning styles. Coloring illustrations reveals gaps in one's understanding of graphic material.

Answers are provided in the back of the study guide for all sections except *Chapter Overview* and *Thought Questions*.

ACKNOWLEDGMENTS

We thank our students who helped us learn what does and does not work, especially our student editors at Orange Coast College who worked through the study guide and informed us of anything that was ambiguous or unclear. We also appreciate the contributions of the reviewers of our previous study guides, and we thank our colleagues and other instructors of biology whose creative ideas and dedication inspired us to prepare this study guide.

Student Editors
Melinda Anne Aaron
Kathleen A. Conroy
Grace S. Kim
Jillian Wong

Ronald S. Daniel, PhD
Professor of Biological Sciences
California State Polytechnic University, Pomona

Sharon Callaway Daniel, MS
Professor of Biological Sciences
Orange Coast College, Costa Mesa, California

Ronald L. Taylor, PhD
Program Manager, Orange County Health Care Agency

CONTENTS

❑

Contents

TO THE STUDENT

❑

We prepared this Study Guide to help you learn biology. We have included the ways of studying and learning that our most successful students find helpful. If you study the textbook, use this study guide regularly, and follow the guidelines provided here, you will learn more about biology and, in the process, you are likely to earn a better grade. It is our sincere desire to provide you with enjoyable experiences during your excursion into the world of life.

Study guide chapters contain the following sections: *Chapter Overview, Chapter Outline and Concept Review, Building Words, Matching, Table, Thought Questions, Multiple Choice,* and *Visual Foundations.* The ways that you can use these sections are described below.

CHAPTER OVERVIEW

Each chapter in this study guide begins with an overview of the chapter. This section, untitled, provides you with a brief synthesis of the chapter *in toto* and demonstrates how major themes are related and how chapter topics are organized. We believe you will find it beneficial to read this section carefully *before* you undertake any of the exercises in the chapter, and again *after* you have completed them. Return to the *Chapter Overview* any time you find yourself losing sight of "the big picture."

CHAPTER OUTLINE AND CONCEPT REVIEW

This section is an outline of the textbook chapter. Within the outline are important concepts presented in the form of fill-in questions. These questions summarize most of the major concepts presented in the chapter.

We suggest that you pay particular attention to these concepts. Be sure that you understand them. Examine the textbook material to be sure that you have correctly filled in all of the blank spaces. Once completed, check your work in the *Answers* section at the back of the study guide.

BUILDING WORDS

The language of biology is fundamental to an understanding of the subject. Many terms used in biology have common word roots. If you know word roots, you can frequently figure out the meaning of a new term. This section provides you with a list of some of the prefixes and suffixes (word roots) that are used in that chapter.

Learn the meanings of these prefixes and suffixes, then use them to construct terms in the blank spaces next to the definitions provided. Once you have filled in the blanks for all of the definitions, check your work in the *Answers* section at the back of the study guide.

Learning the language of biology also means that you can communicate in its language, and effective communication in any language means that you can read, write, and speak the language. Use the glossary in your textbook to learn how to pronounce new terms. Say the words out loud. Listen to them carefully. You will discover that pronouncing words aloud will not only teach you verbal communication, it will also help you remember them.

MATCHING

The matching section tests your knowledge of key terms used in the chapter. Check you answers in the *Answers* sections.

TABLE

The tables relate to key themes in the chapters. Your ability to complete the tables tests your recognition and knowledge of these themes. Check you answers in the *Answers* sections.

THOUGHT QUESTIONS

The questions in this section ask a little more of you. They test your ability to understand and apply material. Many of the questions require analysis, synthesis, and interpretation of information in the chapter. Some questions require linkages to concepts in other chapters. These questions provide more opportunity for you to compare and contrast related processes. Answers are not provided for these questions.

MULTIPLE CHOICE

The questions in this section were selected for two purposes. First, and perhaps most importantly, they help you develop study skills. They provide a means to re-examine, integrate, and analyze selected textbook material. To answer many of them, you must understand how material presented in one part of a chapter relates to material presented in other parts of the chapter. These questions also provide a means to evaluate your mastery of the subject matter, and to identify material that needs additional study.

The questions do not cover everything presented in the textbook. Although they provide a fairly comprehensive sampling of each chapter, the questions may not cover all of the material that your instructor expects you to know.

Once you have studied the textbook material and your notes, close your textbook, put away your notes, and try answering the *Multiple Choice* questions. Check your answers in the *Answers* section. Don't be discouraged if you miss some answers. Take each question that you missed one at a time, go back to the textbook and your notes, and determine what you missed and why. Do not merely copy down the correct answers — that defeats the purpose of the questions. If you dig the answers out of the textbook, you will find that your understanding of the material will improve. Some points that you thought you already knew will become clearer, and you will develop a better understanding of the relationships between different aspects of biology.

VISUAL FOUNDATIONS

Selected visual elements from the textbook are presented in this section. Coloring the parts of biological illustrations causes one to focus on the subject more carefully. Coloring also helps to clearly delineate the parts and reveals the relationship of parts to one another and to the whole.

Begin by coloring the open boxes (□). Colored boxes will provide you with a quick reference guide to the illustration when you come back to study it later. Close your textbook while you are doing this section. After completing the *Visual Foundations* section, use your textbook to check your work. Make notes about anything you missed for future reference. Come back to these figures periodically for review. If an item to be colored is listed in the singular form, and there is more than one of them in the illustration, be sure to color all of them. For example, if we ask you to color the "mitochondrion," be sure to color all of the mitochondria shown in the illustration.

The figures in your textbook and in this study guide have been carefully selected to illustrate principles and help you learn the material. Be sure that you understand the figures and their legends in the textbook.

OTHER STUDY SUGGESTIONS

When possible, read material in the textbook *before* it is covered in class. Don't be concerned if you don't understand all of the material at first, and don't try to memorize every detail in this first reading. Reading before class will contribute greatly to your understanding of the material as it is presented in class, and it will help you take better notes. As soon as possible after the lecture, recopy all of your class notes. You should do this while the classroom presentation is still fresh in your mind. If you are prompt and disciplined about it, you will be able to recall additional details and include them in your notes. Use material in this study guide and your textbook to fill in the gaps.

What is the best way for you to study? That's something you have to figure out for yourself. Reading all of the chapters assigned by your instructor is basic, but it is not all there is to studying.

A technique that many successful students have found beneficial is to organize a study group of 4-6 students. Meet periodically, especially before examinations, and go over the material together. Bring along your textbooks, study guides, and expanded notes. Have each member of the group ask every other member several questions about the material you are reviewing. Discuss the answers as a group until the principles and details are clear to everyone. Use the combined notes of all members to fill in any gaps in your notes. Consider the reasons for correct and incorrect answers to questions in the *Concept Questions* and *Multiple Choice* sections. This study technique can do wonders. There is probably no better way to find out how well you know the material than to verbalize it in front of a critical audience. Just ask any biology instructor!

To the Student

If you follow the guidelines presented here and carefully complete the work provided in this study guide, you will make giant strides toward understanding the principles of biology. You will lay a foundation for a more concentrated study of those areas of the field that you find most interesting. If you decide to specialize in some area of biology, you will find the general background you obtained in this course to be invaluable.

PART I

□

The Organization of Life

CHAPTER 1

A View of Life

Living things are organized entities that share a common set of characteristics that distinguishes them from nonliving things. These characteristics include being composed of one or more cells, growth and development, metabolism, homeostasis, movement, responsiveness, reproduction, and adaptation to environmental change. Organisms transmit information from one generation to the next by means of DNA. They transmit information internally by means of chemical and electrical signals, and from one organism to another by releasing chemicals into the environment, by visual displays, and/or by sound production. Perhaps the most important thing living things do is evolve. Evolution is responsible for the diversity of life on the planet. Living things are organized at a variety of levels, each more complex than the previous, from the chemical level up through the ecosystem level. To study the millions of organisms that have evolved on our planet, scientists have developed a binomial and hierarchical system of nomenclature. In this system, all organisms are assigned to one of five kingdoms — Prokaryotae, Protista, Fungi, Plantae, or Animalia. Living things require a continuous input of energy to maintain the chemical transactions and cellular organization essential to their existence. They are categorized either as producers, consumers, or decomposers, depending on how they obtain and process energy. Living things are studied by a system of observation, hypothesis, experiment, more observation, and revised hypothesis known as the scientific method.

CHAPTER OUTLINE AND CONCEPT REVIEW
Fill in the blanks.

LIFE CAN BE DEFINED IN TERMS OF THE CHARACTERISTICS OF ORGANISMS

1 Nine of the essential characteristics of life include _____
_____
_____.

Organisms are composed of cells

2 The _____ asserts that all living things are composed of cells.

Living organisms grow and develop

3 Living things grow by increasing _____ of cells.

4 _____ is a term that encompasses all the changes that occur during the life of an organism.

Metabolism includes the chemical processes essential to growth, repair, and reproduction

5 Metabolism can be described as the sum of all the _____ that take place in the organism.

6 _____ is the tendency of organisms to maintain a constant internal environment.

Movement is a basic property of cells
Organisms respond to stimuli
Organisms reproduce

7 In general, living things reproduce in one of two ways: _____
_____.

Populations evolve and become adapted to the environment

8 The process by which living things become adapted to their environments, or change, is called
_____.

INFORMATION MUST BE TRANSMITTED WITHIN AND BETWEEN INDIVIDUALS

DNA transmits information from one generation to the next

9 _____ are the units of hereditary material.

Information is transmitted by many types of molecules and by nervous systems

10 Information in living systems is carried by two major categories of chemical messengers,
_____ , and by electrical impulses.

EVOLUTION IS THE PRIMARY UNIFYING CONCEPT OF BIOLOGY

Species adapt in response to changes in the environment
Natural selection is an important mechanism by which evolution proceeds

11 Charles Darwin's famous book, "_____
_____," published in 1859, had and still
has a profound influence on the biological sciences, as well as impacting on the thinking of society
in general.

12 Darwin's theory of natural selection was based on the observations that individuals within species
(a)_____ significantly, and more individuals are produced than can possibly
(b)_____, and individuals with the most advantageous characteristics are the most likely
to (c)_____ and pass those characteristics on to their offspring.

Populations evolve as a result of selective pressures from changes in the environment

13 Natural selection favors organisms with traits that enable them to cope with _____
_____. These organisms are most likely to survive and produce offspring.

14 Well adapted organisms are the products of _____.

Biological organization reflects the course of evolution

15 The basic unit of organization in an organism is the _____.

16 All of the communities of organisms on Earth are collectively referred to as the _____.

Millions of kinds of organisms have evolved on our planet

17 The science of classifying and naming organisms is _____.

18 The scientific name of an organism consists of the _____ names for that organism.

19 The broadest of the taxonomic groups is the _____.

20 The five kingdoms are: (a)_____, consisting of the unicellular bacteria;
(b)_____, unicellular organisms *with* distinct nuclei such as protozoa and algae;
(c)_____, heterotrophic decomposers that include molds and yeasts;
(d)_____, complex multicellular producers that can make food by
photosynthesizing; and (e)_____, complex multicellular organisms that must
obtain food from a "ready made" source.

LIFE DEPENDS ON CONTINUOUS INPUT OF ENERGY

Energy flows through individual cells and organisms

21 _____ is the process by which molecular energy is released for the activities of life.

Energy flows through ecosystems

22 A self-sufficient ecosystem contains three major categories of organisms: _____. It must also have a continuous input of energy.

BIOLOGY IS STUDIED USING THE SCIENTIFIC METHOD

Science is based on systematic thought processes

23 The _____ is a system of observation, hypothesis, experiment, more observation, and revised hypothesis.

24 _____ reasoning are two categories of systematic, scientific thought processes.

Scientists make careful observations and recognize problems
A hypothesis is a proposed explanation

25 A _____ is an educated guess. It is based on presumptive data and, once formulated, it is tested by more observations and carefully controlled experiments.

A prediction is a logical consequence of a hypothesis
Predictions can be tested by experiment
Scientists draw conclusions from the results of experiments
A well-supported hypothesis may lead to a theory

26 A hypothesis becomes a theory only when it is supported by a large body of _____ _____.

Science has ethical dimensions

BUILDING WORDS

Use combinations of prefixes and suffixes to build words for the definitions that follow.

Prefixes	The Meaning	Suffixes	The Meaning
a-	without, not, lacking	-logy	"study of"
auto-	self, same	-stasis	equilibrium caused by opposing forces
bio-	life	-troph	nutrition, growth, "eat"
eco-	home	-zoa	animal
hetero-	different, other		
homeo-	similar, "constant"		
photo-	light		
proto-	first, earliest form of		

Prefix	Suffix	Definition
_____	_____	1. The study of life.
_____	_____	2. The study of groups of organisms interacting with one another and their nonliving environment.

_____	-sexual	3. Reproduction without sex.
_____ _____		4. Maintaining a constant internal environment.
_____	-system	5. A community along with its nonliving environment.
_____	-sphere	6. The planet earth including all living things.
_____ _____		7. An organism that produces its own food.

_____ _____   8. An organism that is dependent upon other organisms for food, energy, and oxygen.

_____ _____   9. Animal-like protists.

_____ -synthesis   10. The conversion of solar (light) energy into stored chemical energy by plants, blue-green algae, and certain bacteria.

MATCHING
For each of these definitions, select the correct matching term from the list that follows.

_____ 1. A group of organisms of the same species that live in the same geographical area at the same time.

_____ 2. The broadest category of classification used.

_____ 3. The science of naming, describing, and classifying organisms.

_____ 4. One of many short, hairlike structures that project from the surface of some cells and are used for locomotion or movement of materials across the cell surface.

_____ 5. A physical or chemical change in the internal or external environment of an organism potentially capable of provoking a response.

_____ 6. A group of closely associated, similar cells that work together to carry out specific functions.

_____ 7. The ability of an organism to adjust to its environment.

_____ 8. A microorganism that breaks down organic material.

_____ 9. The basic structural and functional unit of life, which consists of living material bounded by a membrane.

_____10. A group of organisms with similar structural and functional characteristics that in nature breed only with each other and have a close common ancestry; a group of organisms with a common gene pool.

Terms:

a.	Adaptation	f.	Flagellum	k.	Species
b.	Cell	g.	Kingdom	l.	Stimulus
c.	Cilium	h.	Order	m.	Taxonomy
d.	Consumer	i.	Phylum	n.	Tissue
e.	Decomposer	j.	Population		

TABLE
Fill in the blanks.

Level of Biological Organization	Components of the Organizational Level
Chemical level	Atoms, molecules
#1	Plasma membrane, cytoplasm, nucleus
#2	Group of cells that have similar structure and function
#3	Functional structures made up of tissues
Organ system level	#4
Organism	#5

THOUGHT QUESTIONS
Write your responses to these questions.

1. Describe the characteristics that distinguish living things from nonliving things.
2. Why is evolution considered a unifying principle of biology?
3. Describe the distinguishing characteristics of each of the kingdoms of living organisms.
4. Using examples of each, explain the role of producers, consumers, and decomposers and explain how they interact in nature.
5. State an original hypothesis and design an experiment to test it. Include each of the steps in the scientific method in your example.

MULTIPLE CHOICE
Place your answer(s) in the space provided. Some questions may have more that one correct answer.

_____ 1. Organisms that can use sunlight energy to synthesize complex food molecules are known as
    a. heterotrophs.   d. consumers.
    b. cyanobacteria.   e. producers.
    c. autotrophs.

_____ 2. The kingdom(s) containing organisms that have a true nucleus and are unicellular is/are
    a. Prokaryotae.   d. Plantae.
    b. Protista.   e. Animalia.
    c. Fungi.

_____ 3. A snake is a member of the kingdom/phylum
    a. Prokaryotae/vertebrate.   d. Animalia/chordate.
    b. Fungi/ascomycete.   e. Plantae/angiosperm.
    c. Protista/Aves.

_____ 4. The characteristic(s) common to all living things include(s)
   a. reproduction.
   b. autotrophic synthesis.
   c. heterotrophic synthesis.
   d. growth.
   e. adaptation.

_____ 5. The kingdom(s) containing organisms that are both single-celled and have a "nucleoid" area that is not separated from the cytoplasm by a membrane is/are
   a. Prokaryotae.
   b. Protista.
   c. Fungi.
   d. Plantae.
   e. Animalia.

_____ 6. Maintenance of physiological equilibrium in a changing environment by an organism is known as
   a. response.
   b. adaptation.
   c. homeostasis.
   d. metabolism.
   e. cyclosis.

_____ 7. The Prokaryotae
   a. lack a nuclear membrane.
   b. include bacteria.
   c. contain membrane-bound organelles.
   d. are simple plants.
   e. comprise one kingdom.

_____ 8. Considering the rigors of the scientific method, in which order would the following occur?
   a. hypothesis, data from experiments, law, theory
   b. data from experiments, hypothesis, law, theory
   c. belief, data that fits the belief, law
   d. hypothesis, data from experiments, theory, law
   e. hypothesis, theory, data from experiments, law

_____ 9. To comply with scientific method, a hypothesis must
   a. be true.
   b. withstand rigorous investigations.
   c. be popular and fundable.
   d. be testable.
   e. eventually reach the status of a principle.

The Federal Center for Disease Control tested 600 female prostitutes who regularly use condoms. They found that none of the women had the AIDS virus. Use this study and data to answer questions 10-12.

_____ 10. In terms of determining a relationship between use of condoms by women and the kind of sexual behavior that might lead to contraction of AIDS, this experiment
   a. is useless.
   b. is well-designed.
   c. should be repeated with more subjects.
   d. could be called a theory.
   e. should include male subjects.

_____ 11. To determine the effectiveness of using condoms in preventing AIDS among women, this study was lacking in not having
   a. enough subjects.
   b. male subjects.
   c. controls.
   d. subjects who do not use condoms during sex.
   e. women who are not prostitutes.

_____ 12. The experiment
   a. proves that condoms are effective in preventing AIDS.
   b. suggests that condoms may be effective in preventing AIDS.
   c. indicates that AIDS is not transmitted through sexual activity.
   d. probably means that none of the partners of the female prostitutes had AIDS.
   e. shows that women are less prone to contract AIDS than men.

VISUAL FOUNDATIONS

Questions 1-4 refer to the diagram below. Some questions may have more that one correct answer.

_____ 1. The diagram represents:
    a. developmental theory.
    b. a controlled experiment.
    c. deductive reasoning.
    d. scientific method.
    e. all of the above.

_____ 2. Which of the following statements is/are supported by the diagram?
    a. A theory and a hypothesis are identical.
    b. Conclusions are based on results obtained from carefully designed experiments.
    c. Using results from a single sample can result in sampling error.
    d. A tested hypothesis may lead to new observations.
    e. Science is infallible.

_____ 3. Which of the following statements best describes the activity depicted by the box labeled #3 in the diagram?
    a. Make a prediction that can be tested.
    b. Generate a related hypothesis,
    c. Form generalized inductive conclusions.
    d. Develop a scientific principle.
    e. Test a large number of predicted cases.

_____ 4. Which of the following terms is appropriate for the box labeled #4 in the diagram?
    a. Hypothesis.
    b. Theory.
    c. Principle.
    d. Data.
    e. None of the above.

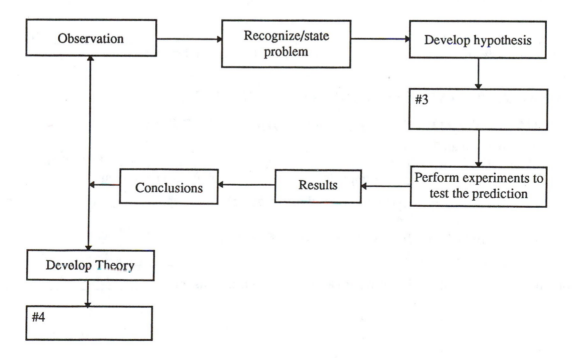

CHAPTER 2

Atoms and Molecules:
The Chemical Basis of Life

Living things, like everything else on our planet, are made of atoms and molecules. Therefore, to understand life is first to understand the interaction and activity of atoms and molecules. Living things share remarkably similar chemical compositions. Although the matter of the universe is composed of 92 chemical elements, about 96% of an organism's mass is made up of just four elements — oxygen, carbon, hydrogen, and nitrogen. The smallest portion of an element that retains its chemical properties is the atom. Atoms, in turn, consist of subatomic particles, the most important of which are protons, neutrons, and electrons. The number and arrangement of electrons determine the chemical behavior of atoms. Molecules and compounds are formed by the joining of atoms. The forces that hold atoms together are called chemical bonds. The unique properties of water are determined by its chemical bonds. Water is the molecule most important to the evolution of life and to the maintenance of life as we know it. Oxidation and reduction are chemical reactions that play critical roles in life's processes. Acids, bases, and salts also play important roles.

CHAPTER OUTLINE AND CONCEPT REVIEW
Fill in the blanks.

ELEMENTS ARE NOT CHANGED IN CHEMICAL REACTIONS

1 An element is a substance that cannot be separated into simpler parts by _____
_____.

2 Of the (a)_____(#?) elements, (b)_____ is the heaviest, and (c)_____ is the lightest. Over 96% of an organism's weight is made up of the four elements (d)_____
_____.

ATOMS ARE THE BASIC PARTICLES OF ELEMENTS

ATOMS CONTAIN PROTONS, NEUTRONS, AND ELECTRONS

3 The three main parts of an atom are the (a)_____
_____. The (b)_____ are in the nucleus, and the (c)_____ occupy energy levels and orbitals around the nucleus.

4 (a)_____ have one unit of positive charge, (b)_____ are neutral, and (c)_____ have a negative charge.

    An atom is uniquely identified by its number of protons

5 The number of protons in an atom is called the _____.

6 The total number of protons and neutrons in an atom is referred to as the _____
_____.

Protons plus neutrons determine atomic mass
Isotopes differ in number of neutrons

7 The (a)_____ of isotopes vary, but the
   (b)_____ remains constant.

Electrons can be involved in chemical changes
Electrons occupy orbitals corresponding to energy levels

8 A neutral atom has (a)_____ net electrical charge because it has the same number of
   (b)_____.

9 Electrons with the same energy comprise an _____.

ATOMS FORM MOLECULES AND COMPOUNDS

10 A _____ forms when two or more atoms are joined by chemical bonds.

11 A chemical compound consists of _____
    _____ that are combined in a fixed ratio.

Chemical formulas describe chemical compounds

12 A (a)_____ formula shows types, numbers and arrangement of atoms in a
    compound, whereas a (b)_____ formula uses chemical symbols to describe the
    chemical composition of a compound.

One mole of any substance contains the same number of units

13 The molecular mass of a compound is the sum of the _____
    _____.

Chemical equations describe chemical reactions

14 (a)_____ are substances that participate in the reaction, (b)_____ are
    substances formed by the reaction

15 Arrows are used to indicate the direction of a reaction; reactants are placed to the (a)_____ of
    arrows, products are placed to the (b)_____ of arrows

ATOMS ARE JOINED BY CHEMICAL BONDS

16 The electrons in the outermost energy level of an atom are called _____ electrons.

17 Energy that is available to do work is potential energy, while energy that is actually doing work is
    referred to as kinetic energy. Chemical bonds represent _____ chemical energy.

Electrons are shared in covalent bonds

18 The _____ of molecules largely determines their functions.

19 Covalent bonds are (a)_____ if the electrons are shared equally between the two atoms; and
    they are (b)_____ if one atom has a greater affinity for electrons than the other.

Ionic bonds form between cations and anions

20 Ionic compounds are comprised of positively charged ions called (a)_____, and negatively
    charged ions called (b)_____.

Hydrogen bonds are weak attractions involving partially charged hydrogen atoms

21 Weak hydrogen bonds form between a hydrogen atom in one molecule and a relatively
    _____ atom in the same or another molecule.

Molecules may interact through van der Waals forces and hydrophobic forces

22 Single van der Waals "bonds" are very weak, but they become significant in large numbers, and when the shape of the _____ involved allow close contact between atoms.

23 Hydrophobic interactions occur between groups of _____ molecules.

OXIDATION INVOLVES THE LOSS OF ELECTRONS; REDUCTION INVOLVES THE GAIN OF ELECTRONS

24 When iron rusts, it loses electrons and is said to be (a)_____, while at the same time oxygen gains electrons and is (b)_____. Reactions involving oxidation _and_ reduction are called (c)_____ reactions.

WATER IS ESSENTIAL TO LIFE

Water molecules are polar

25 Each water molecule can form up to four _____ bonds with adjacent water molecules.

Water is the principal solvent in organisms
Hydrogen bonding makes water cohesive and adhesive

26 Cohesion of water molecules is due to _____ between the molecules.

Water helps maintain a stable temperature

27 Water has a high _____, which helps organisms maintain a relatively constant internal temperature.

ACIDS ARE PROTON DONORS; BASES ARE PROTON ACCEPTORS

28 A base is a proton acceptor that generally dissociates in solution to yield (a)_____ _____. An acid is a proton donor that dissociates in solution to yield

(b)_____.

pH is a convenient measure of acidity

29 The pH scale covers the range (a)_____. Neutrality is at pH (b)_____. A solution with a pH above 7 is a(an) (c)_____. A solution with a pH below 7 is a(an)

(d)_____.

Buffers minimize pH change

30 (a)_____ are comprised of a weak acid and a salt of that acid, or a weak base and a salt of that base. They resist changes in the pH of a solution when (b)_____ are added.

An acid and a base react to form a salt

31 When the (a)_____ of an acid is replaced by some other cation, a salt forms. Salts provide (b)_____ ions that are essential for fluid balance, nerve and muscle function, and many other body functions.

BUILDING WORDS

Use combinations of prefixes and suffixes to build words for the definitions that follow.

Prefixes	The Meaning		Suffixes	The Meaning
equi-	equal		-hedron	face
hydr(o)-	water		-philic	loving, friendly, lover
neutro-	neutral		-phobic	fearing
non-	not			
radio-	beam, ray			
tetra-	four			

Prefix	Suffix	Definition
_____	-n	1. An uncharged subatomic particle.
_____	-nuclide	2. An isotope that emits high-energy radiation when it decays.
_____	-librium	3. The condition of a chemical reaction when the forward and reverse reaction rates are equal.
_____ _____		4. The shape of a molecule when it appears as a three-dimensional pyramid.
_____	-ation	5. The process whereby ionic compounds combine chemically with water and dissolve.
_____ _____		6. Not attracted to water; insoluble in water.
_____	-electrolyte	7. A substance that does not form ions when dissolved in water, and therefore does not conduct an electric current.
_____ _____		8. Having a strong affinity for water; soluble in water.

MATCHING

For each of these definitions, select the correct matching term from the list that follows.

____ 1. The loss of electrons, or the loss of hydrogen atoms from a compound.

____ 2. The total number of protons and neutrons in the atomic nucleus.

____ 3. Substances that can conduct electrical current.

____ 4. The smallest subdivision of an element.

____ 5. Chemical bond that involves one or more shared pairs of electrons.

____ 6. The name of the outermost electrons.

____ 7. An ion with a negative electric charge.

____ 8. An ion bearing an electric charge, either positive or negative.

____ 9. An ion with a positive electric charge.

____ 10. The amount of heat required to raise 1 gram of a substance 1 degree Celsius.

Terms:

a.	Anion	f.	Covalent bond	k.	Ionic bond
b.	Atom	g.	Electrolyte	l.	Oxidation
c.	Atomic mass	h.	Electron	m.	Proton
d.	Calorie	i.	Hydrogen bond	n.	Reduction
e.	Cation	j.	Ion	o.	Valence electrons

TABLE
Fill in the blanks.

Element	Chemical Symbol	Atomic Number	Atomic Weight	Number of Valence Electrons	Bond (Covalent or Ionic) Usually Formed by Atoms of This Element
Hydrogen	H	1	1.00	1	Covalent bond
#1	#2	8	16.00	6	#3
#4	C	6	#5	#6	#7
#8	N	#9	#10	5	Covalent bond
#11	P	#12	30.97	#13	Covalent bond
#14	Na	#15	22.99	1	#16
Chlorine	#17	17	35.45	#18	#19

THOUGHT QUESTIONS
Write your responses to these questions.

1. Define "valence electron" and explain how the term relates to the chemical properties of an atom.
2. Describe the nature and characteristics of the three main types of chemical bonds.
3. What are cations and anions? How do they interact?
4. What are oxidation and reduction? How are "redox" reactions involved in the transfer of energy in living systems?
5. Describe and contrast the properties of acids and bases.

MULTIPLE CHOICE
Place your answer(s) in the space provided. Some questions may have more that one correct answer.

_____ 1. The symbol $_6C$ stands for
    a. calcium with six electrons.
    b. carbon with six protons.
    c. carbon with three protons and three neutrons.
    d. calcium with three protons and three electrons.
    e. an isotope of carbon.

_____ 2. A solution is made acidic by
    a. hydroxyl ions.
    b. water.
    c. hydrogen ions.
    d. protons.
    e. electrons.

_____ 3. Cations are
    a. negative ions.
    b. positive ions.
    c. isotopes.
    d. incomplete products.
    e. negative molecules.

_____ 4. Which of the following is/are among the most common elements found in organisms?

a. carbon

b. oxygen

c. hydrogen

d. nitrogen

e. calcium

_____ 5. A quantum is the amount of energy required to

a. create an isotope.

b. form a covalent bond.

c. form a hydrogen bond.

d. move an electron to a higher energy level.

e. add a neutron to a stable atom.

_____ 6. Adhesion might be described as

a. attraction of atoms of different substances.

b. attraction of atoms within one substance.

c. arrested cohesion.

d. a form of chemical bonding.

e. capillary action.

Use the following chemical equation to answer questions 7-12.

$$2 H_2 + O_2 \longrightarrow 2 H_2O \text{ (water)}$$

_____ 7. How many oxygen atoms are involved in this reaction?

a. one

b. two

c. three

d. four

e. none

_____ 8. The lightest reactant has a mass number of

a. one.

b. two.

c. larger than the product.

d. eight.

e. ten.

_____ 9. The reactant(s) in the formula is/are

a. hydrogen.

b. oxygen.

c. H_2O.

d. water.

e. the arrow.

_____10. The product has a molecular mass of

a. one.

b. twelve.

c. larger than either reactant alone.

d. sixteen.

e. eighteen.

_____11. The product(s) is/are

a. hydrogen.

b. oxygen.

c. H_2O.

d. water.

e. the arrow.

_____12. The direction in which this reaction tends to occur is

a. right to left.

b. from products to reactants.

c. from hydrogen and oxygen to water.

d. left to right.

e. from reactants to products.

_____13. Electrolytes result from ionization of

a. salts.

b. acids.

c. hydrogen.

d. bases.

e. alcohols.

____14. The outer electron energy level of an atom
   a. may contain two or more orbitals.
   b. determines the element.
   c. contains the same number of electrons as protons.
   d. often contains at least one neutron.
   e. determines chemical properties.

____15. If the number of neutrons in an atom is changed, it will also change
   a. atomic mass.
   b. atomic number.
   c. the element.
   d. the number of electrons.
   e. the elements chemical properties.

____16. A form of an element that is progressing toward a more stable state by emitting radiation is called a(n)
   a. Dalton atom.
   b. radioisotope.
   c. enerton.
   d. quantum unitron.
   e. radionuclide.

____17. H–O–H is/are
   a. a structural formula.
   b. water.
   c. a chemical formula.
   d. a molecule.
   e. a structure with the molecular weight 18.

____18. When two atoms combine by sharing one pair of their electrons, the bond is a
   a. divalent, single.
   b. covalent, single.
   c. divalent, double.
   d. covalent, double.
   e. covalent, polar.

____19. Oxidation has occurred when
   a. an atom gains an electron.
   b. a molecule loses an electron.
   c. hydrogen bonds break.
   d. an atom loses an electron.
   e. an ion gains an electron.

____20. A substance containing equal amounts of hydrogen and hydroxyl ions would be
   a. acidic, pH below 7.
   b. alkaline, pH below 7.
   c. acidic, pH above 7.
   d. alkaline, pH above 7.
   e. neutral, pH 7.

VISUAL FOUNDATIONS
Color the parts of the illustration below as indicated.

RED ☐ electron
YELLOW ☐ neutron
BLUE ☐ proton

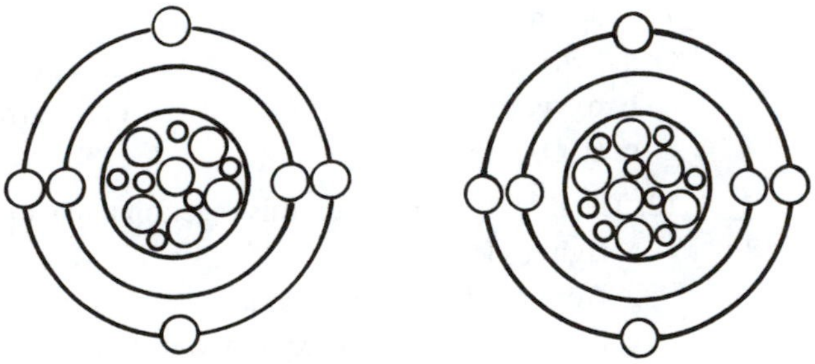

Two atoms are represented in this illustration:
   1. The atom on the left is called (a)_____. The atom on the right is called (b)_____.
____ 2. These atoms:
   a. are ions.
   b. are isotopes.
   c. have the same mass.
   d. are from different elements.
   e. have the same atomic number.

CHAPTER 3

□

The Chemistry of Life: Organic Compounds

Most of the chemical compounds present in organisms are organic compounds. Carbon is the central component of organic compounds, probably because it forms bonds with a greater number of different elements than any other type of atom. The major groups of organic compounds are the carbohydrates, lipids, proteins, and nucleic acids. Each group is characterized by the presence of one or more specific groups of atoms that convey specific properties to the molecules. Many organic molecules are polymers, that is, chains of similar units joined in a row. Carbohydrates serve as energy sources for cells, and as structural components of fungi, plants, and some animals. Carbohydrates are categorized by their degree of complexity into monosaccharides, disaccharides, and polysaccharides. Lipids function as biological fuels, as structural components of cell membranes, and as hormones. Among the biologically important lipids are the neutral fats, phospholipids, steroids, carotenoids, and waxes. Proteins serve as structural components of cells and tissues and as the facilitators of thousands of different chemical reactions. Proteins are made up of amino acid subunits and exhibit four levels of structural organization, each more complex than the previous. The structure of a protein helps determine its biological activity. DNA and RNA are nucleic acids that transmit hereditary information and determine what proteins a cell manufactures. They are made up of subunits called nucleotides. Some single and double nucleotides, such as ATP, cyclic AMP, and NAD+, are central to cell function.

CHAPTER OUTLINE AND CONCEPT REVIEW
Fill in the blanks.

INTRODUCTION

1 _____ are the four major "classes" of macromolecules in living systems.

CARBON ATOMS CAN FORM AN ENORMOUS VARIETY OF STRUCTURES

2 Carbon atoms can form _____(#?) bonds with a large variety of other atoms.

3 Carbon-containing molecules have a three-dimensional structure due to the _____ nature of their bond angles.

ISOMERS HAVE THE SAME MOLECULAR FORMULA, BUT DIFFER IN STRUCTURE

4 Isomers have the same molecular formula but different _____.

FUNCTIONAL GROUPS CHANGE THE PROPERTIES OF ORGANIC MOLECULES

5 Each class of organic compound is characterized by specific _____ groups with distinctive properties.

MANY BIOLOGICAL MOLECULES ARE POLYMERS

6   Macromolecules produced by linking monomers together are called
_____.

CARBOHYDRATES INCLUDE SUGARS, STARCHES, AND CELLULOSE

7   Carbohydrates contain the elements _____
_____ in a ratio of about 1:2:1.

Monosaccharides are simple sugars

8   The simple sugar _____ is an important and nearly ubiquitous fuel molecule in living cells.

Disaccharides consist of two monosaccharide units

9   Disaccharides form when two monosaccharides are bonded by a _____ linkage.

Polysaccharides can store energy or provide structure

10   Long chains of _____ link together to form a polysaccharide.

Some modified and complex carbohydrates have special roles

11   A carbohydrate combined with a protein forms (a)_____; carbohydrates
combined with lipids are called (b)_____.

LIPIDS ARE FATS OR FATLIKE SUBSTANCES

12   Lipids are composed of the elements _____.

Neutral fats contain glycerol and fatty acids

13   A fat consists of one (a)_____ molecule combined with one to three
(b)_____ molecules.

Phospholipids are components of cellular membranes

14   A phospholipid molecule assumes a distinctive configuration in water because of its
_____ property.

Carotenoid plant pigments are derived from isoprene units
Steroids contain four rings of carbon atoms

PROTEINS ARE MACROMOLECULES FORMED FROM AMINO ACIDS

Amino acids are the subunits of proteins

15   Proteins are composed of the elements _____
_____.

16   All amino acids contain (a)_____ groups but vary in the
(b)_____.

Peptide bonds join amino acids

17   Proteins are large, complex molecules made of (a)_____ molecules joined by
(b)_____ bonds.

Proteins have four levels of organization

18   The levels of organization distinguishable in protein molecules are the primary, and the
_____.

Protein structure determines function

19   A mutation that results in a change in (a)_____ sequence or a change in
(b)_____ structure can disrupt the biological activity of a protein.

DNA AND RNA ARE NUCLEIC ACIDS
Nucleic acids consist of nucleotide subunits

20 Each nucleotide subunit in a nucleic acid is composed of a (a)_____
_____ nitrogenous base, a five-carbon (b)_____, and an inorganic
(c)_____ group.

Some nucleotides are important in energy transfers and other cellular functions

21 The energy for life functions is supplied mainly by the nucleotide _____.

BUILDING WORDS
Use combinations of prefixes and suffixes to build words for the definitions that follow.

Prefixes	The Meaning	Suffixes	The Meaning
amphi-	two, both, both sides	-lysis	breaking down, decomposition
amylo-	starch	-mer	part
di-	two, twice, double	-plast	formed, molded, "body"
hex-	six	-saccharide	sugar
hydr(o)-	water		
iso-	equal, "same"		
macro-	large, long, great, excessive		
mono-	alone, single, one		
poly-	much, many		
tri-	three		

Prefix	Suffix	Definition
_____	_____	1. A compound that has the same molecular formula as another compound but a different structure and different properties.
_____	-molecule	2. A large molecule consisting of thousands of atoms.
_____	_____	3. A single organic compound that links with similar compounds in the formation of a polymer.
_____	_____	4. The degradation of a compound by the addition of water.
_____	_____	5. A simple sugar.
_____	-ose	6. A sugar that consists of six carbons.
_____	_____	7. Two monosaccharides covalently bonded to one another.
_____	_____	8. A macromolecule consisting of repeating units of simple sugars.
_____	_____	9. A starch-forming granule.
_____	-glyceride	10. A compound formed of one fatty acid and one glycerol molecule.
_____	-glyceride	11. A compound formed of two fatty acids and one glycerol molecule.
_____	-glyceride	12. A compound formed of three fatty acids and one glycerol molecule.
_____	-pathic	13. Having one hydrophilic end and one hydrophobic end.
_____	-peptide	14. A compound consisting of two amino acids.
_____	-peptide	15. A compound consisting of a long chain of amino acids.

MATCHING

For each of these definitions, select the correct matching term from the list that follows.

_____ 1. A molecule composed of one or more phosphate groups, a 5-carbon sugar, and a nitrogenous base.

_____ 2. A group of yellow to orange pigments in plants.

_____ 3. A large complex organic compound composed of chemically linked amino acid subunits.

_____ 4. Complex lipid molecules containing carbon atoms arranged in four interlocking rings.

_____ 5. An organic compound containing an amino group and a carboxyl group.

_____ 6. Protein structure where two polypeptide chains are bonded together.

_____ 7. An organic compound that is made up of only carbon and hydrogen atoms.

_____ 8. Fatlike substance in which there are two fatty acids and a phosphorus-containing group attached to glycerol.

_____ 9. Polysaccharide used by plants to store energy.

_____10. The principle carbohydrate stored in animal cells.

Terms:

a.   Amino acid
b.   Amino group
c.   Carotenoid
d.   Condensation
e.   Glycogen

f.   Glucose
g.   Hydrocarbon
h.   Lipid
i.   Nucleotide
j.   Phosphate group

k.   Phospholipid
l.   Protein
m.   Starch
n.   Steroid
o.   Tertiary structure

TABLE

Fill in the blanks.

Functional Group	Abbreviated Formula	Organic Compound Characterized by Group	Nature of Interaction With Other Molecules
Carboxyl group	R — COOH	Proteins	Polar, weakly acidic
Sulfhydryl group	R — SH	#1	Form covalent bonds
Amino group	#2	#3	Weak base
Hydroxyl group	#4	#5	#6
#7	R — PO$_4$H$_2$	#8	Weak acid
Methyl group	#9	Component of many organic compounds	#10

THOUGHT QUESTIONS
Write your responses to these questions.

1. What are the principal molecular constituents of each of the four major macromolecules in organisms? Using labelled diagrams, explain how these constituents polymerize to form macromolecules.

Describe the compositions, characteristics, and biological functions of each of the following:
2. Storage polysaccharides and structural polysaccharides.
3. Neutral fats, phospholipids, and steroids.
4. Proteins having different levels of organization.
5. RNA and DNA.

MULTIPLE CHOICE
Place your answer(s) in the space provided. Some questions may have more that one correct answer.

_____ 1. Nucleotides
    a. are polypeptide polymers.
    b. are polymers of simple carbohydrates.
    c. contain a base, phosphate, and sugar.
    d. comprise complex proteins.
    e. store genetic information.

_____ 2. Polymers are
    a. made from monomers.
    b. formed by dehydration synthesis.
    c. found in monomers.
    d. macromolecules.
    e. smaller than monomers.

_____ 3. Proteins
    a. are a major source of energy.
    b. contain carbon, hydrogen, and oxygen.
    c. can combine with carbohydrates.
    d. are found in both plants and animals.
    e. are polypeptide polymers.

_____ 4. The complex carbohydrate storage molecule in plants is
    a. glycogen.
    b. starch.
    c. usually in an extracellular position.
    d. composed of simple carbohydrate subunits.
    e. more soluble than the animal storage molecule.

_____ 5. Proteins contain
    a. glycerol.
    b. polypeptides.
    c. peptide bonds.
    d. amino acids.
    e. nucleotides.

_____ 6. $C_6H_{12}O_6$ is the formula for
    a. glucose.
    b. a type of monosaccharide.
    c. cellulose.
    d. fructose.
    e. a type of disaccharide.

_____ 7. Phospholipids
    a. are found in cell membranes.
    b. form a cubelike structure in water.
    c. contain glycerol, fatty acids, and phosphorus.
    d. are polar.
    e. are polypeptide polymers.

_____ 8. A monomer is to a polymer as a
    a. chain is to a link.
    b. link is to a chain.
    c. monosaccharide is to a polysaccharide.
    d. necklace is to a bead.
    e. bead is a necklace.

____ 9. Hydrolysis results in
   a. smaller molecules.
   b. larger molecules.
   c. reduced ambient water.
   d. increase in polymers.
   e. increase in monomers.

____ 10. Compared to unsaturated fatty acids, saturated fatty acids
   a. contain more hydrogens.
   b. contain more double bonds.
   c. are more like oils.
   d. are more prone to be solids.
   e. contain more carbohydrates.

____ 11. The main structural components of cells and tissues are
   a. carbon-based molecules.
   b. carbohydrates, lipids, and DNA.
   c. basically strings of covalently bonded carbons.
   d. organic compounds.
   e. DNA, RNA, and ATP.

____ 12. Glycogen is
   a. a molecule in animals.
   b. a molecule in plants.
   c. found only in the liver and muscle of mammals.
   d. composed of simple carbohydrate subunits.
   e. more soluble than starch.

____ 13. An isomer can be described as one of a group of molecules with the same atomic composition but different
   a. elements.
   b. compounds.
   c. arrangement of atoms.
   d. arrangement of covalent bonds.
   e. placement of groups.

VISUAL FOUNDATIONS
Color the parts of the illustration below as indicated.

RED   ☐ saturated fatty acid
GREEN   ☐ glycerol
YELLOW   ☐ phosphate group
BLUE   ☐ choline
ORANGE   ☐ unsaturated fatty acid

Questions 1-4 refer to the illustration above. Some questions may have more that one correct answer.

____ 1. The illustration represents
   a. a polypeptide.
   b. cellulose.
   c. glycogen.
   d. lecithin.
   e. starch.

____ 2. The hydrophobic end of this molecule is composed of
   a. amino acids.
   b. fatty acids.
   c. glucose.
   d. phosphate groups.
   e. steroids.

_____ 3. This kind of molecule contributes to the formation of
    a. the lipid bilayer.   d. cellulose.
    b. cell membranes.   e. complex proteins.
    c. RNA.

_____ 4. This molecule is amphipathic, meaning:
    a. it contains polar groups.   d. it is found in all life forms.
    b. it is very large.   e. it can act as an acid.
    c. it contains hydrophobic and hydrophilic regions.

CHAPTER 4

Cellular Organization

The cell is the basic unit of living organisms. Because all cells come from preexisting cells, they have similar needs and therefore share many fundamental features. Most cells are microscopically small because of limitations on the movement of materials across the plasma membrane. Cells are studied by a combination of methods, including microscopy and cell fractionation. Unlike prokaryotic cells, eukaryotic cells are complex, containing a variety of membranous organelles. Although each organelle has its own particular structure and function, all of the organelles of a cell work together in an integrated fashion. Most cells are surrounded by an extracellular matrix that is secreted by the cells themselves.

CHAPTER OUTLINE AND CONCEPT REVIEW
Fill in the blanks.

INTRODUCTION
1 The cell is the smallest unit capable of carrying on _____.

THE CELL IS THE SMALLEST UNIT OF LIFE
2 All (a)_____ are made up of cells, and all cells arise from
   (b)_____.

CELLS SHARE MANY ATTRIBUTES
3 *All* cells are enclosed by a _____.

CELL SIZE IS LIMITED
4 In general, the smaller the cell, the larger the (a)_____, and the
   faster the (b)_____.

CELLS ARE STUDIED BY A COMBINATION OF METHODS
5 Cells are studied by using important techniques and tools; for example, powerful
   (a)_____ microscopes are used to resolve ultrastructure, and cell components are
   separated by a method known as (b)_____.

EUKARYOTIC CELLS ARE MUCH MORE COMPLEX THAN PROKARYOTIC CELLS
6 _____ cells are relatively lacking in complexity and their genetic material is not
   enclosed by membranes.

7 _____ cells are relatively complex and possess both membrane-bound organelles
   and a "true" nucleus.

EUKARYOTIC CELLS CONTAIN SPECIALIZED ORGANELLES

Membranous organelles carry out specific functions

The cell nucleus contains DNA

8 The genetic material in the eukaryotic nucleus consists of hereditary units called _____ that contain codes for producing proteins.

9 Chromosomes are composed of a complex of DNA and protein called _____.

Organelles of the internal membrane system interact extensively

10 The _____ is a complex of intracytoplasmic membranes and cisternae that compartmentalize the cytoplasm.

11 RER is studded with _____ that are involved in protein synthesis.

12 Some proteins constructed on RER are transported by _____ for secretion to the outside or insertion in other membranes.

13 The _____ processes, sorts, and modifies proteins. It consists of stacks of platelike membranes and vesicles, some of which are filled with cellular products.

14 _____ are small sacs containing digestive enzymes that can break down (lyse) complex molecules, foreign substances, and "dead" organelles.

Mitochondria and chloroplasts are energy-converting organelles

15 The chemical reactions that convert food energy to ATP (cellular respiration) take place in organelles called _____.

16 _____ are organelles that contain green pigments that trap light energy for photosynthesis.

17 _____ membranes contain chlorophyll that traps sunlight energy and converts it to chemical energy in ATP.

18 The matrix within chloroplasts, where carbohydrates are synthesized, is called the (a)_____, while the ATP that supplies energy to drive the process is synthesized on membranes called (b)_____

Microbodies are compartments for specialized chemical reactions

All eukaryotic cells contain a cytoskeleton

19 The cytoskeleton is a network of protein filaments that are responsible for both _____ of cells.

20 _____ are both made up of globular protein subunits that constantly polymerize and depolymerize.

21 _____ appear to be the organizing centers for microtubule formation in animal cells.

22 Cell motility is accomplished by two types of movable, whiplike structures that extend from the cell surface called _____.

23 The rapid association and disassociation of (a) _____ microfilaments and (b)_____ filaments results in a sliding motion that generates force and movement.

An extracellular matrix surrounds most cells

24 Plant cell walls are composed primarily of _____ and smaller quantities of other polysaccharides.

25 In some animal cells glycoproteins and glycolipids form a _____ coating.

BUILDING WORDS
Use combinations of prefixes and suffixes to build words for the definitions that follow.

Prefixes	The Meaning
chloro-	green
chromo-	color
cyto-	cell
eu-	good, well, "true"
leuko-	white (without color)
lyso-	loosening, decomposition
micro-	small
myo-	muscle
pro-	"before"

Suffixes	The Meaning
-karyo(te)	nucleus
-plast(id)	formed, molded, "body"
-some	body

Prefix	Suffix	Definition
_____	-phyll	1. A green pigment that traps light for photosynthesis.
_____	_____	2. Organelles containing pigments that give fruits and flowers their characteristic colors.
_____	-plasm	3. Cell contents exclusive of the nucleus.
_____	-skeleton	4. A complex network of protein filaments within the cell.
glyoxy-	_____	5. A microbody containing enzymes used to convert stored fats in plant seeds to sugars.
_____	_____	6. An organelle that is not pigmented and is found primarily in roots and tubers, where it is used to store starch.
_____	_____	7. An organelle containing digestive enzymes.
_____	-filaments	8. Small, solid filaments, 7 nm in diameter, that make up part of the cytoskeleton of eukaryotic cells.
_____	-tubules	9. Small, hollow filaments, 25 nm in diameter, that make up part of the cytoskeleton of eukaryotic cells.
_____	-villi	10. Small, finger-like projections from cell surfaces that increase surface area.
_____	-sin	11. A muscle protein which, together with actin, is responsible for muscle contraction.
peroxi-	_____	12. An organelle containing enzymes that split hydrogen peroxide, rendering it harmless.
_____	_____	13. Precursor organelles.
_____	-karyotes	14. Single-celled organisms, including the bacteria and cyanobacteria; organisms that evolved before organisms with nuclei.
_____	-sol	15. Cell contents exclusive of the nucleus and organelles; the fluid component of the cytoplasm.
_____	_____	16. An organism with a distinct nucleus surrounded by nuclear membranes.

MATCHING
For each of these definitions, select the correct matching term from the list that follows.

____ 1. One of a pair of small, cylindrical organelles lying at right angles to each other near the nucleus.

____ 2. Site of ribosome synthesis.

____ 3. An intracellular organelle that is the site of oxidative phosphorylation.

____ 4. A fluid-filled, membrane-bounded sac found within the cytoplasm; may function in storage, digestion, or water elimination.

____ 5. Any small sac, especially a small spherical membrane-bounded compartment, within the cytoplasm.

____ 6. A chlorophyll-bearing intracellular organelle of some plant cells.

____ 7. A stack of thylakoids within a chloroplast.

____ 8. The fluid region of the chloroplast.

____ 9. An interconnected network of intracellular membranes.

____10. An organelle that is part of the protein synthesis machinery.

Terms:

a. Actin
b. Centriole
c. Chloroplast
d. Endoplasmic reticulum
e. Granum
f. Mitochondrion
g. Nuclear envelope
h. Nucleolus
i. Plasma membrane
j. Ribosome
k. Secretory vesicle
l. Stroma
m. Vacuole
n. Vesicle

TABLE
Fill in the blanks.

Cell Structure	Location of Structure	Function of Structure	Kind of Cell Containing This Structure
Nuclear area	Cytoplasm	Hereditary information	Prokaryotes
Chromosomes	#1	#2	#3
Chloroplasts	#4	#5	Eukaryotes (plants)
Mitochondria	#6	#7	#8
Centrioles	#9	#10	#11
Cell wall	#12	#13	#14
#15	#16	Encloses cellular contents, regulates movement of materials, communicates with other cells	#17

THOUGHT QUESTIONS

Write your responses to these questions.

1. Describe the similarities and differences in the structures and functions of smooth endoplasmic reticulum and rough endoplasmic reticulum.
2. Use words and a labelled diagram of the Golgi complex to describe its structure and functions.
3. Would you expect to find more or fewer lysosomes in a diseased cell? Explain your answer in terms of the structure, contents, and function of lysosomes.
4. Using words and diagrams that show spacial relationships and structural characteristics, explain how chloroplasts and mitochondria synthesize ATP.
5. What are the structural and functional characteristics of the microtrabecular lattice (intracytoplasmic "cytoskeleton")?
6. What are the similarities and differences in prokaryotic and eukaryotic cells? Explain what it is about these cells that causes us to think that one was more likely than the other to be the first to appear in evolutionary history?

MULTIPLE CHOICE

Place your answer(s) in the space provided. Some questions may have more that one correct answer.

_____ 1. Cell membrane functions include
   a. energy transduction.
   b. selective permeability.
   c. isolation of different chemical reactions.
   d. sorting genetic material.
   e. concentration of reactants.

_____ 2. The cell theory states that
   a. new cells come from preexisting cells.
   b. all cells are descended from ancient cells.
   c. cells divide.
   d. living things are composed of cells.
   e. cells contain genetic material.

_____ 3. Chloroplasts and mitochondria both
   a. are found in plant cells.
   b. have two membranes.
   c. contain DNA.
   d. are found in animal cells.
   e. contain a matrix.

_____ 4. The high resolution attained by the electron microscope is attributed to its use of
   a. fixed specimens.
   b. electromagnetic lenses.
   c. short wavelengths.
   d. grid patterns.
   e. electrons instead of light.

_____ 5. Cells are small at least in part because as size increases the surface-to-volume ratio
   a. doubles.
   b. decreases to half.
   c. increases.
   d. decreases.
   e. reduces efficiency of cell activities.

_____ 6. Which of the following cells contain plastids?
   a. animal
   b. plant
   c. some eukaryotic
   d. some prokaryotic
   e. all cells

_____ 7. The membranes that partition the cytoplasm of eukaryotic cells (endomembrane system) include
   a. Golgi complex.
   b. lysosomes.
   c. endoplasmic reticulum.
   d. transport vesicles.
   e. plasma membrane.

_____ 8. The Golgi complex functions to
      a. modify proteins.   d. produce polysaccharides.
      b. process proteins.   e. add carbohydrates to proteins.
      c. form glycoproteins.

_____ 9. The "cytoskeleton" of eukaryotic cells
      a. changes constantly.   d. extends into the nucleus.
      b. includes microfilaments.   e. includes protein.
      c. includes some DNA.

_____10. Which of the following is in the nucleolus, but not normally found in the rest of chromatin?
      a. DNA   d. RNA
      b. protein   e. ribosomes
      c. chromosomes

_____11. The part(s) of a mitochondrion that are rich in enzymes is/are the
      a. cristae.   d. intermembrane space.
      b. outer membrane.   e. inner membrane.
      c. matrix.

_____12. Lysosomes
      a. contain digestive enzymes.   d. break down complex molecules.
      b. possess a membrane.   e. break down organelles.
      c. contain nucleic acids.

_____13. Actin, myosin, and tubulin are
      a. proteins.   d. components of filaments.
      b. in chromatin.   e. components of the plasma membrane.
      c. constructed on ribosomes.

VISUAL FOUNDATIONS

Color the parts of the illustration below as indicated.

    RED   ☐   plasma membrane
   GREEN ☐   nuclear area
  YELLOW ☐   cell wall

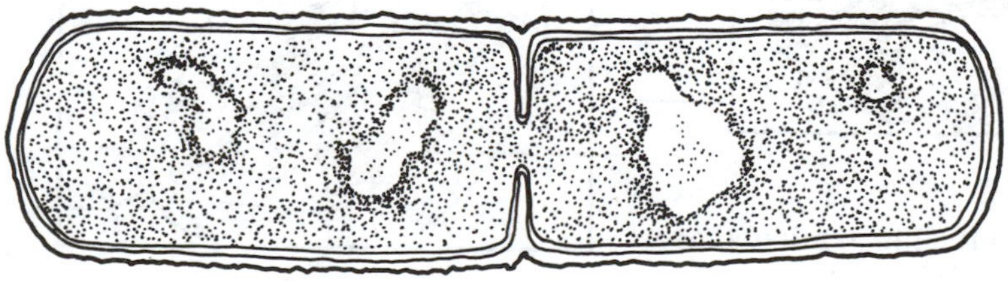

Questions 1 and 2 pertain to the illustration above.

1. Is this a prokaryotic cell or a eukaryotic cell? _____

2. Which of the colored parts is/are unique to this kind of cell? _____

Color the parts of the illustration below as indicated.

RED ☐ plasma membrane

GREEN ☐ nucleus

YELLOW ☐ cell wall

BLUE ☐ prominent vacuole

ORANGE ☐ chloroplast

BROWN ☐ mitochondrion

TAN ☐ internal membrane system (note: The plasma membrane and the outer portion of the nuclear envelope have already been assigned colors. Use tan to identify the remaining structures in this system.)

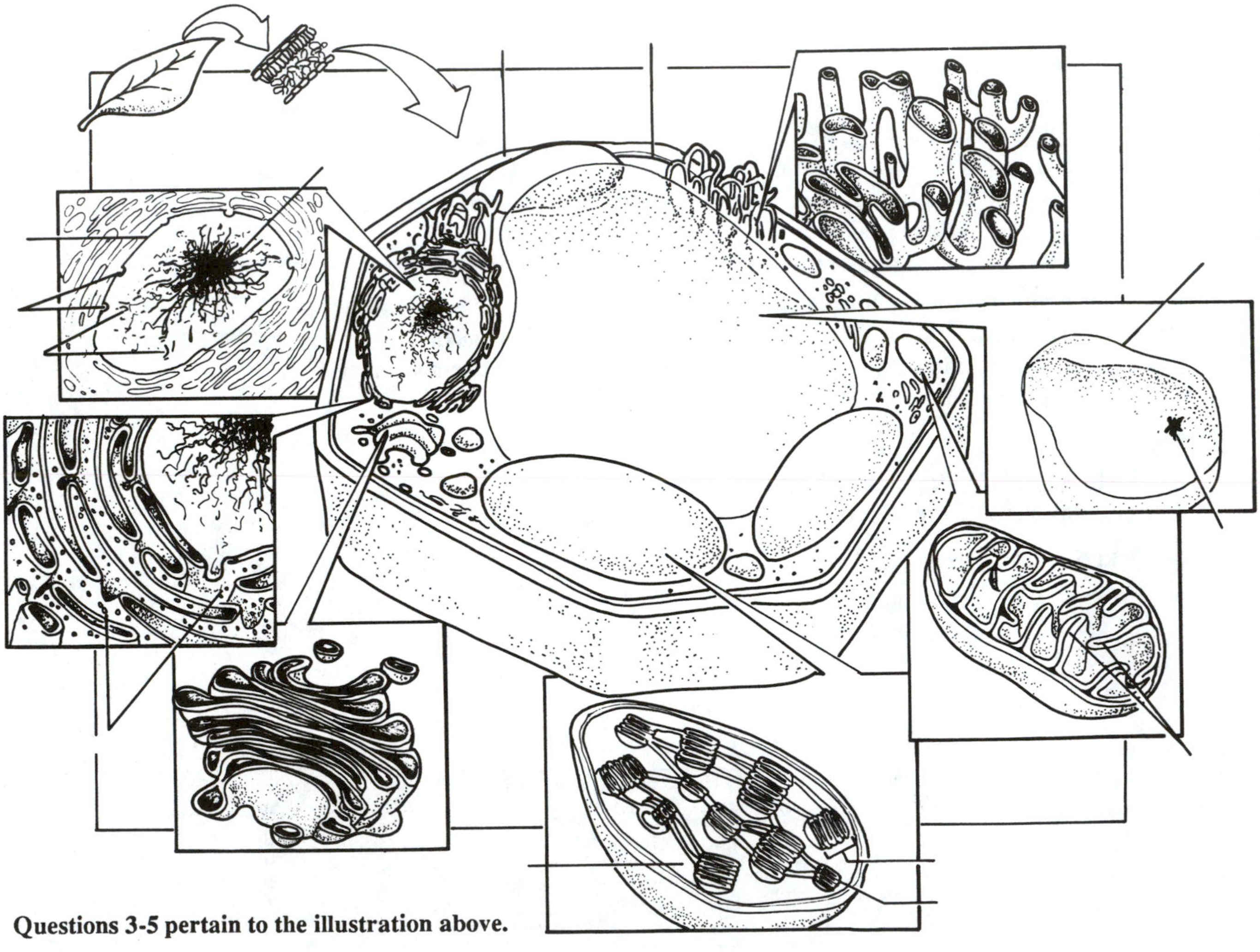

Questions 3-5 pertain to the illustration above.

3. What kind of cell is illustrated by this generalized diagram? _____

4. Which of the colored parts is/are considered components of the cytoplasm of this cell? _____

_____

5. Which of the colored parts is/are characteristic of this kind of cell and not generally associated with other types of cells? _____

Color the parts of the illustration below as indicated.

RED   ❑   plasma membrane

GREEN   ❑   nucleus

YELLOW   ❑   centriole

BLUE   ❑   lysosome

ORANGE   ❑   mitochondrion

BROWN   ❑   internal membrane system (note: The plasma membrane, lysosomes and the outer portion of the nuclear envelope have already been assigned colors. Use tan to identify the remaining structures in this system.)

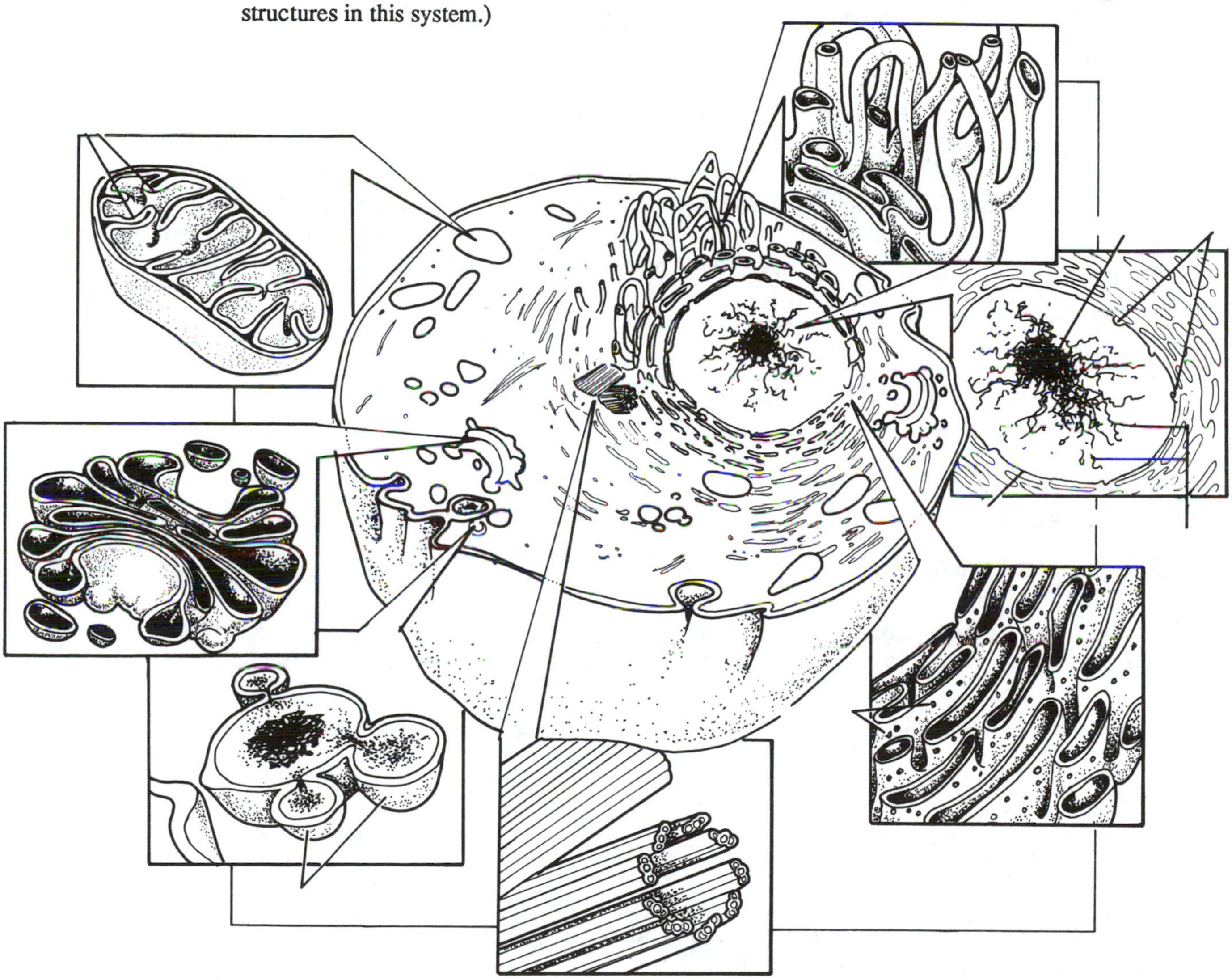

Questions 6-8 pertain to the illustration above.

6. What kind of cell is illustrated by this generalized diagram? _____

7. Which of the colored parts is/are considered components of the cytoplasm of this cell? _____

_____

8. Which of the colored parts is/are characteristic of this kind of cell and not generally associated with other types of cells? _____

Biological Membranes

Biological membranes enclose cells, separating the interior world of the cell from the exterior. In eukaryotes, they also enclose a variety of internal structures and form extensive internal membrane systems. Membranes function to provide work surfaces for many chemical reactions, to regulate movement of materials in and out of the cell, and to transmit signals and information between the environment and the interior of the cell. Membranes are phospholipid bilayers with both internal and peripheral proteins associated with them. Membranes have the ability to seal themselves, round up and form closed vesicles, and fuse with other membranes. Molecules, depending to a large extent on their size and electrical charge, pass through membranes in a variety of ways. In multicellular organisms, cell membranes have specialized structures associated with them that allow neighboring cells to form strong connections with each other or to establish rapid communications between adjacent cells.

CHAPTER OUTLINE AND CONCEPT REVIEW
Fill in the blanks.

INTRODUCTION
1 All cells are physically separated from the external environment by a _____.

BIOLOGICAL MEMBRANES ARE LIPID BILAYERS WITH ASSOCIATED PROTEINS

2 The theory of membrane structure known as the (a)_____ model holds that membranes consist of a dynamic, fluid (b)_____ and embedded globular (c)_____.

3 The physical properties of biological membranes are due primarily to the characteristics of their _____.

Phospholipids form bilayers in water

4 The cylindrical shape and strongly _____ character of phospholipid molecules are features responsible for the formation of the bilayer.

5 The headgroups of phospholipid molecules are said to be (a)_____ because they readily associate with water. On the other hand, the (b)_____ ends turn away from water and associate with each other.

Biological membranes are two-dimensional fluids

6 The crystal-like properties of many phospholipid bilayers is due to the orderly arrangement of (a)_____ on the outside and (b)_____ on the inside.

7 The two-dimensional fluid property of the lipid bilayer is due to the constant motion of _____ _____.

Biological membranes fuse and form closed vesicles

Membrane proteins may be integral or peripheral

8 _____ have hydrophobic stretches of amino acids that contact fatty acid chains of the lipid bilayer.

9 _____ have hydrophobic amino acids that are buried inside the molecule away from water.

Proteins are oriented asymmetrically across the bilayer

10 The differences in characteristics of the outer and inner surfaces of membranes is the result of asymmetrical distribution of constituent _____.

Membrane proteins have specific functions

11 A variety of membrane functions is made possible by the diversity of _____ molecules in membranes.

CELLULAR MEMBRANES ARE SELECTIVELY PERMEABLE

12 Biological membranes are said to be _____ because they allow some substances to pass through while "blocking" others.

Random motion of particles leads to diffusion

13 Diffusion rate is affected by temperature and the _____ _____ of the moving particles.

14 (a)_____ is diffusion of a solute through a selectively permeable membrane. Diffusion of a solvent through a selectively permeable membrane is called (b)_____.

15 The _____ of a solution is determined by the amount of dissolved substances in the solution.

Carrier-mediated transport of solutes requires special integral membrane proteins

16 The two forms of carrier-mediated transport are _____ _____.

17 _____ is the process by which cells expend energy in order to move ions and molecules against a concentration gradient.

Facilitated diffusion occurs down a concentration gradient

Some carrier-mediated active transport systems "pump" substances against their concentration gradients

18 Because of the Sodium-potassium pump, there are (a)_____(fewer or more) potassium ions inside the cell relative to the sodium ions outside, causing the inside to be (b)_____ charged relative to the outside.

Linked cotransport systems indirectly provide energy for active transport

Integrated multiple transport systems use indirect linkages between active transport and facilitated diffusion

Facilitated diffusion is powered by a concentration gradient; active transport requires another energy source

19 The energy for diffusion is provided by a _____.

In exocytosis and endocytosis large particles are transported by vesicles or vacuoles

JUNCTIONS ARE SPECIALIZED CONTACTS BETWEEN CELLS

20 Three types of junctions, the _____
_____, are specialized structures associated with the plasma
membrane of animal cells.

Desmosomes are points of attachment between some animal cells

Tight junctions seal off intercellular spaces between some animal cells

Gap junctions permit transfer of small molecules and ions between some animal cells

21 Gap junctions consist of an array of (a)_____ that form (b)_____ that
connect the cytoplasm of adjacent cells.

Plasmodesmata allow movement of certain molecules and ions between plant cells

22 Plasmodesmata in plant cells are functionally equivalent to _____ in animal
cells.

BUILDING WORDS

Use combinations of prefixes and suffixes to build words for the definitions that follow.

Prefixes	The Meaning		Suffixes	The Meaning
desm(o)-	bond		-cyto(sis)	cell
endo-	within		-desm(a)	bond
exo-	outside, outer, external		-some	body
hyper-	over			
hypo-	under			
iso-	equal, "same"			
phago-	eat, devour			
pino-	drink			

Prefix	Suffix	Definition
_____	_____	1. A process whereby materials are taken into the cell.
_____	_____	2. The process whereby waste or secretion products are ejected from a cell by fusion of a vesicle with the plasma membrane.
_____	-tonic (-osmotic)	3. Having an osmotic pressure or solute concentration that is greater than a standard solution.
_____	-tonic (-osmotic)	4. Having an osmotic pressure or solute concentration that is less than a standard solution.
_____	-tonic (-osmotic)	5. Having an osmotic pressure or solute concentration that is the same as a standard solution.
_____	_____	6. The engulfing of microorganisms, foreign matter, and other cells by a cell.
_____	_____	7. A type of endocytosis whereby fluid is engulfed by vesicles originating at the cell surface.
_____	_____	8. A button-like plaque (body) present on two opposing cell surfaces, that hold (bond) the cells together by means of protein filaments that span the intercellular space.
plasmo-	_____	9. A cytoplasmic channel connecting (bonding) adjacent plant cells and allowing for the movement of small molecules and ions between cells.

MATCHING
For each of these definitions, select the correct matching term from the list that follows.

_____ 1. The modern picture of membranes in which protein molecules float in a phospholipid bilayer.

_____ 2. The diffusion of water across a selectively permeable membrane.

_____ 3. The transport of ions or molecules across a membrane by a specific carrier protein.

_____ 4. Regions in a system of differing concentration, such as exist in a cell and its environment, that cause molecules to move from areas of higher concentration to lower concentration.

_____ 5. Energy-requiring transport of a molecule across a membrane from a region of low concentration to a region of high concentration.

_____ 6. A membrane that allows some substances to cross it more easily than others.

_____ 7. The net movement of molecules from a region of higher concentration to one of lower concentration of that substance.

_____ 8. The internal pressure in a plant cell caused by the diffusion of water into the cell.

_____ 9. A specialized structure between some animal cells, producing a tight seal that prevents materials from passing through the spaces between the cells.

_____10. Structure consisting of specialized regions of the plasma membranes of two adjacent cells containing numerous pores that allow passage of certain small molecules and ions between them.

Terms:

a. Active transport
b. Concentration gradient
c. Cotransport
d. Dialysis
e. Diffusion

f. Facilitated diffusion
g. Fluid-mosaic model
h. Gap junction
i. Osmosis
j. Signal transduction

k. Selectively permeable membrane
l. Tight junction
m. Turgor pressure

TABLE
Fill in the blanks.

Transport Mechanism	Description of the Transport Mechanism
Diffusion	Net movement of a substance from an area of high concentration to an area of low concentration
#1	Diffusion of a solvent across a semipermeable membrane
#2	Diffusion of a solute across a semipermeable membrane
Carrier-mediated transport	#3
Active transport	#4
#5	A form of carrier-mediated transport in which the energy of a concentration gradient is used to move a solute
#6	Transport of food and other materials into the cell
Exocytosis	#7

THOUGHT QUESTIONS
Write your responses to these questions.

1. Construct a labelled diagram that illustrates the "fluid mosaic theory" of membrane structure. Describe the basic functions and relationships between the membrane's molecular components.
2. Using the principles of osmosis and diffusion, explain what will happen to a cell in (a) a hypotonic solution, (b) a hypertonic solution, and (c) an isotonic solution. Describe the differences in reactions that might exist between an animal cell and a plant cell. How do cells in nature compensate for a large difference between cytoplasmic and environmental tonicities?
3. Describe and contrast passive transport of atoms and molecules (e.g., diffusion and osmosis) with transport that requires expenditure of energy by the cell.
4. Explain with a diagram and words how materials sequestered in membranes are transported into, through, and out of a cell.

MULTIPLE CHOICE
Place your answer(s) in the space provided. Some questions may have more that one correct answer.

Red blood cells are placed in three beakers containing the following solutions: beaker A, distilled water; beaker B, isotonic solution; beaker C, 5% salt solution. Use this information to answer questions 1-3.

_____ 1. The cells in beaker A
    a. shriveled.
    b. swelled.
    c. plasmolyzed.
    d. were unaffected.
    e. probably eventually burst.

_____ 2. The cells in beaker B
    a. shriveled.
    b. swelled.
    c. plasmolyzed.
    d. were unaffected.
    e. probably eventually burst.

_____ 3. The cells in beaker C
    a. shriveled.
    b. swelled.
    c. plasmolyzed.
    d. were unaffected.
    e. probably eventually burst.

_____ 4. If membranes were not fluid and dynamic, which of the following might still occur normally?
    a. active transport
    b. facilitated diffusion
    c. simple diffusion
    d. endo- and exocytosis
    e. osmosis

_____ 5. Active transport
    a. moves molecules against a gradient.
    b. does not occur in prokaryotes.
    c. requires use of ATP.
    d. occurs in animal cells, not in plant cells.
    e. moves substances both into and out of cells.

_____ 6. Facilitated diffusion
    a. moves molecules against a gradient.
    b. does not occur in prokaryotes.
    c. requires use of ATP.
    d. involves protein channels.
    e. moves substances both into and out of cells.

_____ 7. A molecule is called amphipathic when it
    a. prevents free passage of substances.
    b. is in a membrane.
    c. has hydrophobic and hydrophilic regions.
    d. is a lipid.
    e. is embedded in a bilipid layer.

____ 8. Cell membranes are
   a. asymmetrical.
   b. 11–12 nm thick.
   c. partly nonpolar, hydrophobic fatty acid chains.
   d. composed mostly of phospholipids and nucleic acids.
   e. all similar in basic structure.

____ 9. Cell walls
   a. are large plasma membranes.
   b. contain carbohydrates.
   c. are thicker than membranes.
   d. are fluid.
   e. contain desmosomes.

____10. Diffusion rate depends on
   a. the flow of water.
   b. concentration gradient.
   c. energy from the cell.
   d. the plasma membrane.
   e. kinetic energy.

____11. Plasma membranes of eukaryotic cells have a large amount of
   a. sterols.
   b. phospholipid.
   c. rigidity.
   d. fluidity.
   e. protein.

____12. Endocytosis may include
   a. secretion vacuoles.
   b. pinocytosis.
   c. phagocytosis.
   d. combination of inbound particles with proteins.
   e. receptor-mediation.

____13. Plasmodesmata
   a. are channels in cytoplasm.
   b. connect plant cells.
   c. plant cell structures equivalent to animal cell desmosomes.
   d. are the same as several plasmodesma.
   e. connect ER of adjacent cells.

____14. Membrane fusion enables
   a. diversity of proteins.
   b. exocytosis.
   c. endocytosis.
   d. fusion of vesicles and plasma membrane.
   e. formation of vesicles.

Visual Foundations on next page ➤

VISUAL FOUNDATIONS

Color the parts of the illustration below as indicated. Label the interior and exterior of the cell.

RED	☐	transmembrane protein
GREEN	☐	hydrophilic heads
YELLOW	☐	hydrophobic tails
BLUE	☐	glycolipid
ORANGE	☐	glycoprotein

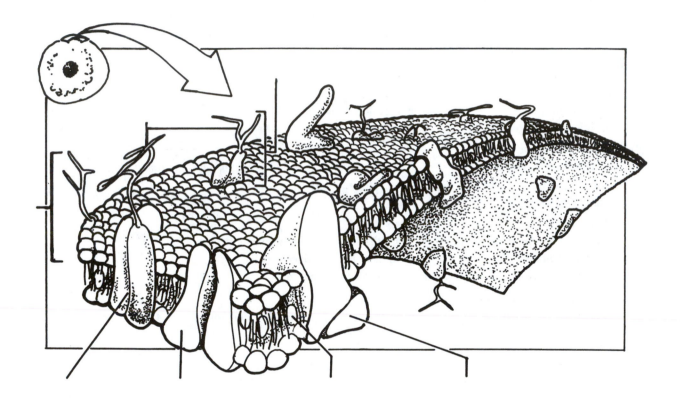

1. What is the name of the model used to generate the illustration above?_____

2. Which of the colored parts have roles in cell recognition, adhesion and mechanical protection? _____

_____

PART II

◻

Energy Transfer Through Living Systems

CHAPTER 6

❑

Energy and Metabolism

Life depends on a continuous input of energy. The myriad chemical reactions of cells that enable them to grow, move, maintain and repair themselves, reproduce, respond to stimuli, etc. involve energy transformations. These transformations are governed by the laws of thermodynamics that explain why organisms cannot produce energy but must continuously capture it from somewhere else, and why in every energy transaction, some energy is dissipated as heat. Some chemical reactions occur spontaneously, releasing free energy which is then available to perform work. Other reactions are not spontaneous, requiring an input of free energy before they can occur. The energy released from spontaneous reactions is stored in the chemical bonds of adenosine triphosphate (ATP), and the energy consumed in energy-requiring reactions is taken from the chemical bonds of ATP. In general, for each energy-requiring reaction occurring in a cell there is an energy-releasing reaction coupled to it, and ATP is what links them. Chemical reactions in organisms are regulated by enzymes, protein catalysts that affect the speed of a chemical reaction without being consumed in the process. Enzymes lower the amount of energy needed to activate reactions. They are highly specific and work by forming temporary chemical compounds with their substrates. Whereas some enzymes consist solely of protein, others consist of a protein and an additional chemical component. Enzymes usually work in teams, with the product of one enzyme-controlled reaction serving as the substrate for the next. Cells regulate enzyme activity by controlling the amount of enzyme produced and the shape of the enzyme. Enzymes work best at specific temperatures and pH levels, and most can be inhibited by certain chemical substances.

CHAPTER OUTLINE AND CONCEPT REVIEW
Fill in the blanks.

INTRODUCTION
1 Life depends on a continuous supply of outside energy. Fortunately, _____ capture outside sources of energy and incorporate it in chemical bonds (food).

BIOLOGICAL WORK REQUIRES ENERGY
2 Energy is the capacity to do ___work___.

Organisms carry out conversions between potential energy and kinetic energy

3 Energy is in one of two forms: (a)_____ is "stored energy" and (b) _____ is "energy of motion."

TWO LAWS OF THERMODYNAMICS GOVERN ENERGY TRANSFORMATIONS
4 The study of energy and its transformations is called _____.

The total energy in the universe does not change

5 The first law of thermodynamics states that energy can be neither (a)_____ _____, however, it can be (b)_____ and changed in form.

The entropy (disorder) of the universe is increasing

6 The term entropy refers to the _____ in the universe.

7 In every energy conversion or transfer some energy is dissipated as _____.

METABOLIC REACTIONS INVOLVE ENERGY TRANSFORMATIONS

Enthalpy (H) is the total potential energy of a system

8 The total potential energy of a chemical reactions, or enthalpy, equals the _____ _____ of the reactants and products.

Free energy is available to do cellular work

Chemical reactions involve changes in free energy

Exergonic reactions do not require outside energy

9 _____ are spontaneous and they release energy that can perform work.

All reactions have a required energy of activation

An endergonic reaction requires an energy source

10 Chemical reactions may release or absorb heat energy. (a)_____ reactions release heat; (b)_____ reactions absorb heat.

Actual free energy changes depend on the concentrations of the reactants and products

11 When a chemical reaction is at equilibrium, the difference in free energy between reactants and products is _____.

A cell maintains its reactions far from equilibrium

Cells drive endergonic reactions by coupling them to exergonic reactions

12 Exergonic reactions (a)_____[release or require input of?] free energy; endergonic reactions (b)_____[release or require input of?] free energy. Endergonic and exergonic reactions are coupled in organisms.

ADENOSINE TRIPHOSPHATE (ATP) IS THE ENERGY CURRENCY OF THE CELL

The ATP molecule has three main parts

13 The three main parts of the ATP molecule are _____, _____, and _____.

ATP can donate energy through the transfer of a phosphate group

14 Phosphate bonds in ATP are broken by the process known as _____.

ATP links exergonic and endergonic reactions

The cell maintains a very high ratio of ATP to ADP

ATP cannot be stockpiled

CELLS TRANSFER ENERGY BY REDOX REACTIONS

Most electron carriers carry hydrogen atoms

Electron carriers transfer energy

ENZYMES ARE CHEMICAL REGULATORS

An enzyme lowers the activation energy needed to initiate a chemical reaction

An enzyme has no effect on free energy change

An enzyme works by forming an enzyme-substrate complex

15 Enzymes lower activation energy by forming an unstable intermediate called the _____ _____.

16 The _____ model and the _____ model are designed to explain the nature of enzyme-substrate binding.

Enzymes are very efficient catalysts

Many enzyme names end in -*ase*

Enzymes are specific

17 Enzymes bind specifically to the _____ of substrates.

Many enzymes require cofactors

18 An organic cofactor is called a _____ .

Enzymes are most effective at optimal conditions

19 Factors that affect enzyme activity include (list three) _____
_____ .

Enzymes are organized into teams in metabolic pathways

20 When enzymes work in teams, the (a)_____ from one enzyme-substrate reaction becomes the (b)_____ for the next enzyme-substrate reaction.

The cell regulates enzymatic activity

21 A cell regulates enzymatic activity by influencing the _____ of the enzyme.

Enzymes can be inhibited by certain chemical agents

22 Inhibition is _____ when the inhibitor-enzyme bond is weak.

23 _____ inhibition occurs when the inhibitor competes with the normal substrate for binding to the active site of the enzyme.

24 _____ inhibition occurs when the inhibitor binds to the enzyme at a site other than the active site.

Enzymes can do damage if they function inappropriately

BUILDING WORDS

Use combinations of prefixes and suffixes to build words for the definitions that follow.

Prefixes	The Meaning		Suffixes	The Meaning
allo-	other, "another"		-calor(ie)	heat
ana-	up		-ergonic	work, "energy"
cata-	down		-steric	"space"
end(o)-	within			
ex(o)-	outside, outer, external			
kilo-	thousand			

Prefix	Suffix	Definition
_____	_____	1. The amount of heat required to raise the temperature of 1000 grams (1 kg) of water from 14.5^0 C to 15.5^0 C.
_____	-bolism	2. In living organisms, the "building up" (synthesis) of more complex substances from simpler ones.
_____	-bolism	3. In living organisms, the "breaking down" of more complex substances into simpler ones.
_____	_____	4. A spontaneous reaction that releases free energy and can therefore perform work.

_____   _____   5. A reaction that requires an input of free energy from the
surroundings.

_____   _____   6. Refers to a receptor site on some region of an enzyme molecule
other than the active site.

MATCHING

For each of these definitions, select the correct matching term from the list that follows.

____ 1. A quantitative measure of the amount of randomness or disorder of a system.

____ 2. An organic substance that is required for a particular enzymatic reaction to occur.

____ 3. A substance on which an enzyme acts.

____ 4. Energy in motion.

____ 5. Principles governing heat or energy transfer.

____ 6. A substance that increases the speed at which a chemical reaction occurs without being used up during the reaction.

____ 7. An organic catalyst that greatly increases the rate of a chemical reaction without being consumed by that reaction.

____ 8. The total potential energy of a system.

____ 9. Stored energy.

____10. The capacity or ability to do work.

Terms:

a.   Catalyst
b.   Coenzyme
c.   Energy
d.   Enthalpy
e.   Entropy

f.   Enzyme
g.   Free energy
h.   Heat energy
i.   Kinetic energy
j.   Potential energy

k.   Substrate
l.   Thermodynamics

TABLE

Fill in the blanks.

Chemical Reaction	Type of Metabolic Pathway	Energy Required (Endergonic) or Released (Exergonic)
Synthesis of ATP	Anabolic	Endergonic
Hydrolysis	#1	#2
Phosphorylation	#3	#4
ATP ——> ADP	#5	#6
Oxidation	#7	#8
Reduction	#9	#10
FAD ——> FADH$_2$	#11	#12

THOUGHT QUESTIONS

Write your responses to these questions.

1. If the law of conservation of energy applies to a biological system, how can it also be true that entropy is increasing in that system? Furthermore, how does the statement that "energy transfers in living systems are never 100% efficient" relate to the laws of thermodynamics?

2. Explain the components of the following formula, then plug into the formula realistic values for a biological situation and solve for "G."

$$G = H - TS$$

3. Illustrate coupled endergonic and exergonic reactions and explain the meaning of these terms.

4. What are the characteristics of ATP that enable it to play a primary role in energy transfers and cellular metabolism?

5. What is energy of activation? Explain how enzymes enable intracellular chemical reactions by lowering energy of activation?

MULTIPLE CHOICE

Place your answer(s) in the space provided. Some questions may have more that one correct answer.

_____ 1. A kilocalorie (kcal) is
    a. a measure of heat energy.   d. the temperature of water.
    b. equal to a Calorie.   e. an essential nutrient.
    c. a way to measure energy generally.

_____ 2. Compliance with the second law of thermodynamics presumes that
    a. disorder is increasing.   d. all energy will eventually be useless to life.
    b. order in a system requires input of energy.   e. heat dissipates in all systems.
    c. entropy will decrease as order in organisms increases.

Use the following formula to answer question 3:

    $ATP + H_2O \longrightarrow ADP + P$ $\Delta G = -7.3$ kcal/mole

_____ 3. This reaction
    a. hydrolyzes ATP.   d. is endergonic.
    b. loses free energy.   e. is exergonic.
    c. produces adenosine triphosphate.

_____ 4. The reaction $HCl + NaOH \longrightarrow H_2O + NaCl + heat$
    a. is exothermic.   d. tends to occur in absence of intervention.
    b. is endothermic.   e. produces products chemically different from reactants.
    c. requires input of energy.

_____ 5. Kinetic energy is
    a. doing work.   d. energy in motion.
    b. stored energy.   e. energy of position or state.
    c. in chemical bonds.

_____ 6. In the formula $\Delta G = \Delta H - T\Delta S$
    a. free energy decreases as entropy decreases.   d. temperature increase decreases free energy.
    b. entropy and free energy are inversely related.   e. increasing enthalpy increases free energy.
    c. change in enthalpy is greater than change in free energy.

Use the following formulae to answer questions 7-9:

$$A + B \longrightarrow C + D$$

$$K = \frac{[C] \times [D]}{[A] \times [B]}$$

____ 7. Which of the following mean(s) "concentration of?"

    a. $A + B \longrightarrow$ d. K

    b. $\longrightarrow C + D$ e. $[C] \times [D]$

    c. brackets

____ 8. The reaction depicted probably has

    a. a small K value.   d. a K value of 10^{-7} or less.

    b. a large K value.   e. a tendency to move toward equilibrium.

    c. substances C and D are different than A and B.

____ 9. The formula for the equilibrium constant indicates that

    a. reactants are divided by products.   d. reactants are multiplied by reactants.

    b. products are divided by reactants.   e. reactants are multiplied by products.

    c. products are multiplied by products.

____10. Enzyme activity may be affected by

    a. cofactors.   d. substrate concentration.

    b. temperature.   e. genes.

    c. pH.

____11. Enzymes

    a. are lipoproteins.   d. become products after complexing with substrates.

    b. lower activation energy.   e. are regulated by genes.

    c. speed up biological chemical reactions.

VISUAL FOUNDATIONS

Color the parts of the illustration below as indicated.

RED   ☐   active sites

GREEN   ☐   substrates

YELLOW   ☐   enzyme

BLUE   ☐   allosteric site

ORANGE   ☐   regulator

BROWN   ☐   cyclic AMP

TAN   ☐   enzyme-substrate complex

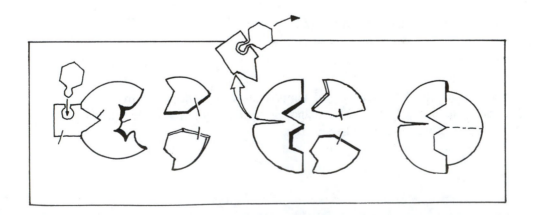

CHAPTER 7

Energy-Releasing Pathways and Biosynthesis

Cells break down nutrients one step at a time. In the process energy is released from chemical bonds in a controlled fashion and transferred to ATP where it is available for cellular work. There are three major pathways by which cells extract energy from nutrients: aerobic respiration, anaerobic respiration, and fermentation. Aerobic respiration involves a series of reactions in which hydrogen is transferred from glucose to oxygen, resulting in the formation of water. Many organisms use nutrients besides glucose, or in addition to it, as a source of energy. These nutrients might be fatty acids or amino acids that become transformed into one of the metabolic intermediates of aerobic respiration. In anaerobic respiration, fuel molecules are broken down in the absence of oxygen, and an inorganic compound such as nitrate or sulfate serves as the final hydrogen (electron) acceptor. Fermentation is a type of anaerobic respiration in which the final electron acceptor is an organic compound derived from the initial nutrient. There is a net gain of only 2 ATPs per glucose molecule in fermentation, compared with 36-38 ATPs produced per glucose molecule by aerobic respiration. Two common types of fermentation are alcohol fermentation and lactate fermentation. Yeast cells carry on alcohol fermentation, in which ethyl alcohol and carbon dioxide are the final products. Lactate fermentation occurs in some fungi and bacteria, and in some animal cells in the absence of sufficient oxygen. In addition to breaking down nutrients, cells also synthesize an array of complex molecules, such as proteins, nucleic acids, lipids, polysaccharides, etc. These biosynthetic reactions are catalyzed by enzymes and require ATP to drive them. At any given time, a cell is in a dynamic state with some molecules being degraded while others are being synthesized.

CHAPTER OUTLINE AND CONCEPT REVIEW
Fill in the blanks.

INTRODUCTION

1 _____ is the endergonic aspect of metabolism involving the synthesis of
   complex molecules.

2 _____ is the exergonic aspect of metabolism involving breakdown of complex
   molecules.

3 Living cells transfer energy from food to the molecule _____
   _____ for later use.

CELLS USE REDOX REACTIONS TO EXTRACT ENERGY FROM NUTRIENTS
Aerobic respiration is a redox process

4 Most cells that live in environments where oxygen is plentiful use the catabolic process called
   _____ to extract free energy from nutrients.

5 In aerobic respiration, a fuel molecule is oxidized, yielding the by-products (a)_____
   _____ with the release of the essential (b)_____ that is required
   for life's activities.

AEROBIC RESPIRATION HAS FOUR STAGES

6 Three principal types of reactions that occur in aerobic respiration are the _____ _____, _____, and the preparation reactions.

In glycolysis, glucose is converted to pyruvate

7 Glycolysis reactions take place in the _____ of the cell.

8 In the first phase of glycolysis, two ATP molecules are consumed and glucose is split into two _____ molecules.

9 In the second phase of glycolysis, each of the molecules resulting from splitting of glucose is oxidized and transformed into a _____ molecule.

10 Glycolysis *nets* _____ (#?) ATPs.

Each pyruvate is converted to acetyl CoA

11 When oxygen is present, pyruvate is converted to acetyl CoA in mitochondria, (a)_____ is reduced, and (b)_____ is released.

The citric acid cycle oxidizes acetyl CoA

12 The eight step citric acid cycle completes the oxidation of glucose. For each acetyl group that enters the citric acid cycle, (a)_____ (#?) NAD^+ are reduced to NADH, (b)_____ (#?) molecules of CO_2 are produced, and (c)_____ (#?) hydrogen atoms are removed.

13 _____(#?) acetyl CoAs are completely degraded with two turns of the citric acid cycle.

The electron transport system is coupled to ATP synthesis

14 The hydrogens removed during glycolysis, acetyl CoA formation, and the citric acid cycle are first transferred to the primary hydrogen acceptors (a)_____, then they are processed through (b)_____.

15 The (a)_____ consists of a chain of electron acceptors embedded in the inner membrane of mitochondria. (b)_____ is the final acceptor in the chain.

16 The electron transport system provides energy to pump protons into the (a)_____ of the mitochondrion, creating an energy gradient. As protons move down the gradient, the energy released is used by the enzyme (b)_____ to produce ATP.

THE AEROBIC RESPIRATION OF ONE GLUCOSE YIELDS A MAXIMUM OF 36 TO 38 ATPs

NUTRIENTS OTHER THAN GLUCOSE ALSO PROVIDE ENERGY

17 Human beings and many other animals usually obtain most of their energy by oxidizing _____ _____. Amino acids are also used.

18 Amino groups are metabolized by a process called _____.

19 The _____ components of neutral fats can be use as fuel.

20 Fatty acids are oxidized and split into acetyl groups by the process of _____ _____. The acetyl molecules enter the citric acid cycle.

CELLS REGULATE AEROBIC RESPIRATION

ANAEROBIC RESPIRATION AND FERMENTATION DO NOT REQUIRE OXYGEN

21 _____ is a means of extracting energy that produces inorganic end products and does not require oxygen.

22 Fermentation is an anaerobic process in which the final acceptor of electrons from NADH is an _____.

Alcohol fermentation and lactate fermentation are inefficient

23 When hydrogens from NAD are transferred to acetaldehyde, _____ is formed.

24 When hydrogens from NAD are transferred to pyruvate, _____ is formed.

25 Fermentation yields a net gain of only (a)_____(#?) ATPs per glucose molecule, compared with about (b)_____(#?) ATPs per glucose molecule in aerobic respiration.

ANABOLIC REACTIONS ARE PART OF BIOSYNTHETIC PROCESSES

BUILDING WORDS
Use combinations of prefixes and suffixes to build words for the definitions that follow.

Prefixes	The Meaning	Suffixes	The Meaning
aero-	air	-be (bios)	life
an-	without, not, lacking	-lysis	breaking down, decomposition
de-	indicates removal, separation		
glyco-	sweet, "sugar"		

Prefix	Suffix	Definition
_____	_____	1. An organism that requires air or free oxygen to live.
_____	-aerobe	2. An organism that does not require air or free oxygen to live.
_____	-hydrogenation	3. A reaction in which hydrogens are removed from the substrate.
_____	-carboxylation	4. A reaction in which a carboxyl group is removed from a substrate.
_____	-amination	5. A reaction in which an amine group is removed from a substrate.
_____	_____	6. A sequence of reactions that breaks down a molecule of glucose (a sugar) to two molecules of pyruvate.

MATCHING
For each of these definitions, select the correct matching term from the list that follows.

____ 1. The process by which a pH gradient drives the formation of ATP.

____ 2. Aerobic series of chemical reactions in which acetyl Co-A is completely degraded to carbon dioxide and water with the release of ATP.

____ 3. An organism that can live in either the presence or absence of oxygen.

____ 4. Anaerobic respiration that utilizes organic compounds both as electron donors and acceptors.

____ 5. Oxygen-requiring pathway by which organic molecules are broken down and energy is released that can be used for biological work.

____ 6. Iron-containing proteins of the electron transport system that are alternately oxidized and reduced.

____ 7. The loss of electrons or hydrogen atoms from a substance.

____ 8. The introduction of a phosphate group into an organic molecule.

____ 9. A series of chemical reactions during which hydrogens or their electrons are passed along from one receptor molecule to another, with the release of energy.

____10. The gain of electrons or hydrogen atoms by a substance.

Terms:

a. Aerobic respiration
b. Anaerobic respiration
c. Chemiosmosis
d. Citric acid cycle
e. Cytochromes

f. Electron transport system
g. Ethyl alcohol
h. Facultative anaerobe
i. Fermentation
j. Lactic acid

k. Oxidation
l. Phosphorylation
m. Pyruvic acid
n. Reduction

TABLE
Fill in the blanks.

Reaction	Hydrogen Acceptor	Energy Yield in Aerobic Respiration	Energy Yield in Anaerobic Respiration
PGAL ——>1,3-bisphosphoglycerate	NAD^+	3 ATP	0 ATP
1,3- bisphosphoglycerate ——> 3-phosphoglycerate	#1	#2	#3
Phosphoenolpyruvate ——> pyruvate	#4	#5	#6
Pyruvate ——> acetyl CoA	#7	#8	#9
Isocitrate ——> alpha-ketoglutarate	#10	#11	#12
Succinate ——> fumarate	#13	#14	#15
Malate ——> oxaloacetate	#16	#17	#18

THOUGHT QUESTIONS
Write your responses to these questions.

1. Use *all* of the components in the list below and chemical formulas as needed to construct an illustration of glycolysis and aerobic respiration. Add any components that are needed to make the illustration clear. Explain what is occurring at each step. The components are: CO_2, H_2O, $C_6H_{12}O_6$, O_2, energy in, energy out, pyruvates, NAD and NADH, acetyl coenzyme A, oxaloacetate, citrate, ATP, FADH and $FADH_2$, electrons, protons, cytosol (cytoplasm), mitochondria, electron transport system, chemiosmosis.
2. How are the pathways for protein and lipid metabolism related to the pathway for glucose oxidation?
3. Could cells survive without anabolism? Why? Without catabolism? Why?

MULTIPLE CHOICE

Place your answer(s) in the space provided. Some questions may have more that one correct answer.

_____ 1. Production of acetyl CoA from pyruvate
   a. is anabolic.
   b. takes place in mitochondria.
   c. takes place in cytoplasm.
   d. takes place in endoplasmic reticulum.
   e. uses derivatives of vitamin B.

_____ 2. Glycolysis
   a. generates two ATPs.
   b. is more efficient that aerobic respiration.
   c. takes place in the cristae of mitochondria.
   d. produces two pyruvates.
   e. reduces glucose to H_2O and CO_2.

_____ 3. During oxidative phosphorylation
   a. ATP converts to ADP.
   b. ATP forms ADP.
   c. three ATPs are produced when two electrons pass from NADH to oxygen.
   d. ATP is coupled to electron flow.
   e. energy is required to make ATPs.

_____ 4. Complete aerobic metabolism of one mole of glucose
   a. yields more than 30 ATPs.
   b. is more than 50% efficient.
   c. yields more ATP than complete metabolism of fatty acid.
   d. involves glycolysis.
   e. releases about 686 kcal.

_____ 5. ATP synthetase
   a. is a cytochrome.
   b. converts ATP to ADP.
   c. forms channels across the inner mitochondrial membrane.
   d. resides in cristae of mitochondria.
   e. couples protons to electrons to form water.

_____ 6. Cells may obtain energy from large molecules by means of
   a. catabolism.
   b. aerobic respiration.
   c. anaerobic respiration.
   d. fermentation.
   e. the Kreb's cycle.

_____ 7. Chemiosmosis involves
   a. flow of protons down an electrical gradient.
   b. flow of protons down a concentration gradient.
   c. pumping of protons into the mitochondrial matrix.
   d. proton channels composed of electrons.
   e. ultimate production of ATP.

_____ 8. In the citric acid cycle
   a. glucose is catabolized to H_2O and CO_2.
   b. NAD+ is a hydrogen acceptor.
   c. FAD is a hydrogen acceptor.
   d. decarboxylation occurs.
   e. oxaloacetate is consumed.

_____ 9. In the electron transport system
   a. water is the final electron acceptor.
   b. cytochromes carry electrons.
   c. the final electron acceptor has a positive redox potential.
   d. glucose is a common carrier molecule.
   e. electrons gain energy with each transfer.

VISUAL FOUNDATIONS

Color the parts of the illustration below as indicated.

RED	☐	ATP synthetase
GREEN	☐	first complex with FMN
YELLOW	☐	cytochrome b-c_1 complex
BLUE	☐	third complex with cytochromes a and a_3
ORANGE	☐	coenzyme Q
BROWN	☐	phospholipid bilayer
TAN	☐	matrix
PINK	☐	intermembrane space
VIOLET	☐	cytochrome C

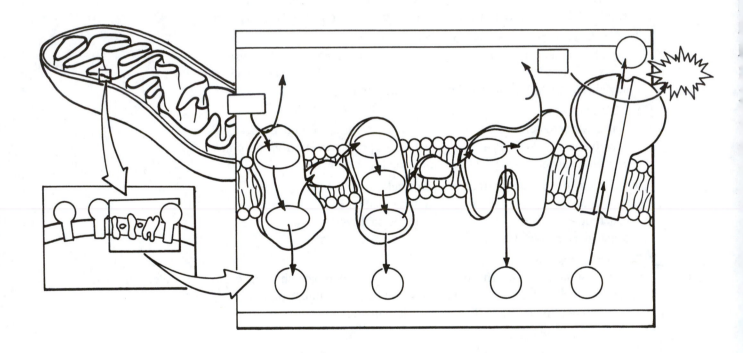

Photosynthesis: Capturing Energy

The existence and survival of all living things depends on chemosynthesis and photosynthesis, processes by which energy is captured from the environment and converted into the chemical bond energy of carbohydrates. This energy is what fuels the metabolic reactions that sustain all life. Photosynthesis, by far the most prevalent and important process of the two, captures solar energy and uses it to manufacture organic compounds from carbon dioxide and water. Photosynthesis occurs in thylakoid membranes. These membranes exist in photosynthetic prokaryotes as extensions of the plasma membrane, and in eukaryotes as organized structures within chloroplasts. Photosynthesis consists of both light-dependent and light-independent reactions. In the light-dependent reactions, light energy is trapped by chlorophyll and temporarily stored in the chemical bonds of ATP and NADPH. Some of this energy is used to split water into hydrogen and oxygen. The oxygen is released to the environment. In the light-independent reactions, additional energy from the ATP and NADPH manufactured earlier is used to make carbohydrate from carbon dioxide and the hydrogen that resulted from the splitting of water. There are two different pathways by which carbon dioxide is assimilated into plants — the C_3 and C_4 pathways. C_3 is the most prevalent. On bright, hot, dry days when plant cells close their stomata and carbon dioxide cannot get into the leaves, C_3 plants consume oxygen and produce carbon dioxide and water, a process called photorespiration. Photosynthetic efficiency is reduced by photorespiration because it removes some of the intermediates that are used in the C_3 cycle.

CHAPTER OUTLINE AND CONCEPT REVIEW
Fill in the blanks.

INTRODUCTION

1 Using the basic raw materials (a) _____,
   photosynthetic autotrophs convert (b) _____ energy to stored
   (c)_____ energy.

2 _____ are organisms that cannot make their own food.

PHOTOSYNTHESIS IN EUKARYOTES TAKES PLACE IN CHLOROPLASTS

3 The (a)_____ is the fluid-filled region within the inner membrane of chloroplasts that
   contains most of the enzymes for photosynthesis. Most chloroplasts are located in the
   (b)_____ cells of leaves.

4 _____ are flat, disk-shaped membranes in chloroplasts that are arranged in stacks
   called grana. Chlorophyll is located in these membranes.

PHOTOSYNTHESIS IS THE CONVERSION OF LIGHT ENERGY TO CHEMICAL BOND ENERGY

5 _____ traps sunlight energy where it is converted to chemical energy in the form of
   the two high energy molecules.

ATP and NADPH are the products of the light-dependent reactions
Carbohydrates are produced during the carbon fixation reactions

6   Light-dependent reactions provide useful chemical energy for synthesis of photosynthetic products. Light-independent reactions transfer the energy from (a)_____ to the bonds in (b)_____ molecules.

THE LIGHT-DEPENDENT REACTIONS CONVERT LIGHT ENERGY TO CHEMICAL BOND ENERGY

Light is composed of particles that travel as waves

7   The lowest energy state for an atom is called the (a)_____. When a molecule absorbs light energy, its electrons are raised (boosted) to higher energy states, and their atoms are said to be (b)_____.

8   (a)_____ occurs when energized electrons return to the ground state, emitting their excess energy in the form of visible light. In photosynthesis, energized electrons leave atoms and pass to an (b)_____ molecule.

Chlorophyll is found in the thylakoid membrane

9   Chlorophyll absorbs light mainly in the _____ portions of the visible spectrum.

10   Of the several types of chlorophyll in plants, (a)_____ is the bright green form that initiates the light-dependent reactions, and the yellowish-green form, (b)_____, is an accessory pigment. Other yellow and orange accessory pigments in plant cells are (c)_____.

Chlorophyll is the main photosynthetic pigment

11   An (a)_____ is a graph that illustrates the relative absorption of different wavelengths of light by a given pigment. It is obtained with an instrument called a (b)_____.

12   The (a)_____ of photosynthesis is a measurement of the actual effectiveness of different wavelengths of light in affecting photosynthesis. It may be greater than can be accounted for by the absorption of chlorophyll alone, the difference accounted for by (b)_____ that transfer energy absorbed from the green wavelengths to chlorophyll.

Photosystems I and II include light-harvesting antenna complexes of pigment molecules
The photosynthetic electron transport system is coupled to ATP synthesis

13   _____ is the process whereby energized electrons provide energy to add a phosphate molecule to ADP.

The chemiosmotic model explains the coupling of ATP synthesis and electron transport

14   Energy spent as electrons cycle along the transport system is used to pump protons across the (a)_____, where their accumulation in the thylakoid lumen creates a strong (b)_____. When "gates" in the thylakoid membrane "open," protons flow out, providing sufficient energy of motion to generate ATP. This phenomenon is known as (c)_____.

15   Protons leak out of the thylakoid lumen through channels in the ATP synthetase molecule called the _____ complex.

Noncyclic photophosphorylation produces ATP and NADPH

16   Both photosystems are utilized during noncyclic photophosphorylation, during which electrons flow in one direction, yielding one (a)_____ molecule and two (b)_____ molecules.

Cyclic photophosphorylation produces ATP only

17 Cyclic photophosphorylation takes place in photosystem I, where excited electrons escape from _____ at the reaction center, pass along an electron transport chain, and ultimately return to their point of origin.

THE CARBON FIXATION REACTIONS REQUIRE ATP AND NADPH

18 ATP and NADPH generated in the light-dependent reactions are used to reduce carbon dioxide to a carbohydrate, a process known as _____.

Most plants use the Calvin (C_3) cycle to fix carbon

19 The light-dependent reactions form the Calvin cycle, six turns of which produce one six carbon (a)_____ molecule. The cycle begins with the combination of one CO_2 molecule and one five carbon sugar, (b)_____, to form a six carbon molecule. This molecule instantly splits into two three carbon molecules called (c)_____, which then are phosphorylated to form (d)_____ _____.

20 To produce one six carbon carbohydrate, the light-independent reactions utilize six molecules of (a)_____, hydrogen obtained from (b)_____, and energy from (c)_____.

The initial carbon fixation step differs in C_4 plants and in CAM plants

21 C_4 plants initially fix CO_2 into the four carbon molecule _____.

Photorespiration reduces photosynthetic efficiency

22 Photorespiration occurs mainly during bright, hot days when plant stomata are closed. It reduces photosynthetic efficiency because CO_2 cannot enter the system, causing the enzyme (a)_____ to bind RuBP to (b)_____ instead of CO_2.

BUILDING WORDS

Use combinations of prefixes and suffixes to build words for the definitions that follow.

Prefixes	The Meaning		Suffixes	The Meaning
auto-	self, same		-lysis	breaking down, decomposition
hetero-	different, other		-plast	formed, molded, "body"
meso-	middle		-troph	nutrition, growth, "eat"
photo-	light			
chloro-	green			

Prefix	Suffix	Definition
_____	-phyll	1. Tissue in the middle of a leaf specialized for photosynthesis.
_____	-synthesis	2. The conversion of solar (light) energy into stored chemical energy by plants, blue-green algae, and certain bacteria.
_____	_____	3. The breakdown (splitting) of water under the influence of light energy trapped by chlorophyll.
_____	-phosphorylation	4. Phosphorylation that uses light as a source of energy.
_____	_____	5. A membranous organelle containing green photosynthetic pigments.
_____	phyll	6. A green photosynthetic pigment.

_____ _____   7. An organism that produces its own food.

_____ _____   8. An organism that is dependent upon other organisms for food, energy, and oxygen.

MATCHING

For each of these definitions, select the correct matching term from the list that follows.

____ 1. The fluid region of the chloroplast surrounding the thylakoids.

____ 2. Interconnected system of flattened sac-like membranous structures inside the chloroplast where light energy is converted into chemical energy.

____ 3. A stack of thylakoids within a chloroplast.

____ 4. A yellow to orange plant pigment.

____ 5. The usual pathway for fixing carbon dioxide in the synthesis reactions of photosynthesis.

____ 6. A cyclic series of reactions occurring during the light-independent phase of photosynthesis.

____ 7. A highly organized cluster of photosynthetic pigments and electron/hydrogen carriers embedded in the thylakoid membranes of chloroplasts.

____ 8. An autotrophic organism that obtains energy and synthesizes organic compounds from inorganic compounds.

____ 9. A particle of electromagnetic radiation; one quantum of radiant energy.

____10. A metabolic pathway that fixes carbon in desert plants.

Terms:

a. C_3 pathway
b. C_4 pathway
c. Calvin cycle
d. CAM
e. Carotenoid

f. Chemoautotroph
g. Chemoheterotroph
h. Granum
i. Photoheterotroph
j. Photon

k. Photosystem
l. Stoma
m. Stroma
n. Thylakoid

TABLE

Fill in the blanks.

Category	Noncyclic Photosynthesis	Aerobic Respiration
Eukaryotic cellular site	Chloroplasts	Cytosol and mitochondria
Type of metabolic pathway	#1	#2
Type of reaction	#3	#4
Site of electron transport chain	#5	#6
#7	NADP	NAD
#8	H_2O	Glucose, or other carbohydrate
Final hydrogen acceptor	#9	#10

THOUGHT QUESTIONS
Write your responses to these questions.

1. Use *all* of the components in the list below and chemical formulas as needed to construct an illustration of photosynthesis. Add any components that are needed to make the illustration clear. Explain what is occurring at each step. The components are: CO_2, H_2O, $C_6H_{12}O_6$, O_2, light (solar) energy in, energy out, NADP and NADPH, ATP, electrons, protons, chloroplast, chlorophylls, photosystems I and II, electron transport system, Calvin cycle, chemiosmosis.
2. Explain how the C_4 pathway functions in certain plants.
3. Explain how the absorption of photons can energize a plant pigment. Illustrate the process using solar energy and electrons in chlorophyll.
4. Contrast the chemical reactions in photosynthesis that require light with the reactions that do not require light.

MULTIPLE CHOICE
Place your answer(s) in the space provided. Some questions may have more that one correct answer.

_____ 1. In photosynthesis, electrons in atoms
    a. are excited.
    b. produce fluorescence.
    c. are accepted by a reducing agent when they escape.
    d. are pushed to higher energy levels by photons of light.
    e. release absorbed energy as another wavelength of light.

_____ 2. Some of the requirements for the light-dependent reactions of photosynthesis include
    a. water.
    b. photons of light.
    c. carbohydrates.
    d. NADP.
    e. oxygen.

_____ 3. The light-independent reactions of photosynthesis
    a. generate PGAL.
    b. generate oxygen.
    c. take place in the stroma.
    d. require ATP.
    e. require NADPH.

_____ 4. Reactions occurring during electron flow in respiration and photosynthesis are
    a. both endergonic.
    b. both exergonic.
    c. exergonic and endergonic respectively.
    d. endergonic and exergonic respectively.
    e. neither endergonic nor exergonic.

_____ 5. Both photosystems I and II
    a. are activated by light.
    b. form NADPH from $NADP^+$.
    c. involve an electron transport system.
    d. occur in higher plants.
    e. occur in photosynthetic bacteria.

_____ 6. Thylakoids
    a. comprise grana.
    b. are continuous with the plasma membrane.
    c. possess the basic fluid mosaic membrane structure.
    d. are located in the stroma.
    e. carry out the light-independent reactions.

_____ 7. Chloroplasts in prokaryotic cells
    a. possess a CF_0–CF_1 complex.
    b. have double membranes.
    c. do not exist.
    d. contain chlorophyll.
    e. carry out photosynthesis II.

_____ 8. The more advanced aspects of photosynthesis, generally found only in higher plants, include
   a. photosystem I.   d. noncyclic phosphorylation.
   b. photosystem II.   e. use of oxygen as a hydrogen acceptor.
   c. cyclic phosphorylation.

_____ 9. Chlorophyll
   a. contains magnesium in a porphyrin ring.   d. is the only pigment found in most plants.
   b. dissolves in water.   e. is found mostly in the stroma.
   c. is green because it absorbs green portions of the light spectrum.

_____ 10. C_4 fixation
   a. replaces C_3 fixation.   d. takes place in bundle sheath cells.
   b. produces PGA.   e. supplements C_3 fixation.
   c. produces oxaloacetate.

_____ 11. The reactions of photosystem II
   a. include splitting water with light photons.   d. use $NADP^+$ as a final electron acceptor.
   b. produce H_2O.   e. produce O_2.
   c. assist in producing an electrochemical gradient across the thylakoid membrane.

_____ 12. The ATP synthetase enzyme
   a. regulates flow of protons.   d. comprises the CF_0–CF_1 complex.
   b. functions in chemiosmosis.   e. removes phosphate from ATP to form free ADP.
   c. forms a channel for passage of protons to the interior of thylakoids.

_____ 13. Photosynthesis involves
   a. enzymes in the stroma.   d. chlorophyll.
   b. enzymes in thylakoids.   e. carotenoids.
   c. a CF_0–CF_1 complex.

_____ 14. Most producers are
   a. chemosynthetic heterotrophs.   d. photosynthetic heterotrophs.
   b. chemosynthetic autotrophs.   e. photosynthetic consumers.
   c. photosynthetic autotrophs.

VISUAL FOUNDATIONS

Color the parts of the illustration below as indicated.

RED   ☐   stoma

GREEN   ☐   mesophyll

YELLOW   ☐   spongy mesophyll

BLUE   ☐   vein

ORANGE   ☐   bundle sheath cell

BROWN   ☐   upper epidermis

TAN   ☐   lower epidermis

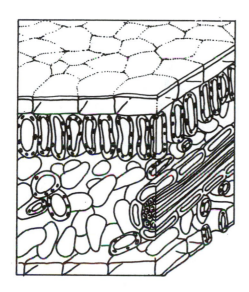
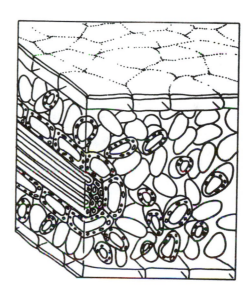

PART III

The Continuity of Life: Genetics

CHAPTER 9

Chromosomes, Mitosis, and Meiosis

Genetic information is transferred from parent to offspring. In prokaryotes, the information is contained in a single circle of DNA. In eukaryotes, it is carried in the chromosomes contained within the cell nucleus. Chromosomes are made up of DNA, protein, and RNA. The DNA is organized into informational units, or genes, that determine the characteristics of the organism. Genes control the structure of all the proteins of the organism, including the enzymes. Each species is unique due to the information specified by its genes. Genes are passed from parent cell to daughter cell by mitosis, a process that ensures that each new nucleus receives the same number and types of chromosomes as were present in the original nucleus. Interphase is the time period between cell divisions when cell growth, DNA replication, and synthesis of chromosomal and other proteins occur. There are two basic types of reproduction — asexual and sexual. In asexual reproduction, a single parent cell usually splits, buds, or fragments into two or more individuals. Genetic material identical to that in the parent is distributed to the new individuals. In sexual reproduction, sex cells, or gametes, are produced by meiosis, a special type of cell division that halves the number of chromosomes in the resulting cells. When two gametes fuse, the resulting cell contains the same number of chromosomes as the parent cells. In animals, gametes are the direct products of meiosis. In plants and some other organisms, the products of meiosis are spores that undergo mitosis one or more times before some of the descendants develop into gametes. Unlike asexual reproduction, the new individual in sexual reproduction is not genetically identical to the cells of the parents because the genetic information from both parents is shuffled during meiosis. Various groups of sexually reproducing eukaryotes differ with respect to the roles of mitosis and meiosis in their life cycles.

CHAPTER OUTLINE AND CONCEPT REVIEW
Fill in the blanks.

EUKARYOTIC CHROMOSOMES CONTAIN DNA, PROTEIN, AND RNA

1 DNA, protein, and RNA form a complex, the _____, that make up chromosomes.

DNA is organized into informational units called genes

2 Genes code for specific (a)_____ molecules, which in turn code for specific (b)_____.

Chromosomes of different species differ in number and informational content

THE CELL CYCLE IS A SEQUENCE OF CELL GROWTH AND DIVISION

3 The period from the beginning of one cell division to the beginning of the next cell division is the _____.

Chromosomes become duplicated during interphase

4 Interphase is divided into the G_1 phase, or (a)_____ phase, the S phase, or (b)_____ phase, and the G_2 phase, or (c)_____ phase.

Mitosis ensures orderly distribution of chromosomes

5 Chromatin threads begin to condense during the _____ stage of mitosis.

6 Chromosomes assemble in the equatorial plane of the cell during the _____ stage of mitosis.

7 Sister chromatids separate and the newly formed chromosomes move toward the poles during the _____ stage of mitosis.

8 Nuclei reform in the daughter cells during the _____ stage of mitosis.

Two separate daughter cells are formed by cytokinesis
Mitosis typically produces two cells genetically identical to the parent cell
Most cytoplasmic organelles are distributed randomly to the daughter cells
The cell cycle is controlled by an internal genetic program interacting with external signals

SEXUAL LIFE CYCLES REQUIRE A MECHANISM TO REDUCE THE CHROMOSOME NUMBER

9 Offspring inherit traits that are virtually identical to those of their single parent when reproduction is _____.

10 In (a)_____ reproduction, offspring receive genetic information from two parents. Haploid (n) gametes from the parents fuse to form a single (b)_____ (2n) cell called the (c)_____.

11 Homologous chromosomes are members of a pair of chromosomes that are similar in _____. They carry genes affecting the same traits.

12 One member of each (a)_____ pair is contributed by each parent to the zygote, thereby restoring (b)_____ in the offspring.

DIPLOID CELLS UNDERGO MEIOSIS TO FORM HAPLOID CELLS

13 Meiosis is a special kind of cell division that produces haploid (a)_____ from diploid (b)_____.

Meiosis produces haploid cells with unique gene combinations
In meiosis homologous chromosomes become distributed into different daughter cells

14 Homologous chromosomes exchange genetic material (crossing over) during the first meiotic (a)_____, providing more (b)_____ among gametes and offspring.

15 The haploid condition is established as the members of each pair of homologous chromosomes separate during the first meiotic _____.

16 The two chromatids of each chromosome separate during the second meiotic _____.

17 Some simple eukaryotes are haploid, producing gametes by (a)_____. The diploid zygote is restored to haploidy by (b)_____.

18 In animals, meiosis reduces diploid _____ cells to haploid _____.

19 Alternation of generations in plants and some algae involves a spore-forming diploid generation, called the (a)_____, alternating with a gamete-producing haploid generation, called the (b)_____.

BUILDING WORDS

Use combinations of prefixes and suffixes to build words for the definitions that follow.

Prefixes	The Meaning		Suffixes	The Meaning
centro-	center		-gen(esis)	production of
chromo-	color		-mere	part
dipl-	double, in pairs		-phyte	plant
gameto-	sex cells, eggs and sperm		-some	body
hapl-	single			
inter-	between			
oo-	egg			
spermato-	seed, "sperm"			
sporo-	spore			

Prefix	Suffix	Definition
_____	_____	1. A dark staining body within the cell nucleus containing genetic information.
_____	-phase	2. The stage in the life cycle of a cell that occurs between successive cell divisions.
_____	-oid	3. An adjective pertaining to a single set of chromosomes.
_____	-oid	4. An adjective pertaining to a double set of chromosomes.
_____	_____	5. The constricted part or region of a chromosome (often near the center) to which a spindle fiber is attached.
_____	_____	6. The process by which gametes (sex cells) are produced.
_____	_____	7. The process whereby sperm are produced.
_____	_____	8. The process whereby eggs are produced.
_____	_____	9. The stage in the life cycle of a plant that produces gametes by mitosis.
_____	_____	10. The stage in the life cycle of a plant that produces spores by meiosis.

MATCHING

For each of these definitions, select the correct matching term from the list that follows.

_____ 1. The last stage of mitosis and meiosis.

_____ 2. The phase in interphase during which DNA and other chromosomal components are synthesized.

_____ 3. Portion of the chromosome centromere to which the mitotic spindle fibers attach.

_____ 4. Process whereby genetic material is exchanged between homologous chromatids during meiosis.

_____ 5. The phase in mitosis during which the chromosomes line up along the equatorial plate.

_____ 6. DNA-protein fibers which condense to form chromosomes during prophase.

_____ 7. Process during which a diploid cell undergoes two successive nuclear divisions resulting in four haploid cells.

_____ 8. Having more than two sets of chromosomes per nucleus.

_____ 9. Division of the cell nucleus resulting in two daughter cells with the identical number of chromosomes as the parental cell.

_____10. Stage of cell division in which the cytoplasm divides into two daughter cells.

Terms:

a.   Anaphase
b.   Chromatin
c.   Crossing over
d.   Cytokinesis
e.   Interkinesis

f.   Kinetochore
g.   Meiosis
h.   Metaphase
i.   Mitosis
j.   Polyploid

k.   Prophase
l.   S phase
m.   Synapsis
n.   Telophase

TABLE
Fill in the blanks.

Event	Mitosis	Meiosis
Condensation of chromosomes	Prophase	Prophase I, prophase II
#1chromosomes line up along an equatorial plane of the cell	Metaphase	Metaphase I, metaphase II
Chromatids separate	#2anaphase	#3anaphase II
Duplication of DNA	#4interphase	#5premeiotic interphase
#6homologous chromosomes move to opposite poles	Does not occur	Anaphase I
Cytokinesis occurs	#7telophase	#8telophase I, telophase II
Tetrads form	#9does not occur	#10prophase I

THOUGHT QUESTIONS
Write your responses to these questions.

1. Make a sketch of each stage in a typical eukaryotic cell cycle emphasizing events that transpire in the nucleus, then describe the major events that occur in each stage.
2. What is sex? Contrast sexual reproduction with asexual reproduction.
3. Describe the major differences in mitosis and meiosis. How do the final products differ?

MULTIPLE CHOICE
Place your answer(s) in the space provided. Some questions may have more that one correct answer.

Use this list to answer questions 1-10 about mitosis:

a. interphase
b. prophase
c. metaphase
d. anaphase

e. telophase
f. T phase
g. G_1 phase

h. G_2 phase
i. S phase
j. M phase

_____ 1. The four stages of mitosis collectively.

_____ 2. Nuclear membranes break down.

_____ 3. Chromosomes begin to condense by coiling.

_____ 4. Active synthesis and growth.

_____ 5. The time between mitosis and start of the synthesis phase.

_____ 6. Centromeres divide.

_____ 7. Chromosomes are lined up in a central plane.

_____ 8. The time between the synthesis phase and prophase.

_____ 9. DNA replicates.

_____10. Condensed chromosomes uncoil.

_____11. In mitosis, cells with 16 chromosomes produce daughter cells with
  a. 32 chromosomes.   d. 8 pairs of chromosomes.
  b. 8 chromosomes.   e. 4 pairs of chromosomes.
  c. 16 chromosomes.

_____12. Gametogenesis typically involves
  a. 2n to 2n.   d. mitosis.
  b. 2n to n.   e. meiosis.
  c. reduction division.

_____13. Eukaryotic chromosomes
  a. contain DNA, RNA, protein.   d. are uncoiled in interphase.
  b. possess asters.   e. align in the equator during prophase.
  c. are distinctly visible in interphase.

_____14. Typical gametes include
  a. somatic cells.   d. ova.
  b. 2n cells.   e. cells resulting from oogenesis.
  c. sperm.

_____15. In meiosis, cells with 16 chromosomes produce daughter cells with
  a. 32 chromosomes.   d. 8 pairs of chromosomes.
  b. 8 chromosomes.   e. 4 pairs of chromosomes.
  c. 16 chromosomes.

Use this list to answer questions 16-25 about meiosis:
  a. interphase   d. anaphase I   g. metaphase II
  b. prophase I   e. telophase I   h. anaphase II
  c. metaphase I   f. prophase II   i. telophase II

_____16. Chromatids separate.

_____17. Pairs of chromosomes align at equatorial plane.

_____18. Crossing over takes place.

_____19. Homologous chromosomes synapse.

_____20. Tetrads form.

_____21. Nuclear membranes break down.

_____22. Members of tetrad separate.

_____23. Single chromosomes align at equatorial plane.

_____24. Diploid to haploid.

_____25. Reduction of chromosome number from 2n to n.

CHAPTER 10

The Basic Principles of Heredity

Experiments conducted in the nineteenth century by Gregor Mendel led to the discovery of the three major principles of heredity: dominance, segregation, and independent assortment. Genes that occupy corresponding loci on homologous chromosomes govern variations of the same characteristic. Such genes, called alleles, exist in pairs in diploid organisms. A monohybrid cross is a cross between two individuals that carry different alleles for a single gene locus. Similarly, a dihybrid cross is a cross between two individuals that carry different alleles at each of two gene loci. The results of monohybrid and dihybrid crosses illustrate the basic principles of genetics. The laws of probability are used to predict the results of a cross between two individuals. The order of genes on a chromosome is determined by calculating the frequency that chromatids exchange segments of chromosomal material during meiosis. The sex of most animal species is determined by special chromosomes called sex chromosomes. The relationship between a single pair of alleles at a gene locus and the characteristic it controls may be simple, or it may be complex. For example, dominance may be complete or incomplete; three or more alleles may exist for a given locus; most genes have many different effects; and the presence of a particular allele of one gene pair may determine whether certain alleles of another gene pair are expressed. Also, many characteristics are not inherited through alleles at a single gene locus, but instead result from multiple independent pairs of genes having similar and additive effects. Selection, inbreeding, and outbreeding are used commercially to develop improved strains of plants and animals.

CHAPTER OUTLINE AND CONCEPT REVIEW
Fill in the blanks.

MENDEL FIRST DEMONSTRATED THE PRINCIPLES OF INHERITANCE

1 Offspring from the P generation cross are heterozygous; they are called the F_1 or
(a)_____ generation. Offspring from the F_1 cross are called the
(b)_____ generation.

The principle of dominance states that one gene can mask the expression of another in a hybrid

2 A (a)_____ gene may mask the expression of a (b)_____ gene, yielding phenotypes different from the genotype.

The principle of segregation states that the genes of a pair separate before gametes are formed

3 During meiosis, members of paired genes at each locus _____ so that each gamete contains only one gene from each locus.

4 During meiosis, each pair of genes separates independently of _____ located in other homologous chromosomes.

ALLELES OCCUPY CORRESPONDING LOCI ON HOMOLOGOUS CHROMOSOMES

5 The site of a gene in a chromosome is called its _____.

A MONOHYBRID CROSS INVOLVES INDIVIDUALS WITH DIFFERENT ALLELES FOR A GIVEN GENE LOCUS

Heterozygotes carry two different alleles for a locus; homozygotes carry identical alleles

The phenotype of an individual does not always reveal its genotype

A Punnett square predicts the ratios of genotypes and phenotypes of the offspring of a cross

A test cross can detect heterozygosity

6 A monohybrid test cross is a cross between _____ _____ individuals.

THE LAWS OF PROBABILITY ARE USED TO PREDICT THE LIKELIHOOD OF GENETIC EVENTS

7 The probability of a specific event occurring is determined by dividing the number of (a)_____ that occurred by the total number of events. Probability can range from (b)_____ (impossible) to (c)_____ (certain).

The product law predicts the combined probabilities of independent events

8 The probability of two independent events occurring together is the _____ of the probabilities of each occurring separately.

The sum law predicts the combined probabilities of mutually exclusive events

9 The probability that either one or the other of two mutually exclusive events will occur is the _____.

The laws of probability can be applied to a variety of calculations

A DIHYBRID CROSS INVOLVES INDIVIDUALS THAT HAVE DIFFERENT ALLELES AT TWO LOCI

10 A dihybrid cross is a cross between homozygous parents (P generation) that differ with respect to their alleles at _____.

The principle of independent assortment states that the alleles of different loci are randomly distributed into gametes

11 Genes in the same chromosome are said to be _____ and do not assort independently.

THE LINEAR ORDER OF LINKED GENES ON A CHROMOSOME CAN BE "MAPPED" BY CALCULATING THE FREQUENCY OF CROSSING OVER

12 Linked genes are recombined when chromatids exchange genetic material, a process known as _____, that occurs during meiotic prophase.

13 A chromosome can be genetically mapped by determining the frequency of _____ among genes.

SEX IS COMMONLY DETERMINED BY SPECIAL SEX CHROMOSOMES

14 The sex or gender of many animals is determined by the X and Y sex chromosomes. The other chromosomes in a given organism's genome are called _____.

The Y chromosome determines male sex in most species of mammals

15 When a Y-bearing sperm fertilizes an ovum, the result is a(n) (a)_____, and fertilization by an X-bearing sperm produces a(n) (b)_____.

X-linked genes have unusual inheritance patterns

Dosage compensation equalizes the expression of X-linked genes in males and females

16 The effect of X-linked genes is made equivalent in males and females by dose compensation, which is accomplished by a (a)_____ X-chromosome in the male or (b)_____ of one X-chromosome in the female.

Sex-influenced genes are autosomal, but their expression is affected by the individual's sex

THE RELATIONSHIP BETWEEN GENOTYPE AND PHENOTYPE IS OFTEN COMPLEX

Dominance is not always complete

Multiple alleles for a locus may exist in a population

17 Multiple alleles are (a)_____(#?) different alleles that can occupy the same (b)_____.

A single gene may affect multiple aspects of the phenotype; alleles of different loci may interact to produce a phenotype

18 _____ refers to the many different effects that can often result from a given gene.

19 _____ is when one allele of a gene pair determines whether alleles of other gene pairs are expressed.

Polygenes act additively to produce a phenotype

20 It is called _____ when two or more independent pairs of genes have similar and additive effects on a phenotype.

SELECTION, INBREEDING, AND OUTBREEDING ARE USED TO DEVELOP IMPROVED STRAINS

21 Inbreeding increases the _____ of recessive genes.

22 Outbreeding, the mating of unrelated individuals, increases (a)_____. (b)_____ is the improvement of offspring as a result of outbreeding.

BUILDING WORDS

Use combinations of prefixes and suffixes to build words for the definitions that follow.

Prefixes	The Meaning
di-	two, twice, double
hemi-	half
hetero-	different, other
homo-	same
mono-	alone, single, one
poly-	much, many

Prefix	Suffix	Definition
_____	-gene	1. Two or more pair of genes that affect the same trait in an additive fashion.
_____	-zygous	2. Having the same (identical) members of a gene pair.
_____	-zygous	3. Having dissimilar (different) members of a gene pair.
_____	-hybrid	4. Pertaining to the mating of individuals differing in two specific pairs of genes.
_____	-hybrid	5. Pertaining to the mating of individuals differing in one pair of genes.
_____	-gametic	6. A condition in which the gametes of the opposite sexes are morphologically different from each other.
_____	-gametic	7. A condition in which the gametes of the opposite sexes are morphologically identical to each other.
_____	-zygous	8. Having only half (one) of a given pair of genes.

MATCHING

For each of these definitions, select the correct matching term from the list that follows.

_____ 1. Condition in which certain alleles at one locus can alter the expression of alleles at a different locus.

_____ 2. An alternative form of a gene.

_____ 3. The physical or chemical expression of an organism's genes.

_____ 4. Condition in which a single gene produces two or more phenotypic effects.

_____ 5. A condensed and inactivated X-chromosome appearing as a distinctive dense spot in the nucleus of certain cells of female mammals.

_____ 6. The allele that is not expressed in the heterozygous state.

_____ 7. The place on a chromosome at which the gene for a given trait occurs.

_____ 8. Condition in which both alleles of a locus are expressed in a heterozygote.

_____ 9. The allele that is always expressed when it is present.

_____10. Mating of genetically similar individuals.

Terms:

a. Allele
b. Barr body
c. Dominant allele
d. Epistasis
e. Genotype

f. Inbreeding
g. Incomplete dominance
h. Linkage
i. Locus
j. Overdominance

k. Phenotype
l. Pleiotropy
m. Recessive allele

TABLE
Fill in the blanks.

	T Y	T y	t Y	t y
T Y	tall plant with yellow seeds (TT YY)	tall plant with yellow seeds (TT Yy)	tall plant with yellow seeds (Tt YY)	tall plant with yellow seeds (Tt Yy)
T y	#1	#2	#3	#4
t Y	#5	#6	#7	#8
t y	#9	#10	#11	#12

THOUGHT QUESTIONS
Write your responses to these questions.

1. Explain the meanings of the following terms and explain the relationship between the pairs of terms:

Compare	With
allele	locus
genotype	phenotype
dominant	recessive
homozygous	heterozygous
test cross	P1 generation
monohybrid cross	dihybrid cross

2. Assume that "R" is a dominant allele on autosome one that produces red flowers and "r" is a recessive allele that produces white flowers, and assume that "T" is an allele on autosome two that produces tall plants and "t" is an allele that produces short plants. What are the possible genotypes and phenotypes in the F1 generation derived from parents with the genotypes RrTT and RrTt? What are the possible genotypes and phenotypes in the F2 generation? How would the ratios of possible genotypes and phenotypes differ if flower color genes and plant height genes are linked?

3. In the above, assume that "R" and "T" are both incompletely dominant to "r" and "t" respectively, the heterozygous conditions producing pink flowers and plants of medium height. Now what are the possible genotypes and phenotypes in the F1 and F2 generations?

4. In the above examples, assume that "R" and "r" exhibit complete dominance and "T" and "t" exhibit incomplete dominance. Now what are the possible genotypes and phenotypes in the F1 and F2 generations?

5. Assume that the alleles A, B, and o all relate to ABO blood types, and that A and B together are incompletely dominant and both A and B are completely dominant over the allele "o." Determine all of the genotypes and phenotypes (ABO blood types) in each of the following "matings:"

Ao x Bo   AB x oo
Ao x oo   AB x AB
blood type A x blood type O   blood type B x blood type AB

6. Suppose the incompletely dominant alleles X, Y, and Z act additively to produce green hair. Which of the following parents would have the greenest hair color — XX YY ZZ or xx yy zz? If this couple produces several children, and those children produce another generation of children, which generation of will have the greatest variety of genotypes and broadest range of hair colors — the P1, or F1, or the F2 generation? Explain your answer.

MULTIPLE CHOICE
Place your answer(s) in the space provided. Some questions may have more that one correct answer.

R and r are genes for flower color. Homozygous dominant and heterozygous genotypes both have red flowers; the homozygous recessive genotype has white flowers. T and t are genes that control plant height. Homozygous dominant plants are tall, heterozygous plants are medium height, and homozygous recessive plants are short. Genes for flower color and height are on different chromosomes. Use these data and the following list to answer questions 1-9. Construct Punnett Squares as needed.

a. RRTT	f. rrtt	k. 1:2:1:2:4:2:1:2:1	p. pink, medium
b. RrTt	g. 1:1	l. red, tall	q. pink, short
c. Rrtt	h. 1:2:1	m. red, medium	r. white, tall
d. rrTT	i. 9:3:3:1	n. red, short	s. white, medium
e. rrTt	j. 1:2:1:1:2:1	o. pink, tall	t. white, short

Plants with the genotypes Rrtt and rrTT are mated. Their offspring are then mated to produce another generation. Questions 1-5 pertain to this last generation.

_____ 1. What are the genotypic and phenotypic ratios among offspring?

_____ 2. What are all of the phenotypes among offspring?

_____ 3. Which of the genotypes listed above are found among offspring?

_____ 4. What are the genotypes found among the offspring that are not in the list?

_____ 5. What would the phenotypic ratio among offspring be if both flower color and height genes had exhibited incomplete dominance?

Questions 6-9 pertain to the following cross: Rrtt x rrTT.

_____ 6. What are the genotypes and phenotypes of the parents?

_____ 7. What are the genotypes and phenotypes of the offspring?

_____ 8. What is the genotypic ratio among offspring?

_____ 9. What is the phenotypic ratio among offspring?

_____ 10. Pleiotropy means that a pair of genes
    a. exhibits incomplete dominance.
    b. affects expression of other genes.
    c. has the same effect as another pair of genes.
    d. has multiple effects.
    e. is sex-influenced.

_____ 11. Given the following information about crossing over, determine the relative positions of three loci (X, Y, Z) on one chromosome: X and Y = 8%, Y and Z = 5%, X and Z = 3 map units.
    a. X, Y, Z
    b. X, Z, Y
    c. Z, Y, X
    d. Y, Z, X
    e. Z, Y, X

_____ 12. Polygenic inheritance means that a pair of genes
    a. exhibits incomplete dominance.
    b. affects expression of other genes.
    c. has the same effect as another pair of genes.
    d. has multiple effects.
    e. is sex-influenced.

_____13. Epistasis means that a pair of genes
    a. exhibits incomplete dominance.   d. has multiple effects.
    b. affects expression of other genes.   e. is sex-influenced.
    c. has the same effect as another pair of genes.

_____14. Which of the following would be true if hairy toes happened to be a recessive X-linked trait?
    a. All men would have hairy toes.   d. More women than men would have hairy toes.
    b. No women would have hairy toes.   e. More men than women would have hairy toes.
    c. Parents with hairy toes could have a child without hairy toes.

_____15. Phenotypic expression
    a. always involves only one pair of genes.   d. refers specifically to the composition of genes.
    b. may involve many pairs of genes.   e. refers to the appearance of a trait.
    c. is partly a function of environmental influences.

_____16. Linked genes are
    a. inseparable.   d. in different chromatids.
    b. generally on the same chromosome.   e. always separated during crossing over.
    c. in separate homologous chromosomes.

_____17. Which of the following is/are true?
    a. XX is usually female.   d. hermaphrodites are XX or XY.
    b. XY is usually male.   e. XXY is usually male.
    c. birds and butterflies do not have sex chromosomes.

_____18. Which of the following is/are consistent with Mendel's principle of dominance?
    a. All F_1 offspring express the dominant trait.   d. Both P generation parents are true breeding.
    b. All F_2 offspring express the dominant trait.   e. Only one P generation parent is true breeding.
    c. Both organisms in the P generation are homozygous.

_____19. Which of the following applies if two pure-breeding P generation plants are used to ultimately produce an F_2 generation with flower colors in a ratio of 1 red: 2 pink: 1 white? (R is a gene for red; r = white)
    a. All F_1 plants were Rr.   d. F_2 genotypic ratio is 3:1.
    b. F_1 plants were red and white.   e. F_2 genotypic ratio is 1:2:1.
    c. At least one P generation parent was pink.

_____20. Mendel's "factors"
    a. are haploid in gametes.   d. are genes.
    b. interact in gametes.   e. affect flower color, but not seed color.
    c. segregate into separate gametes.

VISUAL FOUNDATIONS
Color the parts of the illustration below as indicated.

WHITE ☐ white flowers
RED ☐ red flowers
GREEN ☐ circle parents
BLUE ☐ circle F_1 generation
ORANGE ☐ circle F_2 generation
PINK ☐ pink flowers

Use an "R^1" to indicate the allele for red, and an "R^2" to indicate the allele for white. Label each flower with the correct genotype for its color.

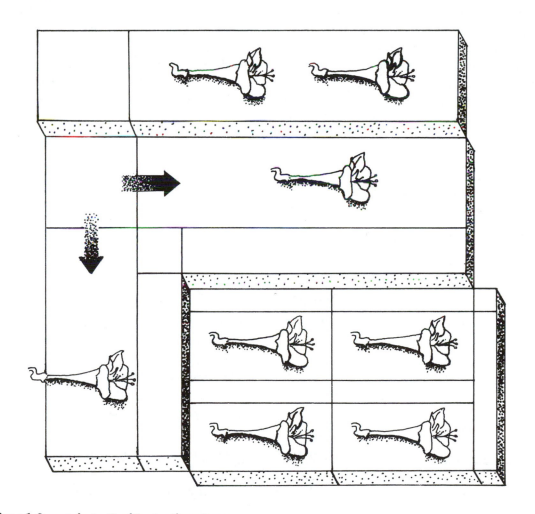

Questions 1-3 pertain to the illustration above.

1. Name the phenomenon illustrated by this figure. _____

2. What color are the flowers of individuals in the F_1 generation? _____

3. What is the phenotypic ratio in the F_2 generation? _____

CHAPTER 11

DNA: The Carrier of Genetic Information

Genes are made of deoxyribonucleic acid (DNA). Each DNA molecule consists of two polynucleotide chains arranged in a coiled double helix. Nucleotides consist of a pentose sugar, a phosphate, and a nitrogenous base. The base is a purine or a pyrimidine. The two chains are joined together by hydrogen bonds between the pyrimidines on one chain and the purines on the other chain. The sequence of bases in DNA provides for the storage of genetic information. When DNA replicates, the hydrogen bonds between the two polynucleotide chains break, and the chains unwind and separate. Each half-helix then pairs with complementary nucleotides, replacing its missing partner. The result is two DNA double helices, each identical to the original and consisting of one original strand from the parent molecule and one newly synthesized complementary strand. DNA replication is complex requiring a number of different enzymes. One polynucleotide chain is synthesized continuously while the other is synthesized discontinuously in short pieces. DNA replication proceeds in both directions from the point at which replication was initiated. DNA is packaged in chromosomes in a highly organized way.

CHAPTER OUTLINE AND CONCEPT REVIEW
Fill in the blanks.

MOST GENES CARRY INFORMATION FOR MAKING PROTEINS

1 Experiments by _____ led to the "one gene one protein" concept.

GENES ARE MADE OF DEOXYRIBONUCLEIC ACID (DNA)

EVIDENCE THAT DNA IS THE HEREDITARY MATERIAL WAS FIRST FOUND IN MICROORGANISMS

2 Evidence that genes are made of nucleic acids was provided by _____ experiments, a process whereby genetic characteristics may pass from one strain of bacteria to another.

3 _____ can reproduce by injecting only their DNA into cells, indicating that DNA is the genetic material.

THE STRUCTURE OF DNA ALLOWS IT TO CARRY INFORMATION AND TO SELF-REPLICATE

Nucleotides can be covalently linked in any order to form long polymers

4 Nucleotides consist of a _____
_____.

DNA is made of two polynucleotide chains intertwined to form a double helix

5 _____ studies by Franklin and Wilkins showed that DNA has a helical structure with nucleotide bases stacked like rungs of a ladder.

6 Watson and Crick devised a DNA model based for the most part on existing data. Their model suggested that DNA was formed from two _____ _____ arranged in a coiled double helix.

In double-stranded DNA, hydrogen bonds form between adenine and thymine and between guanine and cytosine

7 In DNA, the following are found in ratios of about 1: purines to (a)_____, guanine to (b)_____, and adenine to (c)_____.

DNA REPLICATION IS SEMICONSERVATIVE: EACH DOUBLE HELIX CONTAINS AN "OLD" STRAND AND A NEWLY SYNTHESIZED STRAND

8 Replication of DNA is considered semiconservative because each "old" strand serves as a _____ for the formation of a new strand.

DNA replication is complex and has a number of unique features

9 DNA synthesis proceeds in a (a)____' —> (b)____' direction. One strand is copied continuously, the other adds (c)_____ fragments discontinuously.

10 _____ serve as catalysts in the construction of a new DNA strand from deoxyribonucleoside triphosphates.

11 DNA replication usually starts at one or more points known as the "_____ _____," and proceeds in both directions from that/those point(s).

DNA IN CHROMOSOMES IS PACKAGED IN A HIGHLY ORGANIZED WAY

12 Most chromosomes in _____ cells are single strands of circular DNA.

13 The structural unit of eukaryotic chromosomes is the (a)_____. It consists of a segment of DNA coiled around (b)_____ and an adjacent linker DNA/protein complex.

14 The loops that comprise nucleosomes are held together by nonhistone _____ proteins.

BUILDING WORDS
Use combinations of prefixes and suffixes to build words for the definitions that follow.

Prefixes	The Meaning
a-	without, not, lacking
anti-	against, opposite of

Prefix	Suffix	Definition
_____	-virulent	1. Not lethal; lacking in virulence.
_____	-parallel	2. The arrangement of the two polynucleotide chains in a DNA molecule, viz., "running" in opposite directions to each other.

MATCHING
For each of these definitions, select the correct matching term from the list that follows.

____ 1. The incorporation of genetic material by a cell that causes it to change its phenotype.

____ 2. The complex of DNA, protein, and RNA that makes up eukaryotic chromosomes.

____ 3. The enzyme in DNA replication responsible for base pairing.

_____ 4. The region of the DNA where the molecule "unzips" for replication to begin.

_____ 5. The bases adenine and guanine.

_____ 6. Small positively-charged proteins in the nucleus that bind to the negatively-charged DNA.

_____ 7. Random heritable changes in DNA that introduce new alleles into the gene pool.

_____ 8. A molecule composed of one or more phosphate groups, a 5-carbon sugar, and a nitrogenous base.

_____ 9. A pentose sugar lacking a hydroxyl group on carbon-2.

_____10. The shape of the DNA molecule.

Terms:

a.   Chromatin
b.   Deoxyribose
c.   DNA ligase
d.   DNA polymerase
e.   Double helix

f.   Histone
g.   Mutation
h.   Nucleotide
i.   Purine
j.   Pyrimidine

k.   Replication fork
l.   Transformation

TABLE
Fill in the blanks.

Researcher(s)	Contribution to the Body of Knowledge About DNA
Beadle and Tatum	The one gene, one enzyme hypothesis
#1	A = T and G = C
Messelson and Stahl	#2
Watson and Crick	#3
#4	Only the DNA of a bacteriophage is necessary for the reproduction of new viruses
#5	Chemically identified the transforming principle described by Griffith as DNA
#6	Inferred from x-ray crystallographic films of DNA patterns that nucleotide bases are stacked like rungs in a ladder

THOUGHT QUESTIONS
Write your responses to these questions.

1. Sketch the four nucleotides that comprise DNA, then use them to illustrate the structure of a DNA molecule, all the while observing the rules of base-pairing, appropriate hydrogen bonding between bases, and other physical and biochemical characteristics of DNA that make it unique.

2. What is the evidence that supports the semiconservative view of DNA replication? Explain why the preponderance of evidence does not support either a conservative or dispersive theory of DNA replication.

3. Explain when, why, and *how* DNA replicates.

MULTIPLE CHOICE

Place your answer(s) in the space provided. Some questions may have more that one correct answer.

_____ 1. Which of the following base pairs is/are correct?
a. A – T
b. C – A
c. G – C
d. T – A
e. T – C

_____ 2. A large quantity of homogentistic acid in urine
a. is associated with alkaptonuria.
b. indicates a kidney disease.
c. is caused by a block in the metabolic reactions that break down phenylalanine and tyrosine.
d. is due to absence of an oxidizing enzyme.
e. indicates that a diuretic has been consumed.

_____ 3. Chargaff's rules state or infer that
a. [A] = [T].
b. [G] = [C].
c. ratio of purines to pyrimidines = 1.
d. ratio of T to A = 1.
e. ratio of G to C = 1.

_____ 4. The investigators credited with elucidating the structure of DNA are
a. Hershey and Chase.
b. Meselson and Stahl.
c. Avery, MacLeod, and McCarty.
d. Watson and Crick.
e. Franklin and Wilkens.

_____ 5. Which of the following is/are true of linkages in DNA?
a. 5' phosphate to 3' sugar carbon.
b. The backbone has 5', 3' phosphodiester bonds.
c. The two strands are joined by covalent bonds.
d. One strand ends with a 5' phosphate.
e. Both strands end with a 3' hydroxyl group.

_____ 6. *Neurospora* was a good choice for the Beadle and Tatum studies because
a. it is diploid.
b. its sexual phase enables genetic analysis.
c. it can be cultivated on minimal medium.
d. it had no known mutant forms.
e. the control strain could not use arginine.

_____ 7. Deoxyribose and phosphate are joined in the DNA backbone by
a. one of four bases.
b. purines.
c. pyrimidines.
d. phosphodiester bonds.
e. the 1' carbon of the sugar.

_____ 8. A mutation is
a. a change in an enzyme.
b. a change in a gene.
c. a change in DNA.
d. an alteration of mRNA.
e. sometimes inheritable.

_____ 9. Beadle and Tatum concluded that
a. one gene affects one polypeptide.
b. one gene affects one protein.
c. a mutation directly affects an enzyme.
d. mutations have no effect on enzymes.
e. proteins mutate when exposed to radiations.

_____ 10. The 3' end of one DNA fragment is linked to the 5' end of another fragment by means of
a. DNA ligase.
b. helicase enzymes.
c. a DNA polymerase.
d. hydrogen bonds.
e. complementary base pairing.

____11. The molecule(s) thought to be responsible for untying knots in replicating DNA is/are
    a. histones.
    d. topoisomerases.
    b. chromatin.
    e. protosomes.
    c. scaffolding proteins.

____12. The following can be said about the leading strand and lagging strand, respectively:
    a. Both form continuously.
    d. Forms in short pieces, forms continuously.
    b. Both form in short pieces.
    e. Forms continuously, assembles Okazaki fragments.
    c. Forms continuously, forms in short pieces.

____13. Nucleosomes are held together in large coiled loops by
    a. histones.
    d. topoisomerases.
    b. chromatin.
    e. protosomes.
    c. scaffolding proteins.

____14. The double helix structure of DNA was suggested as a result of X-ray diffraction data collected by
    a. Hershey and Chase.
    d. Watson and Crick.
    b. Griffith.
    e. Franklin and Wilkens.
    c. Avery, MacLeod, and McCarty.

____15. The direction for synthesis of DNA is
    a. $5' \longrightarrow 3'$.
    d. $3' \longrightarrow 3'$.
    b. $3' \longrightarrow 5'$.
    e. variable.
    c. $5' \longrightarrow 5'$.

____16. The proteins intimately associated with DNA in eukaryotic chromosomes are
    a. nucleosomes.
    d. absent in prokaryotes.
    b. chromatin.
    e. histones.
    c. topoisomerases.

____17. The genetic code is carried in the
    a. DNA backbone.
    d. Okazaki fragments.
    b. sequence of bases.
    e. histones.
    c. arrangement of 5', 3' phosphodiester bonds.

____18. In one molecule of DNA one would expect the composition of the two strands to be
    a. both either old or new.
    d. one old, one new.
    b. both all new.
    e. unpredictable.
    c. both partly new fragments and partly old parental fragments.

VISUAL FOUNDATIONS
Color the parts of the illustration below as indicated.

RED ❑ thymine

GREEN ❑ adenine

YELLOW ❑ cytosine

BLUE ❑ guanine

ORANGE ❑ sugar

BROWN ❑ phosphate

TAN ❑ circle the 5', 3' phosphodiester bonds

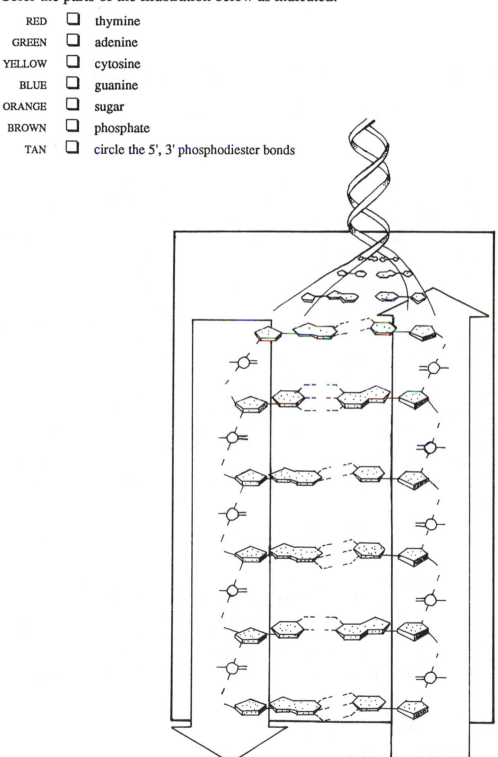

CHAPTER 12

RNA and Protein Synthesis: The Expression of Genetic Information

The manufacture of proteins by cells involves two major steps: the transcription of base sequences in DNA into the base sequences in RNA, and the translation of the base sequences in RNA into amino acid sequences in proteins. Messenger RNA (mRNA) is synthesized by DNA-dependent RNA polymerase enzymes using nucleoside triphosphate precursors. Transcription begins when an RNA polymerase recognizes a specific base sequence at the beginning of a gene. mRNA contains coding sequences that directly code for specific proteins and noncoding sequences that do not directly code for protein. Protein synthesis involves the enzymatic linkage of transfer RNA (tRNA) molecules to their respective amino acids. Ribosomes then couple these tRNAs to their codons on mRNA, catalyze peptide bonding of the amino acids, and move the mRNAs so the next codon can be read. The first stage of protein synthesis involves the formation of the protein-synthesizing complex. The second stage involves the addition of amino acids to a growing polypeptide chain. The third stage involves the release of the polypeptide and the dissociation of the ribosomal subunits which, sooner or later, are used with another mRNA molecule. Whereas prokaryotic mRNAs are translated as they are transcribed, eukaryotic mRNAs undergo posttranscriptional modification and processing prior to translation. Many eukaryotic genes differ from prokaryotic genes by having their protein-coding regions interrupted by noncoding regions. Of 64 possible codons, 61 code for amino acids and three serve as signals that specify the end of the coding sequence for a polypeptide chain. The genetic code is nearly universal, suggesting that it was established early in evolutionary history. A gene is a transcribed nucleotide sequence that yields a product with a specific cellular function. A mutation is a change in a DNA nucleotide sequence. Genes can be altered by mutation in a number of ways.

CHAPTER OUTLINE AND CONCEPT REVIEW
Fill in the blanks.

BASE SEQUENCES IN DNA ARE TRANSCRIBED AS BASE SEQUENCES IN RNA, THEN TRANSLATED INTO AMINO ACID SEQUENCES IN PROTEINS

1 Sequencing of amino acids in a protein involves two major steps: first is (a)_____ _____, wherein DNA codes are read into a special messenger RNA, and second is (b)_____, involving the actual construction of polypeptide chains.

2 Each _____ in mRNA consists of three-bases that specify one amino acid in a polypeptide chain.

3 Translation of messages in mRNA is accomplished by (a)_____ in tRNA. Each tRNA is specific for only one (b)_____.

TRANSCRIPTION IS THE SYNTHESIS OF RNA FROM A DNA TEMPLATE

4 _____ are the three principal types of RNA transcribed from DNA that are involved directly in protein synthesis.

5 mRNA synthesis is catalyzed by DNA-dependent (a)_____ enzymes, and it is formed from (b)_____ precursors.

Messenger RNA contains base sequences that code for protein

6 Transcription is initiated at sites containing specific base sequences called the _____ regions of DNA.

7 Any point closer to the 3' end of transcribed DNA (and therefore toward the 5' end of mRNA) is said to be (a)_____ relative to a given reference point. Conversely, areas toward the 5' end of DNA and the 3' end of mRNA are (b)_____.

Messenger RNA contains additional base sequences that do not directly code for protein

8 Well before the protein-coding sequences at the 5' end of mRNA, a segment called the (a)_____ contains recognition signals for ribosome binding. In addition, coding sequences are followed by (b)_____ that specify the end of the protein.

THE NUCLEIC ACID MESSAGE IS DECODED DURING TRANSLATION

An amino acid must be attached to its specific transfer RNA prior to becoming incorporated into a polypeptide

9 The specific enzymes that catalyze the formation of covalent bonds between amino acids and their respective tRNAs are the _____.

Transfer RNA molecules have specialized regions with specific functions
Ribosomes bring together all the components of the translational machinery

10 Ribosomes couple the (a)_____ in tRNAs to their appropriate codons on mRNA. They also catalyze the formation of the (b)_____ bonds that link amino acids, and they move mRNA so the next codon can be read.

Translation includes initiation, elongation, and termination

11 The codon _____ initiates the formation of the protein-synthesizing complex.

12 The addition of amino acids to a growing polypeptide chain is called _____.

13 Protein synthesis proceeds from the (a)_____ end to the (b)_____ terminal end of the growing peptide chain.

14 Newly formed polypeptides are released and ribosome subunits dissociate as the result of sequences in mRNA called _____.

A polyribosome is a complex of one mRNA and many ribosomes

TRANSCRIPTION AND TRANSLATION ARE MORE COMPLEX IN EUKARYOTES THAN IN PROKARYOTES

15 _____ serves as a protective cap on the 5' end of some eukaryotic mRNA chains.

Both noncoding nucleotide sequences (introns) and coding sequences (exons) are transcribed from eukaryotic genes

16 Intron is an abbreviation for (a)_____, and exon stands for (b)_____. Introns must be removed in order to form a continuous string of "readable" exons.

THE GENETIC CODE IS READ AS A SERIES OF CODONS
The genetic code is redundant
17 There are (a)_____(#?) codons for about (b) _____(#?) different tRNAs.

A GENE IS DEFINED AS A FUNCTIONAL UNIT
18 A gene consists of regulatory sequences and the sequence of _____ that can be transcribed to yield a product with a specific cellular function.

MUTATIONS ARE CHANGES IN DNA
19 A mutation is a change in the _____ in DNA.

20 (a)_____ mutations involve a change in only one pair of nucleotides. These can lead to the substitution of one amino acid for another, a so called (b)_____ mutation or a (c)_____ mutation involving the conversion of an amino acid specifying codon to a termination codon.

21 _____ are regions of DNA that are particularly susceptible to mutations.

BUILDING WORDS
Use combinations of prefixes and suffixes to build words for the definitions that follow.

Prefixes	The Meaning	Suffixes	The Meaning
anti-	against, opposite of	-gen	production of
poly-	much, many	-some	body

Prefix	Suffix	Definition
_____	-codon	1. A sequence of three nucleotides in tRNA that is complementary to ("opposite of"), and combines with, the three-nucleotide codon on mRNA.
ribo-	_____	2. An organelle (microbody) composed of RNA and proteins that functions in protein synthesis.
_____	_____	3. A submicroscopic complex consisting of many ribosomes attached to an mRNA molecule during translation.
muta-	_____	4. A substance capable of producing mutations.
carcino-	_____	5. A substance capable of producing cancer.

MATCHING
For each of these definitions, select the correct matching term from the list that follows.

____ 1. Site on DNA to which RNA polymerase attaches to begin transcription.

____ 2. The making of mRNA from a DNA template.

____ 3. An RNA virus that produces a DNA intermediate in its host cell.

____ 4. A pyrimidine found in RNA.

____ 5. A triplet of mRNA bases that specifies an amino acid or a signal to terminate the polypeptide.

____ 6. The stage of translation during which amino acid chains are built.

____ 7. Chromosome abnormality in which part of one chromosome has become attached to another.

____ 8. The RNA responsible for base pairing during protein synthesis.

____ 9. In eukaryotes, the coding region of DNA.

____10. RNA that has been transcribed from DNA that specifies the amino acid sequence of a protein.

Terms:

a. Adenine
b. Codon
c. Elongation
d. Exon
e. Intron

f. mRNA
g. Promoter
h. Retrovirus
i. rRNA
j. Transcription

k. tRNA
l. Translocation
m. Uracil

TABLE
Fill in the blanks.

mRNA Codon	DNA Base Sequence	Amino Acid Specified	tRNA Anticodon
5' — UUU — 3'	3' — AAA — 5'	Phenylaline	3' — AAA — 5'
5' — AAA — 3'	#1	#2	#3
#4	#5	Tryptophan	#6
5' — GGA — 3'	#7	#8	3' — CCU — 5'
5' — GGU — 3'	#9	#10	3' — CCA — 5'
5' — AUG — 3'	3' — TAC — 5'	#11	3' — UTC — 5'
#12	#13	None (stop codon)	#14

THOUGHT QUESTIONS
Write your responses to these questions.

1. Describe each step in protein synthesis beginning with transcription and ending with a complete protein.
2. How are transcription and replication similar? How are they different?
3. How do ribosomes function in protein synthesis?
4. Diagram and explain the terms upstream, downstream, codes, codons, anticodons, initiation, chain elongation, and chain termination as they relate to transcription, translation, and protein synthesis.
5. Explain the nature of a point mutation and frameshift mutation, then describe how each might affect a protein that is coded for by the mutated portion of DNA.

MULTIPLE CHOICE
Place your answer(s) in the space provided. Some questions may have more that one correct answer.

_____ 1. The A site accepts
   a. aminoacyl-tRNA.   d. peptidyl transferase.
   b. coding sequence.   e. leader sequence.
   c. initiator.

_____ 2. The elongation stage of protein synthesis involves
   a. the code AUG.   d. adding amino acids to a growing polypeptide chain.
   b. initiation tRNA.   e. GTP.
   c. loading initiation tRNA on the large ribosomal subunit.

_____ 3. Ribosomes attach to
   a. 5' end of mRNA.   d. tRNA.
   b. 3' end of mRNA.   e. RNA polymerase.
   c. an mRNA recognition sequence.

_____ 4. Translation involves
   a. hnRNA.   d. copying codes into codons.
   b. decoding of codons.   e. copying codons into codes.
   c. adapter molecules.

_____ 5. Energy for transfer of a peptide chain from the A site to the P site is provided by
   a. ATP.   d. ADP.
   b. enzymes.   e. guanosine triphosphate.
   c. GTP.

_____ 6. Transcription involves
   a. mRNA synthesis.   d. copying codes into codons.
   b. decoding of codons.   e. peptide bonding.
   c. copying DNA information.

_____ 7. DNA-dependent RNA polymerases are enzymes that
   a. add groups of polynucleotides to mRNA.   d. recognize promoters on a gene.
   b. require a primer.   e. are present in all cells.
   c. use nucleoside triphosphates as substrates.

_____ 8. A recognition sequence for ribosome binding to mRNA is/are
   a. aminoacyl-tRNA.   d. upstream from the coding sequence.
   b. coding sequence.   e. leader sequence.
   c. initiator.

_____ 9. Which of the following is/are termination codons?
   a. UAG   d. GUA
   b. UUA   e. UGA
   c. UAA

_____10. An "adapter" molecule containing an anticodon and a region to which an amino acid is bonded is a/an
   a. aminoacyl-tRNA.   d. peptidyl transferase.
   b. coding sequence.   e. leader sequence.
   c. initiator.

_____11. Post-transcriptional modification and processing of mRNA takes place in
   a. cytoplasm.
   b. the nucleus.
   c. prokaryotes only.
   d. eukaryotes only.
   e. both prokaryotes and eukaryotes.

_____12. A string of dozens to hundreds of adenine nucleotides thought to prevent mRNA degradation is a description of a
   a. cap on the 5' end of mRNA.
   b. poly-A tail.
   c. sequence added to mRNA about one minute after transcription.
   d. portion of the activated ribosomal complex.
   e. cap on the 3' end of mRNA.

_____13. Introns probably function to
   a. separate structural domains of proteins.
   b. integrate special coding sequences.
   c. provide for recombination.
   d. organize assembly of amino acids in proteins.
   e. indirectly provide an evolutionary advantage.

_____14. Initiation of protein synthesis involves
   a. the code AUG.
   b. initiation tRNA.
   c. loading initiation tRNA on the large ribosomal subunit.
   d. adding amino acids to a growing polypeptide chain.
   e. peptidyl transferase.

VISUAL FOUNDATIONS
Color the parts of the illustration below as indicated.

RED	❑	formylated methionine tRNA
GREEN	❑	small ribosome subunit
YELLOW	❑	large ribosome subunit
BLUE	❑	P site
ORANGE	❑	A site
BROWN	❑	mRNA binding site
TAN	❑	circle the initiation complex

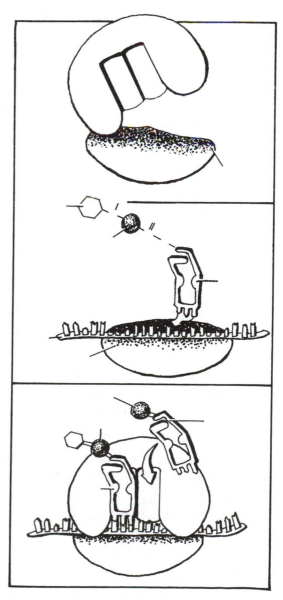

CHAPTER 13

Gene Regulation: The Control of Gene Expression

Genes are regulated and only part of the genetic information in any given cell is expressed. In prokaryotes, most regulated genes are organized into units called operons which may encode several proteins. The transcription of an operon is initiated at its promoter site, a sequence of DNA bases located upstream from the protein-coding region. Protein synthesis by the operon is controlled by the operator, a sequence of DNA bases also located upstream from the operon and overlapping part of the promoter site. Transcription is blocked when a repressor protein binds to the operator, thereby preventing RNA polymerase from binding to the promoter site. Inducible operons are normally turned off and are only turned on when a metabolite binds to the repressor protein so it cannot bind to the operator. Repressible operons are normally turned on and are only turned off when a metabolite binds to the operator and turns the operon off. Groups of operons may be organized into multigene systems called regulons that are controlled by a single regulatory gene. Constitutive genes are neither inducible nor repressible, but are active at all times. DNA-binding regulatory proteins recognize and bind to specific sequences of DNA bases, thereby inactivating functional groups of base pairs. The control of gene expression in prokaryotes also occurs during and after translation. Regulation of gene expression in eukaryotes occurs at the level of transcription, mRNA processing, translation, and the protein product. Eukaryotic transcription of both constitutive and inducible genes requires a promoter consisting of an RNA polymerase-binding site and short DNA sequences that affect the strength of the promoter. The rate of RNA synthesis of an inducible gene is increased by an enhancer, a regulatory element that can be located large distances away from the actual coding region of a gene. Some genes whose products are required in large amounts exist as multiple copies in the chromosome. Some genes are inactivated by changes in chromosome structure.

CHAPTER OUTLINE AND CONCEPT REVIEW
Fill in the blanks.

GENE REGULATION IN PROKARYOTES EMPHASIZES ECONOMY

1 _____ are genes that encode essential proteins that are in constant use.

Operons in prokaryotes permit coordinated control of functionally related genes

2 A group of functionally related genes may be controlled by one _____
_____ that is located upstream from the protein-coding region.

3 The _____ is a sequence of bases that switches mRNA synthesis "on" or "off."

4 Transcription of an operon is blocked when a _____ binds to the operator sequence and covers part of the promoter.

5 An (a)_____ inactivates a repressor in order to turn "on" a gene or operon. Such a system is called an (b)_____ system.

6 Repressible operons are turned "off" when a repressor is activated by binding to a product of the system called the _____.

7 A _____ is a set of operons that is controlled by one regulatory gene.

Some posttranscriptional regulation occurs in prokaryotes

8 _____ is when a product blocks its own production by binding to an enzyme that is required to generate that product.

GENE REGULATION IN EUKARYOTES IS MULTIFACETED

Eukaryotic transcription is controlled at many sites and by many different regulatory molecules

9 RNA polymerase in multicellular eukaryotes binds to a portion of the promoter known as the _____.

10 The efficiency of a eukaryotic promoter depends largely on the number and type of _____.

11 The structure of eukaryotic chromosomes varies with the degree of gene activity. Chromosomes containing inactive genes are called (a)_____, while chromosomes containing genes in an active state are referred to as (b)_____.

The long-lived, highly processed mRNAs of eukaryotes provide many opportunities for posttranscriptional control

12 As a result of _____ _____, the same gene that produces calcitonin in the thyroid gland can produce a neurotransmitter in the brain.

The activity of eukaryotic proteins may be altered by posttranslational chemical modifications

13 The addition or removal of phosphate groups is an example of _____, a mechanism used by eukaryotic cells to regulate protein activity.

BUILDING WORDS

Use combinations of prefixes and suffixes to build words for the definitions that follow.

Prefixes	The Meaning
co-	with, together, in association
eu-	good, well, "true"
hetero-	different, other
homo-	same

Prefix	Suffix	Definition
_____	-chromatin	1. Chromatin that appears loosely coiled. Because it is the only chromatin capable of transcription, it is considered to be the "good" or "true" chromatin.
_____	-chromatin	2. The inactive chromatin that appears highly coiled and compacted, and is generally not capable of transcription. (Chromatin that is "different from" or "other than" the euchromatin.)

_____   -repressor   3. A substance that, together with a repressor, represses protein
                                    synthesis in a specific gene.

_____   -dimer   4. A dimer in which the two component polypeptides are different.

MATCHING

For each of these definitions, select the correct matching term from the list that follows.

____ 1. One of the control regions of an operon.

____ 2. A site located on an enzyme that enables a substance other than the normal substrate to bind to the molecule, and to change the shape of the molecule and the activity of the enzyme.

____ 3. A regulatory protein that represses expression of a specific gene.

____ 4. A group of operons that are coordinately controlled.

____ 5. A regulatory mechanism that controls the rate at which a particular mRNA molecule is translated.

____ 6. Process by which multiple copies of a gene are produced by selective replication, thus allowing for increased synthesis of the gene product.

____ 7. Genes that are constantly transcribed.

____ 8. In prokaryotes, a group of structural genes that are coordinately controlled and transcribed as a single message, plus their adjacent regulatory elements.

____ 9. Regulatory elements that can be located long distances away from the actual coding regions of a gene.

____10. Control in which the presence of a substrate induces the synthesis of an enzyme.

Terms:

a. Allosteric binding site   f. Operator   k. Transcriptional control
b. Constitutive gene   g. Operon   l. Translational control
c. Enhancer   h. Regulon
d. Gene amplification   i. Repressible system
e. Inducible system   j. Repressor

TABLE
Fill in the blanks.

Gene Regulation Strategy	Prokaryotic or Eukaryotic Strategy
Operons	Prokaryotic
Emphasis on the control of transcription	#1
Use of preformed enzymes	#2
Emphasis on specificity of cell form and function	#3
Dominant theme of gene regulation is economy	#4
Rapid turnover of mRNA molecules	#5
Many genes are regulated after transcription	#6

THOUGHT QUESTIONS
Write your responses to these questions.

1. Cite examples of an inducible operon and a repressible operon and explain how each works.
2. Illustrate and describe translation of an inducible operon.
3. What are positive and negative controls? How do they regulate an operon?
4. Explain the functions of DNA-binding proteins in eukaryotes.
5. How is it that a given gene can sometimes cause different products to be produced in different cells?

MULTIPLE CHOICE
Place your answer(s) in the space provided. Some questions may have more that one correct answer.

_____ 1. The basic way(s) that cells control their metabolic activities is/are by
    a. regulating enzyme activity.   d. controlling the number of enzyme molecules.
    b. mutations.   e. transduction.
    c. developing different genes for different purposes.

_____ 2. Codes for repressor and activator proteins are
    a. "off" usually.   d. a constitutive gene.
    b. always "on".   e. sometimes "off," sometimes "on".
    c. facultative.

_____ 3. Feedback inhibition is an example of
    a. transcriptional control.   d. an inducible system.
    b. pretranscriptional control.   e. a repressible system.
    c. a control mechanism affecting events after translation.

_____ 4. The lactose repressor
    a. can convert to an operator.   d. becomes an activator when lactose is present.
    b. is several bases upstream from the operator.   e. is always "on".
    c. is downstream from RNA polymerase coding sequences.

_____ 5. The genes and genetic information in different cells of multicellular organisms are
    a. all slightly different.   d. identical.
    b. distinctly different.   e. almost all identical.
    c. identical in the same tissues, different in different tissues.

_____ 6. Duplication of specific genes in cells that need more of them is an example of
    a. magnification.   d. a positive control system.
    b. amplification.   e. a regulon.
    c. induction.

_____ 7. The lactose operon promoter
    a. contains a translation termination codon.   d. is the recognition site for three genes.
    b. is several bases downstream from the operator.   e. is sometimes covered by a repressor.
    c. is downstream from RNA polymerase coding sequences.

_____ 8. One would expect a gene involved in the manufacture of ATP to be
    a. "off" usually.   d. a constitutive gene.
    b. always "on".   e. sometimes "off," sometimes "on".
    c. facultative.

____ 9. The tryptophan operon in *E. coli* is
    a. usually on.
    b. on in the absence of corepressor.
    c. repressed only when tryptophan binds to repressor.
    d. an inducible system.
    e. a repressible system.

____10. UPEs
    a. are in DNA near promoters.
    b. are amino acids.
    c. seem to determine the strength of a repressor.
    d. are sequences of DNA bases.
    e. affect promoter activity.

____11. The lactose operon in *E. coli*
    a. contains three structural genes.
    b. is turned on by a repressor.
    c. contains an operator that overlaps a promoter.
    d. is transcribed as part of a single RNA molecule.
    e. is an inducible system.

____12. "Zinc fingers" seems to be involved in
    a. unwinding DNA.
    b. activating transcription.
    c. insertion of a regulator domain into grooves of DNA.
    d. binding a regulatory protein to DNA.
    e. blocking posttranscriptional events.

____13. Transcription of a gene is turned "on" instead of "off" in/when
    a. positive control systems.
    b. negative control systems.
    c. systems that sense a metabolic substrate.
    d. regulators stimulate transcription.
    e. regulons.

____14. A group of operons controlled by one regulator may be
    a. embedded in the lactose operon.
    b. a regulon.
    c. involved in nitrogen metabolism.
    d. responding to changing environments.
    e. a lactose repressor.

____15. A tightly coiled portion of DNA that contains inactive genes is
    a. called heterochromatin.
    b. called euchromatin.
    c. found in most eukaryotes and few prokaryotes.
    d. found only in prokaryotes.
    e. found only in eukaryotes.

VISUAL FOUNDATIONS

Color the parts of the illustration below as indicated.

RED	❑	operator region
GREEN	❑	promoter region
YELLOW	❑	repressor gene
BLUE	❑	mRNA
ORANGE	❑	inducer
BROWN	❑	inactivated repressor protein
TAN	❑	structural gene
PINK	❑	RNA polymerase

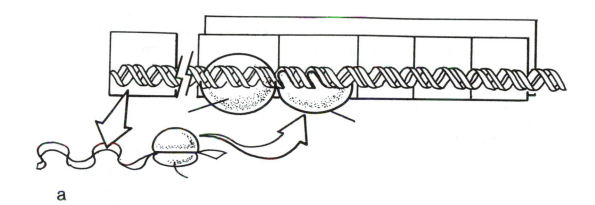

a

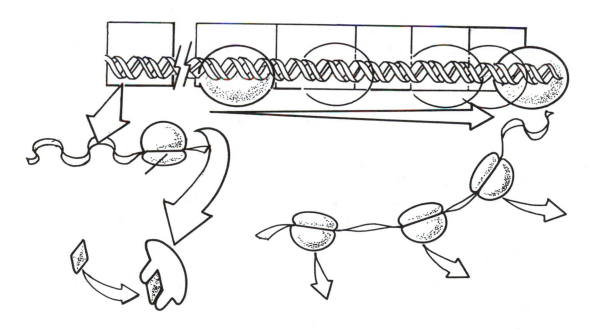

b

CHAPTER 14

Genetic Engineering

Recombinant DNA technology involves introducing foreign DNA into cells of microorganisms where it replicates and is transmitted to daughter cells. In this way, a particular DNA sequence can be amplified to provide millions of identical copies that can be isolated in pure form. Bacteria manufacture specific enzymes that cut DNA molecules at specific base sequences. The resulting fragments are incorporated into vector molecules, forming recombinant DNA molecules. In this manner, an entire library of recombinant molecules can be formed which contains all of the fragments of an organism's genome. In eukaryotes, libraries of recombinant molecules can also be formed by making copies of mRNA which can then be incorporated into vector molecules. The recombinant molecules can be cloned by introduction into, and multiplication of, host cells. Once a piece of DNA is cloned, its functions can be studied and possibly engineered for a particular application. DNA can also be amplified in vitro by means of the polymerase chain reaction. Once a DNA fragment is cloned, a restriction map of it is constructed which provides the information needed to isolate (subclone) smaller DNA fragments. The subcloned regions of the fragment can then be DNA sequenced or used as DNA probes for analytical purposes. Sequencing DNA enables identification of the parts that contain protein-coding and regulatory regions. Recombinant DNA technology has led to genetic engineering — the modification of the DNA of an organism to produce new genes with new characteristics. Engineered genes now produce improved pharmaceutical and agricultural products. Expression of eukaryotic genes in bacteria is often difficult because the gene must be linked to regulatory elements that the bacterium may not have. The expression of such genes in eukaryotic organisms is promising since eukaryotes already have the machinery to process and modify eukaryotic proteins. Transgenic organisms, i.e., plants and animals that have incorporated foreign genes, are useful in research and commerce.

CHAPTER OUTLINE AND CONCEPT REVIEW
Fill in the blanks.

RECOMBINANT DNA METHODS GREW OUT OF RESEARCH IN MICROBIAL GENETICS
Restriction enzymes are "molecular scissors" that cleave DNA reproducibly
1 A _____ sequence reads the same as its complement, but in the opposite direction.

2 _____ cut isolated DNA at specific loci, producing precise fragments that can be incorporated into vector molecules.

Recombinant DNA is formed when DNA is spliced into a vector (DNA carrier)
3 Recombinant DNA vectors are usually constructed from _____

Cloning techniques provide the means for replicating and isolating many copies of a specific recombinant DNA molecule
4 _____ are radioactive strands of RNA or DNA that are complementary to specific targeted cloned fragments. They are used to identify a particular fragment in a large population of clones.

5 A _____ is a population of recombinant plasmids that collectively contain a fragmented genome.

6 In order to avoid cloning introns, the enzyme _____ is used to create complementary DNA (cDNA) on a mRNA template.

The polymerase chain reaction is a technique for amplifying DNA in vitro

7 The polymerase chain reaction (PCR) technique for amplifying DNA is advantageous because it side-steps the cumbersome, time-consuming process of cloning DNA. PCR is an *in vitro* process that alternately uses (a)_____ to replicate DNA with the use of (b)_____ to dissociate the replicated DNA strands for further replications.

A cloned gene sequence is usually first analyzed by restriction mapping, followed by DNA sequencing

8 _____ are detailed "charts" that help identify the specific sites cut by selected restriction enzymes.

9 "Riflips," an acronym for _____ _____, are used to elucidate the degree of variability among genes in a population. "Riflips" are especially useful to study genetic relatedness of individuals.

GENETIC ENGINEERING HAS MANY APPLICATIONS

Additional engineering is required for a recombinant eukaryotic gene to be expressed in bacteria

10 Once a eukaryotic gene is incorporated into a bacterial genome, in order to produce the object product (an encoded protein), the gene must be linked to _____ _____ sequences that bacterial RNA polymerase can recognize.

Transgenic organisms have incorporated foreign DNA into their cells

11 Transgenic organisms (plants or animals) are generally produced by injecting the DNA of a gene into a recipient's (a)_____, or by using (b)_____ as recombinant DNA vectors.

SAFETY GUIDELINES HAVE BEEN DEVELOPED FOR RECOMBINANT DNA TECHNOLOGY

BUILDING WORDS
Use combinations of prefixes and suffixes to build words for the definitions that follow.

Prefixes	The Meaning
retro-	backward
trans-	across

Prefix	Suffix	Definition
_____	-genic	1. Pertains to organisms that have had their genome altered by recombinant DNA technology. (Genes are taken from one organism and transferred "across" into another organism.)
_____	-virus	2. An RNA virus that makes DNA copies of itself by reverse transcription.

MATCHING
For each of these definitions, select the correct matching term from the list that follows.

_____ 1. Enzyme produced by retroviruses to enable the transcription of DNA from the viral RNA in the host cell.

_____ 2. Small, circular DNA molecules that carry genes separate from the main bacterial chromosome.

_____ 3. Techniques that involve introducing normal copies of a gene into some of the cells of the body of a person afflicted with a genetic disorder.

_____ 4. A cluster of genetically-identical cells.

_____ 5. Bacterial enzymes used to cut DNA molecules at specific base sequences.

_____ 6. Method used to produce large amounts of DNA from tiny amounts.

_____ 7. An agent, such as a plasmid or virus, that transfers genetic information.

_____ 8. Any DNA molecule made by combining genes from different organisms.

_____ 9. A radioactively-labeled segment of RNA or single-stranded DNA that is complementary to a target gene.

_____10. A unit of measurement applied to the size of DNA fragments.

Terms:

a. Colony
b. DNA ligase
c. Gene therapy
d. Genetic probe
e. Kilobase

f. Plasmid
g. PCR
h. Recombinant DNA
i. Restriction enzyme
j. Restriction map

k. Reverse transcriptase
l. Vector

TABLE
Fill in the blanks.

Technique	Application
DNA blotting methods	Used in diagnosis and to identify carriers of some genetic diseases
#1	Can be used to estimate the degree of genetic relationship among individuals in a population
Genetic engineering	#2
#3	Method for replicating and isolating many copies of a specific recombinant DNA molecule
#4	Method used to detect a specific DNA sequence
#5	Used to separate DNA fragments on the basis of size
Restriction mapping	#6

THOUGHT QUESTIONS
Write your responses to these questions.

1. Explain how restriction enzymes are used to create recombinant DNA.
2. How are plasmids used to clone DNA?
3. What are DNA hybridization probes and what are some of their practical uses?
4. What are restriction maps and how are they used?
5. Explain why transgenic plants or animals might be preferred over *E. coli* when producing proteins coded for by eukaryotic DNA.
6. Sketch and explain a common DNA sequencing technique.

MULTIPLE CHOICE
Place your answer(s) in the space provided. Some questions may have more that one correct answer.

_____ 1. Creating a eukaryotic gene bank involves the use of
   a. eukaryotic DNA fragments.
   b. bacteriophages.
   c. plasmids treated with restriction enzymes.
   d. clones.
   e. identical plasmid and eukaryotic fragments.

_____ 2. Restriction enzymes are normally used by bacteria to
   a. defend against viruses.
   b. remove extraneous introns.
   c. denature antibiotics.
   d. cut plasmids from the large chromosome.
   e. insert operators in active DNA.

_____ 3. A restriction map of a DNA fragment involves
   a. plasmid markers.
   b. isolating "subfragments."
   c. reconstructing a gene from its fragments.
   d. identifying sites attacked by restriction enzymes.
   e. sizing "subfragments."

_____ 4. The Ti used to insert genes into plant cells is
   a. an RNA fragment.
   b. only effective in dicots.
   c. useful to increase yields of grain foods.
   d. a plasmid.
   e. a vector.

_____ 5. A base sequence that is palindromic to 3'-AGCTTAA-5' would read
   a. 3'-AGCTTAA-5'.
   b. 5'-TCGAATT-3'.
   c. 3'-TCGAATT-5'.
   d. 5'-UCGAAUU-3'.
   e. 3'-UCGAAUU-5'.

_____ 6. Restriction fragment length polymorphisms are
   a. complementary.
   b. RNA variants.
   c. the result of changes in DNA.
   d. possibly due to mutations.
   e. variants paired in order to obtain different clones.

_____ 7. The vector(s) commonly used to incorporate a DNA fragment into a carrier is/are
   a. prophages.
   b. *E. coli*.
   c. plasmids.
   d. translocation microphages.
   e. bacteriophages.

_____ 8. Palindromic sequences are base sequences from one DNA strand that
   a. produce RNA polymerase.
   b. form RNA complementary to DNA.
   c. can be cut by restriction enzymes.
   d. cut specific fragments from DNA.
   e. read in the opposite direction as its complement.

_____ 9. cDNA libraries are compiled
    a. from introns.   d. from introns and exons.
    b. DNA copies of mRNA.   e. with the use of reverse transcriptase.
    c. from multiple copies of DNA fragments.

_____ 10. When a virus is used as a vector in mammalian cells, it is first
    a. transformed.   d. disabled.
    b. stripped of foreign DNA.   e. coated with activating protein.
    c. incorporated into calcium phosphate crystals.

_____ 11. After "sticky ends" pair with other DNA molecules cut by the same enzyme, the two fragments can be covalently linked to form recombinant DNA by treating them with
    a. plasmozymes.   d. DNA ligase.
    b. restriction enzymes.   e. vector enzymes.
    c. DNA polymerase.

_____ 12. DNA fragments for recombination are produced by cutting DNA
    a. and introducing a cleaved plasmid.   d. with bacteriophages.
    b. with restriction enzymes.   e. with DNA ligase.
    c. at specific base sequences.

_____ 13. Obstacles that stand in the way of developing gene products of higher organisms in bacteria include
    a. obtaining eukaryotic fragments.   d. prokaryotic processing of eukaryotic introns.
    b. identifying a vector.   e. association of engineered gene with a promoter.
    c. association of engineered gene with a regulator.

_____ 14. RNA viruses that use reverse transcriptase to make DNA copies of themselves
    a. include the AIDS virus.   d. are the choice for plant vectors.
    b. are bacteriophages.   e. are retroviruses.
    c. can be used to eliminate troublesome introns.

_____ 15. Transgenic organisms are
    a. dead animals.   d. animals with foreign genes.
    b. vectors for retroviruses.   e. plants with foreign genes.
    c. used to produce recombinant proteins.

_____ 16. A radioactive strand of RNA or DNA that is used to locate a cloned DNA fragment is
    a. a genetic probe.   d. a restriction enzyme.
    b. the genome.   e. complementary to the ligase used.
    c. complementary to the targeted gene.

VISUAL FOUNDATIONS

Color the parts of the illustration below as indicated.

RED ☐ exon

GREEN ☐ intron

YELLOW ☐ reverse transcriptase

BLUE ☐ cDNA copy of mRNA

ORANGE ☐ complementary DNA copy of cDNA

PINK ☐ DNA polymerase

Also label DNA, pre-mRNA, and mRNA.

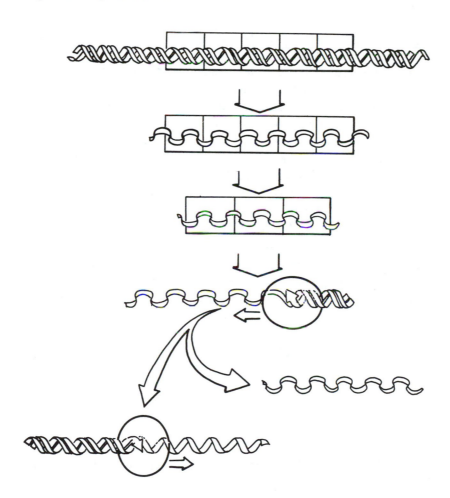

CHAPTER 15

Human Genetics

Human geneticists, unable to make specific crosses of pure strains, must rely on studies of populations, analyses of family pedigrees, and molecular studies. The development of human characteristics is regulated by a large number of interacting genes. Some birth defects are inherited; others are produced by environmental factors. Studies of the number and kinds of chromosomes in the nucleus permit detection of various chromosomal abnormalities, most of which are lethal or cause serious defects. Autosomal abnormalities are usually more severe than sex chromosome abnormalities. Abnormal alleles at a number of loci cause many inherited diseases. Most human genetic diseases with simple inheritance patterns are transmitted as autosomal recessive traits. Autosomal dominant disorders and x-linked recessive disorders also occur. Some genetic diseases can be diagnosed before birth by amniocentesis or chorionic villus sampling. Genetic counselors advise prospective parents with a history of genetic disease about the probability of their having an affected child. Gene replacement therapy is currently being actively researched. A great deal of the natural variation that exists in human beings has a genetic basis. Much of our knowledge of human variation is based on studies of ABO blood group and Rh alleles. Many human characteristics, such as height and intelligence, show continuous variation because a number of genes are involved in the expression of the characteristic. A major research project is underway to determine the total informational content of the human genome. Many people are misinformed about genetic diseases and their effect on society. A high frequency of one abnormal allele in a given group does not mean that the group has a higher frequency of abnormal alleles in general. Virtually everyone is heterozygous for several abnormal alleles. Alleles that cause a genetic disease when homozygous may confer an advantage when heterozygous. Matings of close relatives produce a greater frequency of abnormal offspring.

CHAPTER OUTLINE AND CONCEPT REVIEW
Fill in the blanks.

INTRODUCTION
1 Human geneticists can rarely make specific crosses of _____; that is, pure strains of individuals that are homozygous at nearly all loci.

ANALYSIS OF INHERITANCE PATTERNS IN HUMANS REQUIRES ALTERNATIVE METHODS

MOST HUMAN TRAITS RESULT FROM COMPLEX GENETIC AND ENVIRONMENTAL INTERACTIONS

SOME BIRTH DEFECTS ARE INHERITED

2 Defects that are present at birth are called birth defects, or _____ defects.

CHROMOSOMAL ABNORMALITIES ARE RESPONSIBLE FOR SOME BIRTH DEFECTS

3 _____ is the study of chromosomes and their role in inheritance.

Karyotyping is the analysis of chromosomes

4 The karyotype of an individual is a composite _____ that shows his/her chromosome composition.

Most chromosome abnormalities are lethal or cause serious defects

5 _____, the presence of multiple sets of chromosomes, is usually lethal in humans. ·

6 The abnormal presence or absence of one chromosome in a set is called (a)_____.
For the affected chromosome, the normal condition, (b)_____, becomes a
(c)_____ condition with the addition of an extra chromosome, and a
(d)_____ condition when one member of a pair is missing.

7 (a)_____ occurs when chromosomes do not separate at anaphase.
(b)_____ is an example of a condition that results from such an occurrence in chromosome 21.

Sex chromosome abnormalities are usually less severe than autosomal abnormalities

8 (a)_____ attempts to determine the gender of an individual based on the presence (female) or absence (male) of a (b)_____.

9 Persons with _____ are nearly normal males, except for an extra X chromosome and Barr bodies.

10 Persons with _____ are sterile females without Barr bodies.

Chromosome abnormalities are relatively common at conception but usually result in prenatal death

MOST GENETIC DISEASES ARE INHERITED AS AUTOSOMAL RECESSIVE TRAITS

11 PKU and akaptonuria are examples of disorders involving enzyme defects, collectively known as
_____.

Phenylketonuria (PKU) is due to an enzyme deficiency and can be treated with a special diet

Sickle cell anemia results from a hemoglobin defect

12 An infliction called _____ expresses itself in the form abnormal hemoglobin in RBCs.

Cystic fibrosis results from defective ion transport

13 _____ results from a defective protein, the one that evidently controls the transport of chloride ions across cell membranes.

Tay-Sachs disease is a result of abnormal lipid metabolism in the brain

HUNTINGTON'S DISEASE IS AN AUTOSOMAL DOMINANT DISORDER THAT AFFECTS THE NERVOUS SYSTEM

HEMOPHILIA A IS AN X-LINKED RECESSIVE DISORDER THAT AFFECTS BLOOD CLOTTING

SOME GENETIC ABNORMALITIES AND OTHER BIRTH DEFECTS CAN BE DETECTED BEFORE BIRTH

14 _____ may be used to diagnose prenatal genetic diseases. It analyzes cells in amniotic fluid withdrawn from the uterus of a pregnant woman.

15 One technique designed to detect prenatal genetic defects is _____
_____, which involves inspecting fetal cells involved in forming the placenta.

GENETIC COUNSELORS EDUCATE PEOPLE ABOUT GENETIC DISEASES AND GIVE THEM INFORMATION NEEDED TO MAKE REPRODUCTIVE DECISIONS

GENE REPLACEMENT THERAPY IS BEING DEVELOPED FOR SEVERAL GENETIC DISEASES

16 Some victims of _____, a disorder affecting the immune system, have been successfully treated by introducing normal bone marrow cells or injecting the enzyme ADA.

A GREAT DEAL OF NATURAL VARIATION EXISTS IN THE HUMAN POPULATION

Studies on blood contribute to our understanding of genetic diversity in humans

17 Genotypes (a)_____ produce blood type A, (b)_____ produce type B, (c)_____ produces AB, and (d)_____ results in type O.

18 _____ may result when a pregnant Rh-negative woman's sensitized WBCs are producing large quantities of anti-D antibodies.

Quantitative traits are controlled by polygenes

Many common physical characteristics are inherited

THE HUMAN GENOME INITIATIVE IS A SYSTEMATIC STUDY OF ALL HUMAN GENES

BOTH HUMAN GENETICS AND OUR BELIEFS ABOUT GENETICS HAVE AN IMPACT ON SOCIETY

19 Autosomal recessive genetic diseases appear with greater frequency in the offspring resulting from _____ than from random matings among individuals that are not related.

BUILDING WORDS

Use combinations of prefixes and suffixes to build words for the definitions that follow.

Prefixes	The Meaning
cyto-	cell
iso-	equal, "same"
poly-	much, many
trans-	across, beyond, through
tri-	three

Prefix	Suffix	Definition
_____	-genic	1. Pertaining to homozygosity (having the same alleles) at all loci.
_____	-genetics	2. The branch of biology that uses the methods of cytology to study genetics.
_____	-ploidy	3. The presence of multiples of complete chromosome sets.
_____	-somy	4. Condition in which a chromosome is present in triplicate instead of duplicate (i.e., the normal pair).
_____	-location	5. An abnormality in which a part of a chromosome breaks off and attaches to another chromosome.

MATCHING
For each of these definitions, select the correct matching term from the list that follows.

_____ 1. A technique used to detect birth defects by sampling the fluid surrounding the fetus.

_____ 2. Abnormal separation of homologous chromosomes or sister chromatids caused by their failure to disjoin properly during cell division.

_____ 3. Genetic disease that produces mental deterioration, painful paralysis, sensory loss and ultimately death.

_____ 4. Syndrome in which afflicted individuals have only 47 chromosomes.

_____ 5. Traits that represent some measurable quantity, such as height.

_____ 6. Birth defect caused by an extra copy of human chromosome 21.

_____ 7. Normal condition in which there are two of each kind of chromosome.

_____ 8. General term for the abnormal situation in which one member of a pair of chromosomes is missing.

_____ 9. Matings of close relatives.

_____10. Abnormality involving the presence of an extra chromosome or the absence of a chromosome.

Terms:

a. Amniocentesis	f. Huntington's Disease	k. Nuclear sexing
b. Aneuploidy	g. Karyotype	l. Quantitative trait
c. Consanguineous mating	h. Klinefelter Syndrome	
d. Disomic	i. Monosomic	
e. Down Syndrome	j. Nondisjunction	

TABLE
Fill in the blanks.

Genetic Disease	Method of Transmission	Abnormality	Clinical Description
Sickle cell anemia	Autosomal recessive trait	Valine substitutes for glutamic acid in hemoglobin	Abnormal RBCs block small blood vessels.
#1	#2	Membrane proteins that transport chloride ion malfunction	Characterized by abnormal secretions in respiratory and digestive systems
Tay-Sachs	#3	#4	Blindness, severe retardation
Huntington's disease	#5	#6	Ultimately causes insanity
#7	Autosomal trisomy	#8	Retardation, abnormalities
#9	#10	Lack of blood clotting Factor VIII	Severe bleeding from even slight wounds
PKU	Autosomal recessive trait	#11	Retardation caused by toxic phenylketones

THOUGHT QUESTIONS
Write your responses to these questions.

1. What is nondisjunction? How can it lead to abnormalities in offspring? Are such abnormalities considered to be inherited or environmentally induced? Explain your answer.
2. How are chorionic villus sampling and amniocentesis used in prenatal evaluations?
3. If a pregnant mother already has two children, both less than two years old, what are the possible consequences to the next child under the following circumstances:

Father	Mother
Rh negative	Rh negative
Rh positive	Rh negative
Rh negative	Rh positive
Rh positive	Rh positive

4. What benefits have already been realized from the "Human Genome Initiative?" What further benefits might accrue?

MULTIPLE CHOICE
Place your answer(s) in the space provided. Some questions may have more that one correct answer.

_____ 1. An ideal organism for genetic studies would be one that
    a. is polygenic.
    b. is heterozygous.
    c. has a short generation time.
    d. produces one offspring per mating.
    e. can be controlled.

_____ 2. Consanguineous matings produce more abnormal offspring because of the increased possibility of
    a. heterozygosity.
    b. dominant homozygosity.
    c. recessive homozygosity.
    d. nondisjunctions.
    e. monosomy.

_____ 3. If a couple has one child with cystic fibrosis, their chance of having another affected child is
    a. 10%.
    b. 25%.
    c. 50%.
    d. 75%.
    e. almost 100%.

_____ 4. A person with Down syndrome will likely
    a. have 47 chromosomes.
    b. have trisomy 21.
    c. be mentally retarded.
    d. be born of a mother in her teens.
    e. have a father with Down syndrome.

_____ 5. The study of human genetics relies mainly on
    a. pedigrees.
    b. Punnett squares.
    c. fetal karyotypes.
    d. distribution of a trait in a population.
    e. nondisjunction.

_____ 6. Amniocentesis involves
    a. examination of maternal blood.
    b. insertion of a needle into the uterus.
    c. examination of fetal cells.
    d. examination of karyotypes.
    e. appraisal of paternal abnormalities.

_____ 7. A person with Klinefelter syndrome will
    a. have small testes.
    b. be female.
    c. have 45 chromosomes.
    d. have two X chromosomes.
    e. have a Barr body.

_____ 8. Congenital defects
- a. are acquired in infancy.
- b. may be inherited.
- c. may be produced by the environment.
- d. are present at birth.
- e. are only expressed when alleles are homozygous.

_____ 9. Translocations may involve
- a. loss of genes.
- b. acquisition of extra genes.
- c. duplication of genes.
- d. loss of a chromosome.
- e. acquisition of an extra chromosome.

_____10. A trisomic individual
- a. is missing one of a pair of chromosomes.
- b. may result from nondisjunction.
- c. has at least three X chromosomes.
- d. is a male.
- e. has an extra chromosome.

_____11. Nondisjunction of one chromosome in the second meiotic division would result in which of the following combination(s) of numbers of chromosomes and ratios in human gametes?
- a. 1-24:1-22
- b. 1-22:2-23:1-24
- c. all 23
- d. all 22
- e. 1-23:1-22

_____12. The most accurate statement about the frequency of abnormal genes is that they are found in
- a. certain ethnic groups.
- b. certain cultural groups.
- c. college professors.
- d. college students.
- e. everyone.

Visual Foundations on next page ➤

VISUAL FOUNDATIONS
Questions 1-3 show pedigrees of the type that are used to determine patterns of inheritance. Use the following list to answer questions 1-3:

        a. x-linked recessive
        b. x-linked dominant
        c. autosomal dominant
        d. autosomal recessive
        e. cannot determine the pattern of inheritance

1. What is the pattern of inheritance for the following pedigree?

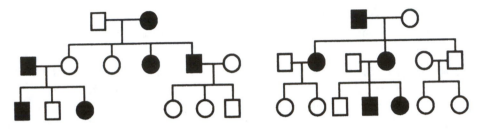

2. What is the pattern of inheritance for the following pedigree?

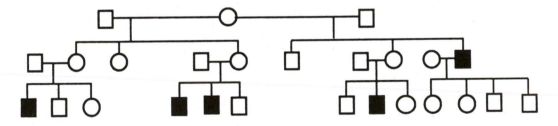

3. What is the pattern of inheritance for the following pedigree?

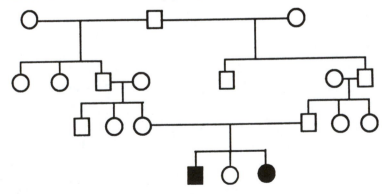

CHAPTER 16

Genes and Development

Development is the process by which cells specialize and organize into a complex organism. Groups of cells become gradually committed to specific patterns of gene activity and progressively organized into recognizable structures. Cellular specialization is not due to the loss of genes during development, but rather to differential gene activity. A variety of organisms and methodologies are used today to identify genes that control development and to determine how those genes work. The earliest stages of fruit fly development are controlled by maternal genes. Then, following 13 cell divisions, the genes in the developing embryo begin to be expressed, extending the developmental program beyond the pattern established by the maternal genome. Still later-acting genes are responsible for specifying the identity of each body segment. Many of these genes contain a sequence of 180 DNA base pairs; homologous sequences have been found in a wide range of other organisms, including humans, and are thought to control development. Early development in the nematode worm is very rigid; if a particular cell is destroyed or removed, the structures that would normally develop from that cell are missing. Genes have been identified in the nematode that control developmental processes, such as interactions with neighboring cells, programmed cell death, and developmental timing. Mouse development, in contrast to that of the nematode, is not rigid. An embryo with extra or missing cells will still develop normally. Foreign DNA injected into fertilized mouse eggs can be incorporated into the chromosomes and expressed. The resulting mice provide information about gene activation during development. Genes affecting development have also been identified in certain plants. Although uncommon, one type of developmental regulation involves structural changes in the DNA that result in new coding sequences, an important mechanism for the development of the immune system. Another uncommon type of regulation involves making multiple copies of DNA sequences that code for products in high demand.

CHAPTER OUTLINE AND CONCEPT REVIEW
Fill in the blanks.

INTRODUCTION

1 Development involves cell specialization, a developmental process involving a gradual commitment by each cell to a specific pattern of gene activity. This process is called
(a)_____, and the final step in the process is referred to as
(b)_____.

2 The distinctive organizational pattern, or form, that characterizes an organism is acquired through a process known as (a)_____. It encompasses a series of steps known as
(b)_____, by which groups of cells organize into identifiable structures.

CELLULAR DIFFERENTIATION USUALLY DOES NOT INVOLVE CHANGES IN DNA

3 A body of data supports the assertion that, in at least some organisms, almost all nuclei of differentiated cells are identical to each other and to the nucleus of the single cell from which they are descended, which has given rise to the concept of _____.

4 _____ cells are fully differentiated, yet, since they can still develop, they apparently still contain all of the original genetic equipment of their common single ancestral cell.

A totipotent nucleus contains all the information required to direct normal development

Most differences among cells are due to differential gene expression

MOLECULAR GENETICS IS REVOLUTIONIZING THE STUDY OF DEVELOPMENT

Certain organisms are particularly well suited for studies on the genetic control of development

DROSOPHILA MELANOGASTER PROVIDES RESEARCHERS WITH A WEALTH OF DEVELOPMENTAL MUTANTS

5 Large, multistranded interphase chromosomes called _____ chromosomes result when DNA replicates repeatedly without mitosis occurring.

6 _____ are "decondensed" segments of multistranded chromosomes that are actively synthesizing RNA.

The *Drosophila* life cycle includes egg, larval, pupal, and adult stages

Many *Drosophila* developmental mutants affect the body plan

7 The initial stages of development in *Drosophila* are under the control of _____ genes.

8 Once maternal genes have initiated pattern formation in the embryonic fruit fly, a set of (a)_____ begin to act on development. Among these are the (b)_____ that begin to organize segments and the (c)_____ _____ that specify the ultimate identity of each imaginal disc and segment.

9 The _____ is a short DNA sequence that is consistently found in association with known developmental genes. It is useful as a probe for locating and cloning developmental mutants.

CAENORHABDITIS ELEGANS IS A ROUNDWORM WITH A VERY RIGID EARLY DEVELOPMENTAL PATTERN

10 The fates of cells in adult roundworms are predetermined by a few _____ _____ that form in the early embryo.

11 Development in *Caenorhabditis elegans* is rigidly fixed in a predetermined pattern; that is, specific cells throughout the early embryo are already irrevocably "programmed" to develop in a particular way. An embryo with this pattern of development is said to be _____.

THE MOUSE IS A MODEL FOR MAMMALIAN DEVELOPMENT

Cells of very early mouse embryos are totipotent

12 Early mice embryos can be fused to produce a _____, a term for offspring that exhibit a diversity of characteristics derived from two or more kinds of genetically dissimilar cells from different zygotes.

13 A self-regulating embryo exhibits a form of development referred to as _____. Such an embryo can still develop normally when it has extra cells or missing cells .

Transgenic mice are used in studies on developmental regulation

HOMEOTIC-LIKE MUTATIONS OCCUR IN PLANTS

SOME EXCEPTIONS TO THE PRINCIPLE OF NUCLEAR EQUIVALENCE HAVE BEEN FOUND

Genomic rearrangements involve structural changes in the DNA

14 _____ are physical changes in gene structure that affect gene activity as, for example, in cases of recombination of coding sequences or replacement of an active gene with a "silent" gene.

Gene amplification increases the number of copies of specific genes

THE STUDY OF DEVELOPMENTAL BIOLOGY PRESENTS MANY FUTURE CHALLENGES

BUILDING WORDS
Use combinations of prefixes and suffixes to build words for the definitions that follow.

Prefixes	The Meaning		Suffixes	The Meaning
chrono-	time		-gen(esis)	production of
morpho-	form			
poly-	much, many			

Prefix	Suffix	Definition
_____	_____	1. The development and differentiation (production) of the form and structures of the body during embryonic development.
_____	-tene	2. Pertains to a large, "many-stranded" chromosome.
_____	-gene	3. A gene that is involved in the timing of development.

MATCHING
For each of these definitions, select the correct matching term from the list that follows.

____ 1. An organism that has incorporated foreign DNA into its genome.

____ 2. Mass of tissue that grows in an uncontrolled manner.

____ 3. Development toward a more mature state; a process changing a young, relatively unspecialized cell to a more specialized cell.

____ 4. The ability of a cell (or nucleus) to provide the information necessary for the development of an entire organism.

____ 5. A phenomenon in which differentiation of a cell is influenced by interactions with particular neighboring cells.

____ 6. A gene that controls the formation of specific structures during development.

____ 7. Specific DNA sequence found in many genes that are involved in controlling the development of the body plan.

____ 8. The progressive limitation of a cell line's potential fate during development.

____ 9. Any of a number of genes that usually play an essential role in cell growth or division, and that cause the formation of a cancer cell when mutated.

____10. A body cell not involved in reproduction.

Terms:

a. Determination
b. Differentiation
c. Homeobox
d. Homeotic gene
e. Induction

f. Oncogene
g. Somatic cell
h. Stem cell
i. Totipotent
j. Transgenic

k. Tumor
l. Zygotic gene

TABLE
Fill in the blanks.

Germ Layer	Tissue	Specialized Cell
Ectoderm	Epidermis	Skin cell
#1	#2	Neuron
#3	#4	Egg and sperm
#5	Gonad	Cells other than those involved in gamete production
#6	#7	Striated muscle cell
Endoderm	Pancreas	#8
#9	Urinary bladder	#10
Mesoderm	#11	Red blood cell

THOUGHT QUESTIONS
Write your responses to these questions.

1. Explain the relationship between totipotency and nuclear equivalence.
2. How are transgenic organisms useful in studies of the genetic control of development?
3. What do cellular determination, cellular differentiation, and pattern formation mean and how are they related?
4. Explain the differences between maternal effect genes, zygotic genes, and homeotic genes.
5. What is programmed cell death and what role, if any, does it play in development?

MULTIPLE CHOICE
Place your answer(s) in the space provided. Some questions may have more that one correct answer.

_____ 1. Morphogens are
    a. hypothetical chemicals.
    b. derived from a homeobox.
    c. determined by embryonic segmentation genes.
    d. thought to exist in *Drosophila* eggs.
    e. unraveled portions of active DNA.

_____ 2. If destruction of a single cell early in development results in the absence of a structure in the adult, the embryo of that organism
a. is mosaic.
b. is a fruit fly.
c. contains predetermined cells.
d. contained founder cells.
e. is controlled to some extent by homeotic genes.

_____ 3. A group of cells grown in liquid medium from a single root cell is
a. formed by a predetermined pattern.
b. a mosaic.
c. totipotent.
d. predetermined.
e. an embryoid.

_____ 4. A parasitic organism that can defeat its host's immune system by switching glycoproteins in the host cell surfaces probably has
a. genetically engineered genes.
b. mosaic development.
c. a regulative development pattern.
d. founder cells that give rise to adult parts.
e. many genomic rearrangements.

_____ 5. An embryo with an invariant developmental pattern in which the fate of cells is predetermined is said to be
a. induced.
b. mosaic.
c. homeotic.
d. transgenic.
e. differentiated.

_____ 6. Embryos that act as a self-regulating whole and can accommodate missing parts
a. are transgenic.
b. are mosaics.
c. have regulative development patterns.
d. have founder cells that give rise to adult parts.
e. contain many genomic rearrangements.

_____ 7. Gap genes in *Drosophila*
a. are a type of segmentation gene.
b. act on all embryonic segments.
c. are the first zygotic genes to act.
d. are complementary to maternal genes.
e. determine segment polarity.

_____ 8. Genes in *Drosophila* that designate the adult structure formed by an imaginal disc are
a. gap genes.
b. segmentation genes.
c. maternal genes.
d. homeotic genes.
e. morphogens.

_____ 9. The concept of nuclear equivalence holds that
a. adult cells are identical to one another.
b. embryonic cells are identical to one another.
c. adult somatic nuclei are identical to the nucleus of the fertilized egg.
d. adult cells are identical to egg cells.
e. adult somatic nuclei are identical.

_____10. The process by which developing cells become committed to a particular pattern of gene activity is
a. determination.
b. differentiation.
c. transgenesis.
d. morphogenesis.
e. the first step in cell specialization.

_____11. If differentiated cells can be induced to act like embryonic cells, the cells are
a. formed by a predetermined pattern.
b. mosaics.
c. totipotent.
d. predetermined.
e. embryoids.

_____12. The process by which the differentiation of an embryonic cell is controlled or influenced by other embryonic cells is
a. transformation.
b. induction.
c. mosaic development.
d. an example of programmed cell death.
e. determinate morphogenesis.

_____13. Pattern formation is a series of steps leading to
a. determination.
b. differentiation.
c. predetermination.
d. morphogenesis.
e. the final step in cell specialization.

_____14. Differences in the molecular composition of different cells in a multicellular organism are due to
a. loss of genes during development.
b. transformation.
c. regulation of the activities of different genes.
d. nuclear equivalency.
e. differential gene activity.

_____15. Polytene chromosomes
a. form imaginal discs.
b. are formed during interphase.
c. are found in the nematode _Caenorhabditis elegans_.
d. form by DNA replication without mitosis.
e. result from fusion of two or more chromosomes.

PART IV

Evolution

CHAPTER 17

Evolution: Mechanisms and Evidence

All organisms that exist today evolved from earlier organisms. Evolution is a genetic change in a population over many generations. The concept of evolution is the cornerstone of biology; biologists do not question its occurrence. Although ideas about evolution existed before Charles Darwin, it remained for Darwin to discover natural selection — the actual mechanism of evolution. According to this mechanism, each species produces more offspring than will survive to maturity, genetic variation exists among the offspring, organisms compete for resources, and individuals with the most favorable characteristics are most likely to survive and reproduce. Favorable characteristics are thus passed to succeeding generations. Over time, changes occur in the gene pools of geographically separated populations that may lead to the evolution of new species. A subsequent modification of Darwin's theory, the synthetic theory of evolution, explains Darwin's observation of variation among offspring in terms of mutation and recombination. The concept of evolution is supported by an enormous body of scientific observations and experiments. The most direct evidence supporting evolution comes from the fossil record. Additional evidence comes from studies of comparative anatomy, embryology, and biogeography. Especially compelling evidence supporting evolution comes from studies of biochemistry and molecular biology.

CHAPTER OUTLINE AND CONCEPT REVIEW
Fill in the blanks.

INTRODUCTION

1 Evolution means change, or as Darwin described it with respect to organisms, in broad, general terms it is "_____."

2 The genetic changes that bring about evolution do not occur in individual organisms, but rather in _____.

3 All of the genes in a population are referred to as the _____ for that population.

IDEAS ABOUT EVOLUTION ORIGINATED BEFORE DARWIN

4 _____ proposed a theory of evolution in 1809, coincidentally in the year of Darwin's birth. His theory held that traits *acquired* during an organism's lifetime could be passed along to their offspring.

DARWIN WAS INFLUENCED BY HIS CONTEMPORARIES

5 Charles Darwin served as the naturalist on the ship (a)_____ during its five year cruise around the world. Central to the theory Darwin would propose were data about the similarities and differences among organisms in the (b)_____, a chain of islands west of mainland Ecuador.

6 In the mid-1800s, most believed that organisms did not change, but some understood the significance of variations among domestic animals that were intentionally induced by breeders, a process called _____.

7 _____ made the important mathematical observation that populations increase in size geometrically until checked by factors in the environment.

DARWIN PROPOSED THAT EVOLUTION OCCURS BY NATURAL SELECTION

8 In 1858, both Darwin and _____ arrived at the conclusion that evolution occurred by natural selection.

9 Darwin published his monumental book, _____
_____ in 1859.

10 Darwin's mechanism of natural selection consists of four premises based on observations about the natural world. In summary, these are: _____
_____

THE SYNTHETIC THEORY OF EVOLUTION COMBINES DARWIN'S THEORY WITH GENETICS

11 The synthetic theory of evolution explains variation in terms of _____
_____, both of which are inheritable and therefore potential explanations for *how* traits are passed from one generation to another.

MANY TYPES OF SCIENTIFIC EVIDENCE SUPPORT EVOLUTION
The fossil record indicates that evolution occurred in the past

12 _____ are the remains or traces of organisms that were preserved in large numbers over a relatively short geological time. They are used to identify specific sedimentary layers and to arrange strata in chronological order.

Comparative anatomy of related species demonstrates similarities in their structures

13 _____ organs have basic structural similarities, even though the organs may be used in different ways. They indicate evolutionary ties between the organisms possessing them.

14 _____ organs have similar functions, whether or not structurally similar. They do not indicate close evolutionary ties.

15 The occasional presence of "remnant organs" in organisms is to be expected as ancestral species evolve and adapt to different modes of life. Such organs or parts of organs, called _____
_____, are degenerate and have no apparent function.

Related species have similar patterns of embryological development
The distribution of plants and animals supports evolution

16 The study of the distribution of plants and animals is _____.

17 It is generally assumed that each species originated only once at a location called its _____
_____. New species spread out from that point until halted by a barrier of some kind.

Molecular comparisons among organisms provide evidence for evolution

BUILDING WORDS
Use combinations of prefixes and suffixes to build words for the definitions that follow.

Prefixes	The Meaning
bio-	life
homo-	same

Prefix	Suffix	Definition
_____	-logous	1. Pertains to having a similarity of form due to having the same evolutionary origin.
_____	-geography	2. The study of the geographical distribution of living things.

MATCHING

For each of these definitions, select the correct matching term from the list that follows.

____ 1. The ability of an organism to adjust to its environment.

____ 2. The independent evolution of structural or functional similarity in two or more organisms of widely-different, unrelated ancestry.

____ 3. Parts or traces of an ancient organism preserved in rock.

____ 4. The period of time required for one-half of the atoms of a radioisotope to change into a different atom.

____ 5. A unified explanation of evolution combining Darwin's theory and Mendelian genetics.

____ 6. Rudimentary; an evolutionary remnant of a formerly functional structure.

____ 7. The portion of the Earth in which a particular species occurs.

____ 8. A fossil that is restricted to a narrow unit of time and is found in the same sedimentary layers in different geographical areas.

____ 9. The genetic change in a population of organisms that occurs over time.

____ 10. Similar in function or appearance, but not in origin or development.

Terms:

a. Adaptation
b. Analogous
c. Artificial selection
d. Convergent evolution
e. Evolution

f. Fossil
g. Gene pool
h. Half-life
i. Index fossil
j. Natural selection

k. Range
l. Vestigial

TABLE

Fill in the blanks.

Evidence for Evolution	How the Evidence Supports Evolution
Fossils	Direct evidence in the form of remains of ancient organisms
#1	Indicate evolutionary ties between organisms that have basic structural similarities, even though the structures may be used in different ways
#2	Indicate that organisms with different ancestries can adapt in similar ways to similar environmental demands
Vestigial organs	#3
#4	Areas of the world that have been separated from the rest of the world for a long time have organisms unique to that area
#5	Evidence that all life is related

THOUGHT QUESTIONS
Write your responses to these questions.

1. What specific data does the fossil record contribute to an understanding of evolution? Comparative anatomy? Developmental biology? Molecular biology? Biogeography?
2. What does the word evolution mean literally? Describe three theories (or notions) that attempt to explain how evolution takes place. What are the basic differences in the theories about organic evolution?
3. Compare and contrast Darwin's proposal for evolution with "synthetic evolution."
4. Summarize the theory of evolution as proposed by Darwin.

MULTIPLE CHOICE
Place your answer(s) in the space provided. Some questions may have more that one correct answer.

____ 1. If phylogenetic trees were to be constructed independently from molecular characteristics and from morphological characteristics, the two trees would likely be
a. basically dissimilar.   d. identical.
b. totally dissimilar.   e. misleading.
c. basically similar.

____ 2. The first comprehensive theory of evolution, which proposed that the mechanism was an inner drive for improvement, was proposed by
a. Aristotle.   d. Wallace.
b. Lamarck.   e. da Vinci.
c. Darwin.

____ 3. The first to formally propose that organisms change over time as the result of natural phenomena rather than divine intervention was
a. Aristotle.   d. Wallace.
b. Lamarck.   e. Lyell.
c. Darwin.

____ 4. That tulip trees are found only in the eastern United States, Japan, and China is probably due to
a. a past land bridge.   d. long range dispersal.
b. similar climates.   e. two centers of origin.
c. migrations.

____ 5. The person(s) who suggested that populations increase geometrically and that their numbers must eventually be checked was
a. Malthus.   d. Wallace.
b. Lamarck.   e. Lyell.
c. Darwin.

____ 6. The individuals who simultaneously concluded that organisms evolve as nature selects the fittest were
a. Aristotle.   d. Wallace.
b. Lamarck.   e. Lyell.
c. Darwin.

____ 7. Organs of different organisms that have a similar form due to a common origin are
a. analogous.   d. usually vestigial.
b. artifacts.   e. found in fossils but never in contemporary organisms.
c. homologous.

_____ 8. That Earth's physical features were formed over long periods of time by a series of gradual changes was proposed by

a. Aristotle.

b. Lamarck.

c. Darwin.

d. Wallace.

e. Lyell.

_____ 9. The humerus in a bird and human are

a. analogous but not homologous.

b. homologous but not analogous.

c. neither analogous nor homologous.

d. both analogous and homologous.

e. probably derived from a common ancestor.

_____10. The first to propose that the mechanism for the evolution of organisms is by natural selection was

a. Malthus.

b. Lamarck.

c. Darwin.

d. Wallace.

e. Lyell.

_____11. The wings of a butterfly and a bat are

a. analogous but not homologous.

b. homologous but not analogous.

c. neither analogous nor homologous.

d. both analogous and homologous.

e. probably derived from a common ancestor.

_____12. Evolution of organisms occurs by means of

a. changes in individual organisms.

b. uniformitarianism.

c. changes in gene frequencies in the gene pool.

d. extinctions.

e. changes in populations.

_____13. If Darwin had known about and employed current biochemical techniques, it most likely would have resulted in

a. a different theory of evolution.

b. no theory of evolution.

c. his conviction of immutable species.

d. confirmation of his theories.

e. an argument with Wallace.

Visual Foundations on next page ➤

VISUAL FOUNDATIONS
Color the parts of the illustration below as indicated.

RED ☐ humerus

GREEN ☐ phalanges

YELLOW ☐ radius

BLUE ☐ carpels

ORANGE ☐ ulna

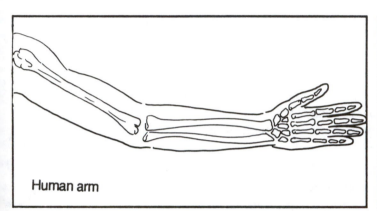

Human arm

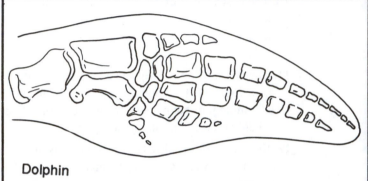

Dolphin

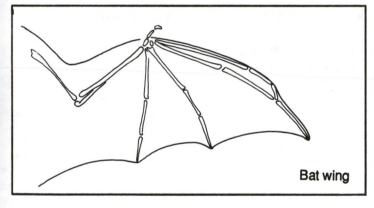

Bat wing

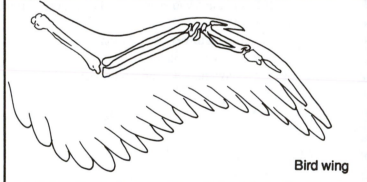

Bird wing

Population Genetics

The gene pool of a population of organisms consists of all possible alleles at each locus of each chromosome in the population. Evolution has occurred when changes occur in allele frequencies in the gene pool over successive generations. Allele frequencies may change by mutation, genetic drift, gene flow, and natural selection. Mutations contribute new genes to the gene pool. Genetic drift causes changes in allele frequencies due to chance events that result in a marked decrease in the number of individuals. Genetic drift also occurs when one or a few individuals extend beyond the population's normal range and establish a new colony. Gene flow, caused by migration of individuals between different populations of the same species, causes a corresponding movement of alleles. Natural selection checks the random effects of mutation, genetic drift, and gene flow, and leads to a change in allele frequencies that bring the population into harmony with the environment. Natural selection is the most significant factor in changing allele frequencies in a population. It involves stabilizing, directional, and disruptive selection — processes that cause changes in the normal distribution of phenotypes in a population. Genetic variation in a gene pool is the raw material for evolutionary change. The gene pools of most populations contain a large reservoir of variability. When the heterozygote has a higher degree of fitness than either homozygote, both alleles tend to be maintained in the population. Often genetic variation is maintained in populations of prey by frequency-dependent selection. The predator catches and consumes the commoner phenotype, while ignoring the rarer phenotypes. Thus, frequency-dependent selection acts to decrease the frequency of the more common phenotypes and increase the frequency of the less common types. Some of the genetic variation seen in a population confers no selective advantage or disadvantage to the individuals possessing it.

CHAPTER OUTLINE AND CONCEPT REVIEW
Fill in the blanks.

INTRODUCTION

1 The type of evolution involving changes *within* a population is referred to as

_____.

THE HARDY-WEINBERG PRINCIPLE DEMONSTRATES GENETIC EQUILIBRIUM

2 Mendelian genetics predicts the frequency of genotypes expected among offspring from one mating, whereas the Hardy-Weinberg law describes the frequencies of genotypes in an entire _____. The Hardy-Weinberg law states that gene frequencies in a population tend to remain constant in successive generations unless certain factors are operating.

3 When the distribution of genotypes in a population conforms to the binomial equation
(a)_____, the population is in a genetic equilibrium and is not evolving. This equilibrium is called the (b)_____.

4 The proportion of alleles in successive generations does not change in a population when certain conditions are met. Among them, briefly stated, are the following five: _____
_____
_____.

EVOLUTION OCCURS WHEN ALLELE FREQUENCIES UNDERGO CHANGE IN A GENE POOL

Mutation increases variation within a gene pool

5 _____ is the ultimate source of all new alleles.

Genetic drift causes changes in allele frequencies by random, or chance events

6 Genetic drift is the random change in _____ of a small breeding population. The changes are usually not adaptive.

7 Genetic drift may occur when the size of a population is suddenly reduced as a function of such temporary causes as a depleted food supply or disease. Such an event is called a _____.

8 Genetic drift may occur among a few individuals that have established a colony outside the usual range of their population. This is known as the _____.

Gene flow, caused by migration, generally increases variation in the gene pool

9 Occasionally, individuals from one local population migrate to another localized population, resulting in a movement of alleles between populations, a phenomenon called _____.

Natural selection changes allele frequencies in a way that increases adaptation

10 The most important reason for adaptive changes in gene frequencies is _____.

SELECTION OPERATES ON AN ORGANISM'S PHENOTYPE

Stabilizing selection favors intermediate phenotypes
Directional selection favors one phenotype over another
Disruptive selection selects for phenotypic extremes

GENETIC VARIATION IS NECESSARY IF EVOLUTION IS TO OCCUR

Genetic polymorphism is an example of variation

11 _____ is the presence of more than one allele for a given locus.

Balanced polymorphism can exist for long periods of time

12 Frequency-dependent selection decreases the frequency of (a)_____ phenotypes and their respective genotypes and increases the frequency of (b)_____ types.

Neutral mutations give no selective advantage or disadvantage

BUILDING WORDS
Use combinations of prefixes and suffixes to build words for the definitions that follow.

Prefixes	The Meaning
micro-	small

Prefix	Suffix	Definition
_____	-evolution	1. Changes in allele frequencies over successive generations; involves small changes within a population.

MATCHING
For each of these definitions, select the correct matching term from the list that follows.

_____ 1. Natural selection that acts against extreme phenotypes and favors intermediate variants.

_____ 2. A phenomenon in which the heterozygous condition confers some special advantage on an individual that either homozygous condition does not.

_____ 3. A random change in gene frequency in a small, isolated population.

_____ 4. The movement of alleles between local populations, or demes, due to migration and subsequent interbreeding.

_____ 5. The gradual replacement of one phenotype with another due to environmental change.

_____ 6. The presence in a population of two or more genetic variants that are maintained in a stable frequency over several generations.

_____ 7. Genetic drift that results from a small number of individuals colonizing a new area.

_____ 8. All the genes present in a freely-interbreeding population.

_____ 9. The principle that in a randomly-mating large population, regardless of dominance or recessiveness, the relative frequencies of allelic genes do not change from generation to generation.

_____10. A group of organisms of the same species that live in the same geographical area at the same time.

Terms:

a. Balanced polymorphism
b. Directional selection
c. Disruptive selection
d. Founder effect
e. Gene flow

f. Gene pool
g. Genetic drift
h. Hardy-Weinberg principle
i. Heterozygote advantage

j. Natural selection
k. Population
l. Stabilizing selection

TABLE
Fill in the blanks.

Category	Mendelian Genetics	Hardy-Weinberg Principle
Basic unit studied	Individual	Populations
Frequencies of genotypes studied	In the offspring of a single mating	#1
Duration of study	#2	#3
Genetic description used	#4	#5
#6	No more than two different alleles for each locus	Substantial polymorphism involving many loci
Ability of basic unit of study to evolve	#7	#8

THOUGHT QUESTIONS
Write your responses to these questions.

1. Which is most influential in bringing about evolution — changes in a gene pool or changes in a genotype? Justify your answer.
2. List the conditions that are needed for genetic equilibrium in a population and explain how each operates to stabilize and/or to "destabilize" a specific gene pool.
3. How do genetic drift, gene flow, mutation, and natural selection affect allele frequencies in a population?
4. Create a "normal" bell-shaped curve for a given trait in a population. Now show how that curve changes when the population is subjected to directional selection, and to disruptive selection, and to stabilizing selection.
5. The Hardy-Weinberg formula asserts that $p^2 + 2pg + q^2 = 1.00$. Which part of this formula refers to each of the following phrases:

frequency of recessive alleles in a population	homozygous dominant alleles
frequency of dominant alleles in a population	heterozygous frequency
homozygous recessive alleles	

6. What percentage of a population will have a given recessive gene if the frequency of the dominant allele is 0.80?
7. What is the frequency of a dominant allele if $p^2 = 0.25$? Of a recessive allele? Of the heterozygotes?

MULTIPLE CHOICE
Place your answer(s) in the space provided. Some questions may have more that one correct answer.

_____ 1. A relatively quick, extreme environmental change that favors several phenotypes at the expense of the mean phenotype will likely result in
   a. stabilizing selection.   d. disruptive selection.
   b. no selection.   e. directional selection.
   c. a shift in gene frequencies.

_____ 2. The source of all new alleles is
   a. gene flow.   d. genetic drift.
   b. natural selection.   e. the heterozygote advantage.
   c. mutations.

_____ 3. Natural selection associated with a well-adapted population is
   a. stabilizing selection.   d. disruptive selection.
   b. no selection.   e. directional selection.
   c. characterized by radical shifts in gene frequencies.

_____ 4. If undisturbed by other forces, random sexual reproduction among members of a large population in nature leads to
   a. new species.   d. new traits.
   b. changes in gene frequencies.   e. a change in the frequency of 2pq, but not p^2 or q^2.
   c. generations of unchanged gene frequencies.

_____ 5. When gradual environmental changes favor phenotypes at the extreme of the normal distribution curve, the result in time is likely to be
   a. stabilizing selection.   d. disruptive selection.
   b. no selection.   e. directional selection.
   c. a shift in gene frequencies.

____ 6. Changes in allele frequencies within a population are referred to as

    a. evolution.   d. p^2 shifts.

    b. macroevolution.   e. q^2 shifts.

    c. microevolution.

____ 7. The proportion of alleles in a population does not change if there is/are

    a. random mating.   d. mating with individuals in a similar population.

    b. natural selection of advantageous alleles.   e. a large number of individuals in the population.

    c. no more than ten percent of the alleles mutating in each generation.

____ 8. Random evolutionary changes in a small breeding population resulting from random changes in gene frequencies are referred to as

    a. gene flow.   d. genetic drift.

    b. natural selection.   e. the heterozygote advantage.

    c. mutations.

____ 9. The gene pool for a given species is

    a. fixed and unchanging.   d. isolated from other species.

    b. unique to that species.   e. the sum of all genes in the species.

    c. found in each individual in the species.

Use the following information, the Hardy-Weinberg equation, and the list below to answer questions 10-15. The phenotype coded for by the genotype "tt" is found among 400 individuals in a population of 10,000 randomly mating individuals.

a. 0.01	f. 0.16	k. 0.80	p. 4.00	u. 20.0
b. 0.02	g. 0.20	l. 1.00	q. 6.40	v. 32.0
c. 0.04	h. 0.32	m. 1.60	r. 9.60	w. 40.0
d. 0.08	i. 0.40	n. 2.00	s. 10.0	x. 64.0
e. 0.10	j. 0.64	o. 3.20	t. 16.0	y. 96.0

____ 10. Frequency of q.

____ 11. Frequency of p.

____ 12. Frequency of heterozygous individuals.

____ 13. Frequency of homozygously recessive individuals.

____ 14. Frequency of homozygously dominant individuals.

____ 15. Percent of individuals containing one or more of the dominant alleles.

Visual Foundations on next page ➤

VISUAL FOUNDATIONS
Color the parts of the illustration below as indicated.

RED   ❑   axis indicating number of individuals

GREEN   ❑   axis indicating phenotype variation

YELLOW   ❑   normal distribution

BLUE   ❑   unsuitable phenotypes

ORANGE   ❑   more suitable phenotypes

Also label directional selection, disruptive selection, and stabilizing selection.

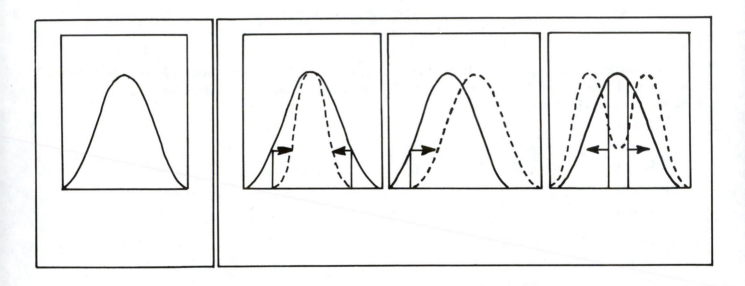

CHAPTER 19

Speciation and Macroevolution

A species is a group of organisms with a common gene pool that is reproductively isolated from other organisms. Most species have two or more mechanisms that serve to preserve the integrity of the gene pool. Some mechanisms prevent fertilization from occurring in the first place. Others ensure reproductive failure should fertilization occur. Speciation, or the development of a new species, most commonly occurs when one population becomes geographically separated from the rest of the species and subsequently evolves. A new species can also evolve, however, within the same geographical region as its parent species. This latter type of speciation is especially common in plants and occurs as a result of hybridization and polyploidy. There are two theories about the pace of evolution. One says that there is a slow, steady change in species over time, and the other says there are long periods of little evolutionary change followed by short periods of rapid speciation. Macroevolution, or evolutionary change above the level of species, includes the origin of unusual features (e.g., wings with feathers), evolutionary trends (e.g., increase in body size), the evolution of many species from a single species, and the extinction of species. Although the synthetic theory of evolution (Chapter 17) appears adequate to explain macroevolution, it is essentially unproved.

CHAPTER OUTLINE AND CONCEPT REVIEW
Fill in the blanks.

INTRODUCTION

1 The 18th century biologist, _____, is generally considered the founder of modern taxonomy. His system for separating plants into different species based on their morphological characteristics formed the basis for modern methods of systematics.

2 Although still very important in describing species, morphological features alone are not enough to delineate species. Biologists now define a species as a group of (a)_____ isolated organisms with a common (b)_____ _____.

SPECIES ACHIEVE REPRODUCTIVE ISOLATION IN VARIOUS WAYS

Prezygotic barriers interfere with mating

3 Prezygotic isolating mechanisms prevent _____.

4 Prezygotic isolating mechanisms include (a)_____, which occurs because two groups reproduce at different times; (b)_____, involving different, incompatible courtship patterns; (c)_____, isolation due to anatomical differences that thwart successful matings; and (d)_____, in which chemical differences between gametes prevents interspecific fertilization.

Postzygotic barriers prevent successful reproduction if mating occurs

5 Embryos are aborted due to _____.

6 _____ occurs when the gametes produced by interspecific hybrids are abnormal and nonfunctional.

THE KEY TO SPECIATION IS REPRODUCTIVE ISOLATION

7 _____ is the evolution of a new species.

Long physical isolation and different selective pressures result in allopatric speciation

8 Allopatric speciation occurs when one population becomes _____ _____ from the rest of the species and subsequently evolves.

Two populations diverge in the same physical location by sympatric speciation

9 Sympatric speciation usually occurs *within* a geographical region as a result of _____ _____.

10 (a)_____ is the possession of more than two sets of chromosomes, a common phenomenon in plants. When it results from the pairing of chromosomes from different species, it is known as (b)_____.

EVOLUTIONARY CHANGE CAN OCCUR RAPIDLY OR GRADUALLY

11 According to proponents of the theory of _____, evolution proceeds in spurts; that is, short periods of active evolution are followed by long periods of inactivity or stasis.

12 According to proponents of the theory of _____, populations slowly and steadily diverge from one another by the accumulation of adaptive characteristics within a population.

MACROEVOLUTION INVOLVES MAJOR EVOLUTIONARY EVENTS

Evolutionary novelties originate from mutations that alter development

13 Evolutionary "novelties" originate from mutations that alter developmental pathways. For example, (a)_____ occurs when developing body parts grow at different rates, and (b)_____ results from differences in the *timing* of development.

Adaptive radiation is the diversification of an ancestral species into many species

14 (a)_____ are new ecological roles made possible by an adaptive advancement. When an organism with a newly acquired evolutionary advancement assumes a new ecological role made possible by its advancement(s), its diversification is known as (b)_____.

Extinction is an important aspect of evolution

15 The geological record seems to indicate that two types of extinction have occurred: (a)_____ extinction, which is a continuous, ongoing, relatively low frequency form, and (b)_____ extinction, a relatively rapid and widespread loss of numerous species.

IS MICROEVOLUTION RELATED TO SPECIATION AND MACROEVOLUTION?

BUILDING WORDS
Use combinations of prefixes and suffixes to build words for the definitions that follow.

Prefixes	The Meaning
allo-	other
macro-	large, long, great, excessive
poly-	much, many

Prefix	Suffix	Definition
_____	-ploidy	1. The presence of multiples of complete chromosome sets.
_____	-patric	2. Originating in or occupying different (other) geographical areas.
_____	-polyploidy	3. Polyploidy following the joining of chromosomes from two different species (half of the chromosomes come from one parent and half from the other parent).
_____	-evolution	4. Large-scale evolutionary change.
_____	-metric	5. Varied rates of growth for different parts of the body during development (some parts grow at rates different from other parts).

MATCHING
For each of these definitions, select the correct matching term from the list that follows.

_____ 1. A postzygotic isolating mechanism in which the embryonic development of an interspecific hybrid is aborted.

_____ 2. The concept that evolution proceeds with periods of inactivity followed by very active phases, so that major adaptations or clusters of adaptations appear suddenly in the fossil record.

_____ 3. The evolution of a new species within the same geographical region as the parent species.

_____ 4. The evolution of several to many species from an unspecialized ancestor.

_____ 5. A prezygotic isolating mechanism in which sexual reproduction between two individuals cannot occur because of chemical differences between the gametes of the two species.

_____ 6. A model of evolution in which the evolutionary change of a species is due to a slow, steady transformation over time.

_____ 7. Interbreeding between members of different taxa.

_____ 8. Evolution of a new species.

_____ 9. An isolating mechanism in which gamete exchange between two groups is prevented because each group possesses its own characteristic courtship behavior.

_____10. The end of a lineage, occurring when the last individual of a species dies.

Terms:

a.	Adaptive radiation	f.	Hybrid breakdown	k.	Punctuated equilibrium
b.	Behavioral isolation	g.	Hybrid inviability	l.	Reproductive isolation
c.	Extinction	h.	Hybrid sterility	m.	Speciation
d.	Gametic isolation	i.	Hybridization	n.	Sympatric speciation
e.	Gradualism	j.	Mechanical isolation		

TABLE
Fill in the blanks.

Reproductive Isolating Mechanism	Example From the Text	Description of the Mechanism (Use the Example From the Text)
Mechanical isolation	Black and white sage	The floral structures are different, assuring pollination by different insects
#1	#2	One plant flowers in early spring, the other flowers in late spring and early summer
#3	Eggs of a bullfrog fertilized by sperm from a leopard frog	#4
#5	#6	Specific proteins on the surface of the egg will only bind to complementary molecules on the surface of sperm cells from the same species
#7	Bowerbird	#8
#9	#10	One species is sexually active only in the morning, the other species is sexually active only in the afternoon
#11	Mules	Two parental species have different chromosome numbers
#12	#13	Isolation in spring fed pools after glacial lakes dried up in Death Valley

THOUGHT QUESTIONS
Write your responses to these questions.

1. Which definition of species do you favor? Why? Why is it that definitions of species have exceptions?
2. Compare and contrast allopatric speciation and sympatric speciation.
3. Which concept regarding the pace of evolution do you favor and why — gradualism or punctuated equilibrium? Could it be that a combination of these two concepts applies in nature?
4. What are the basic similarities and differences between microevolution, macroevolution, and speciation?
5. Define the following terms and explain how they relate to concepts about species: preadaptation, allometric growth, paedomorphosis, evolutionary novelty, adaptive radiation, premating and postmating isolating mechanisms, extinction.

MULTIPLE CHOICE
Place your answer(s) in the space provided. Some questions may have more that one correct answer.

_____ 1. A population that evolves within the same range as its parent species is an example of
     a. cladogenesis.   d. sympatric speciation.
     b. character displacement.   e. allopatric speciation.
     c. anagenesis.

_____ 2. If two distinct species produce a fertile hybrid, the hybrid resulted from
     a. anagenesis.   d. diversifying evolution.
     b. phyletic evolution.   e. allopolyploidy.
     c. cladogenesis.

_____ 3. The fact that dogs come in many sizes, shapes, and colors proves that
     a. dogs have one gene pool.   d. reproduction within a species produces sterile offspring.
     b. dogs can produce hybrids.   e. all dogs belong to one species.
     c. morphology alone is not enough to define a species.

_____ 4. If two distinct populations of organisms occasionally interbreed in the wild, they are considered to be
     a. separate species.   d. one gene pool.
     b. one species.   e. evolving.
     c. reproductively isolated.

_____ 5. If a taxonomist discovered a species with two distinct populations, each population with its own characteristic gene complex, she/he would likely consider them to be
     a. two species.   d. varieties of one species.
     b. subspecies.   e. demes.
     c. two populations.

_____ 6. If flower-loving female botany majors were required to smell a flower before male botany majors would mate with them, and if male zoology majors would only mate with females that did not sniff flowers, zoology majors and botany majors would likely become
     a. separate species.   d. one gene pool.
     b. one species.   e. behaviorally isolated.
     c. reproductively isolated.

_____ 7. The largest unit within which gene flow can occur is the
     a. species.   d. individual organism.
     b. subspecies.   e. deme.
     c. population.

_____ 8. Hybrid breakdown affects the
     a. P_1 generation.   d. F_3 generation.
     b. F_1 generation.   e. course of evolution.
     c. F_2 generation.

_____ 9. A population that evolves as the result of its separation from the rest of the species is an example of
     a. cladogenesis.   d. sympatric speciation.
     b. character displacement.   e. allopatric speciation.
     c. anagenesis.

____10. If college-educated persons mated only at night and noncollege-educated persons mated only at the lunch hour, and no amount of laboratory experimentation could change these habits, these two populations would be considered

a. separate species.   d. members of one gene pool.

b. one species.   e. temporally isolated.

c. reproductively isolated.

____11. A species typically has

a. a common gene pool.   d. members with different morphological characteristics.

b. an isolated gene pool.   e. become reproductively isolated.

c. the capacity to reproduce with other species in the laboratory.

VISUAL FOUNDATIONS

Color the parts of the illustration below as indicated.

RED ❑ original species

GREEN ❑ second species to evolve

YELLOW ❑ time

BLUE ❑ first species to evolve

ORANGE ❑ evolutionary change

Also label gradualism and punctuated equilibrium.

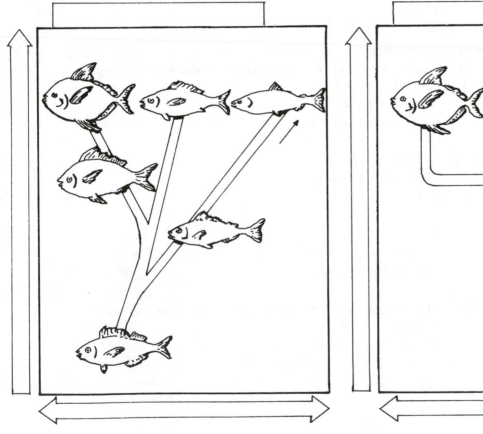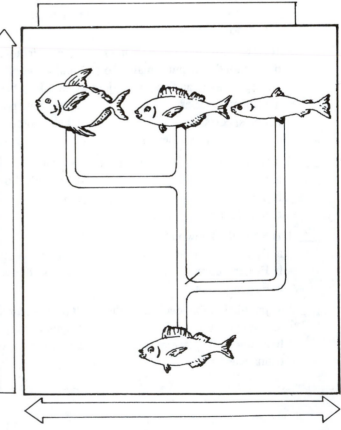

CHAPTER 20

□

The Origin and Evolutionary History of Life

Life on earth developed from nonliving matter. Small organic molecules formed, accumulated, and became organized into complicated assemblages. Cells evolved from these macromolecular assemblages. This process was enabled by the absence of oxygen and the presence of energy, the chemical building blocks of organic molecules, and sufficient time for the molecules to accumulate and react. The first cells to evolve were anaerobic and prokaryotic. Two crucial events in the evolution of cells was the origin of molecular reproduction and the development of metabolism. The earliest cells are believed to have obtained their energy from organic compounds in the environment. Later, cells evolved that could obtain energy from sunlight. Photosynthesis produced enough oxygen to change significantly the atmosphere and thereby alter the evolution of early life. Organisms evolved that had the ability to use oxygen in cell respiration. Eukaryotes are believed to have evolved from prokaryotes. Mitochondria, chloroplasts, and certain other organelles may have originated from symbiotic relationships between two prokaryotic organisms. In Precambrian times, life began and evolved into bacteria, protists, fungi, and animals. During the Paleozoic era, all major groups of land plants evolved except for flowering plants, and fish and amphibians flourished. The Mesozoic era was characterized by the evolution of flowering plants and reptiles; insects flourished, and birds and mammals appeared. The Cenozoic era has been characterized by the diversification of flowering plants and mammals, including humans and their ancestors.

CHAPTER OUTLINE AND CONCEPT REVIEW
Fill in the blanks.

EARLY EARTH PROVIDED THE CONDITIONS FOR CHEMICAL EVOLUTION

1 The four basic requirements for the chemical evolution of life are: _____
_____.

Organic molecules formed on primitive Earth

2 The idea that organic molecules could form from nonliving components on primitive Earth originated in the early 20th century with the Russian biochemist (a)_____ and the Scottish biologist (b)_____.

3 In the 1950s, (a)_____ constructed conditions in the laboratory that were thought to prevail on primitive Earth. They started with an atmosphere of water (H_2O) and the three gases (b)_____.
These and other experiments have produced a variety of organic molecules.

4 _____ are assemblages of organic polymers that form spontaneously, are organized and to some extent resemble living cells.

The first cells probably assembled from organic molecules

5 (a)_____ may have been the first polynucleotide to carry "hereditary" information. Interestingly, some forms of this molecule, called (b)_____, can catalyze the formation of more forms.

133

The first cells were probably heterotrophs, not autotrophs

6 The first cells were (aerobic or anaerobic?) (a)_____ and (prokaryotic or eukaryotic?) (b)_____ .

7 Organisms like ourselves that require preformed organic molecules (food), because they cannot synthesize them *de novo*, are called (a)_____. On the other hand, (b)_____ possess the chemical apparatus required to synthesize food.

8 The evolution of photosynthesis ultimately changed early life and afforded vastly new opportunities for diversity because it generated _____.

Aerobes appeared after oxygen increased in the atmosphere
Eukaryotic cells descended from prokaryotic cells

9 The _____ theory proposes that mitochondria, chloroplasts, and possibly centrioles and flagella evolved from prokaryotes engulfed by other cells.

THE FOSSIL RECORD PROVIDES US WITH CLUES TO THE HISTORY OF LIFE

Evidence of living cells is found in Precambrian times
A considerable diversity of organisms evolved during the Paleozoic era

10 The Paleozoic era began about (a)_____ years ago with a profound burst of evolution. Later, in the (b)_____ period, the first vertebrate, the jawless, bony-armored fish called (c)_____ appeared.

The dinosaurs and other reptiles dominated the Mesozoic era

11 The Mesozoic era began about (a)_____ years ago. It consisted of the three periods: (b)_____. Mammals appeared in the (c)_____ period and birds appeared in the (d)_____ period.

The Cenozoic era is known as the "Age of Mammals"

12 The Cenozoic era began about _____ years ago and extends into the present.

BUILDING WORDS
Use combinations of prefixes and suffixes to build words for the definitions that follow.

Prefixes	The Meaning		Suffixes	The Meaning
aero-	air		-be (bios)	life
an-	without, not, lacking		-bio(nt)	life
auto-	self, same		-troph	nutrition, growth, "eat"
Pre-	before, prior to			
hetero-	different, other			
proto-	first, earliest form of			

Prefix	Suffix	Definition
_____	_____	1. An organism that produces its own food.
_____	_____	2. Spontaneous assemblages of organic polymers that may have been involved in the evolution of the earliest forms of life.
_____	_____	3. An organism that is dependent upon other organisms for food, energy, and oxygen.

_____   -cambrian   4. The geologic time period prior to the Cambrian.

_____ _____   5. An organism that requires air or free oxygen to live.

_____   -aerobe   6. An organism that does not require air or free oxygen to live.

MATCHING
For each of these definitions, select the correct matching term from the list that follows.

____ 1. A geological time period that is a subdivision of an era, and is divided into epochs.

____ 2. Term applied to divisions of geological time that are divided into periods.

____ 3. The "Age of Mammals."

____ 4. A column-like rock that is composed of many minute layers of prokaryotic cells, usually cyanobacteria.

____ 5. A type of protobiont formed by adding water to abiotically-formed polypeptides.

____ 6. A major interval of geological time; a subdivision of a period.

____ 7. Blue-green algae.

____ 8. An organism that lives symbiotically inside a host cell.

____ 9. The remains of a microscopic organism.

____10. A molecule of RNA that has catalytic ability.

Terms:

a. Cenozoic era
b. Cyanobacteria
c. Endosymbiont
d. Epoch
e. Era

f. Mesozoic era
g. Microfossil
h. Microsphere
i. Paleozoic era
j. Period

k. Ribozyme
l. Stromatolite

TABLE
Fill in the blanks.

Relative Time	Land Mass (-es)	Period	Dominant Life Form (s) Associated With the Period
Present	Antarctica, Australia, Africa and Eurasia are joined, the Americas are joined	Quaternary	*Homo sapiens*
#1	Laurasia, Gondwana	#2	#3
260 million years ago	#4	Late Permian	Seed plants, therapsids
505 million years ago	#5	#6	#7
240 million years ago	Pangaea	#8	#9
#10	The Americas are separated, Australia is on the equator	#11	#12

THOUGHT QUESTIONS
Write your responses to these questions.

1. Describe how conditions on the primitive earth, which enabled life to originate, were different from conditions on earth today, which do not support the *de novo* creation of life.
2. Describe the similarities and essential differences between a "protobiont," the cell precursor that resembles a living cell, and a primitive living cell.
3. What differences, if any, do you think there might be between an extant prokaryote and the first living cell on earth?
4. What compelling circumstances on the primitive earth virtually mandated that the first cells were anaerobic heterotrophs as opposed to autotrophs and/or aerobic organisms?
5. Create a table that shows the climate, principal geological features, characteristic organisms, and approximate dates for each of the four geological eras.

MULTIPLE CHOICE
Place your answer(s) in the space provided. Some questions may have more that one correct answer.

_____ 1. The atmosphere of early Earth contained large quantities of
    a. oxygen.
    b. water vapor.
    c. hydrogen.
    d. carbon monoxide.
    e. helium.

_____ 2. The basic requirements for chemical evolution include
    a. DNA.
    b. oxygen.
    c. energy.
    d. water.
    e. time.

_____ 3. Initially, Earth's atmosphere became oxygenated through the photosynthetic activity of
    a. cyanobacteria.
    b. purple sulfur bacteria.
    c. green sulfur bacteria.
    d. water-splitting autotrophs.
    e. hydrogen-sulfide bacteria.

_____ 4. Oparin's coacervates were
    a. proteinoids.
    b. advanced liposomes.
    c. capable of replicating and dividing.
    d. cells.
    e. protobionts.

_____ 5. In a sequence of simplest (primitive) to most complex (advanced), which one of the following items is out of order: monomer, coacervate, proteinoid, prokaryote, eukaryote?
    a. microsphere
    b. proteinoid
    c. prokaryote
    d. coacervate
    e. monomer

_____ 6. According to the endosymbiont theory of the origin of eukaryotic cells, mitochondria were likely once
    a. free-living prokaryotes.
    b. coacervates.
    c. viruses.
    d. cells or proteinoids.
    e. eukaryotes.

_____ 7. Aerobes are generally better competitors than anaerobes because aerobic respiration
    a. is more efficient.
    b. splits water.
    c. is confined to symbiotic mitochondria.
    d. has had a longer period of evolution.
    e. extracts more energy from a molecule.

_____ 8. Experimenters that simulated conditions thought to have existed on early Earth found that these conditions could produce
   a. DNA.
   b. RNA.
   c. nucleotide bases.
   d. proteins.
   e. amino acids.

_____ 9. The consensus among scientists is that the first cells were
   a. aerobic autotrophic prokaryotes.
   b. anaerobic autotrophic prokaryotes.
   c. aerobic heterotrophic prokaryotes.
   d. anaerobic heterotrophic prokaryotes.
   e. anaerobic heterotrophic eukaryotes.

_____10. Strong support for the endosymbiotic theory of eukaryotic cell origins is derived from the fact that mitochondria and chloroplasts have
   a. unique DNA.
   b. two membranes.
   c. ribosomes.
   d. tRNA.
   e. endosymbionts.

_____11. Which of the following era/period combinations is/are mismatched?
   a. Cenozoic/Tertiary
   b. Paleozoic/Miocene
   c. Mesozoic/Jurassic
   d. Cenozoic/Cambrian
   e. Paleozoic/Devonian

_____12. The first large organic polymers (e.g., proteins, nucleic acids) most likely formed
   a. in the atmosphere.
   b. in shallow seas.
   c. in coacervates.
   d. from monomers.
   e. on clay surfaces.

_____13. The Earth is thought to be about how many years old?
   a. 10-20 billion.
   b. 4,000.
   c. 3.5 billion.
   d. 4.6 billion.
   e. 10-20 million.

_____14. About how many years passed between the Earth's formation and the appearance of recognizable cells?
   a. 10 million
   b. 100 million
   c. one billion
   d. five billion
   e. 10 billion

Visual Foundations on next page ➤

VISUAL FOUNDATIONS
Color the parts of the illustration below as indicated.

RED   ☐   DNA
GREEN   ☐   chloroplast
YELLOW   ☐   aerobic bacterium
BLUE   ☐   nuclear envelope
ORANGE   ☐   mitochondrion
BROWN   ☐   endoplasmic reticulum
TAN   ☐   photosynthetic bacterium

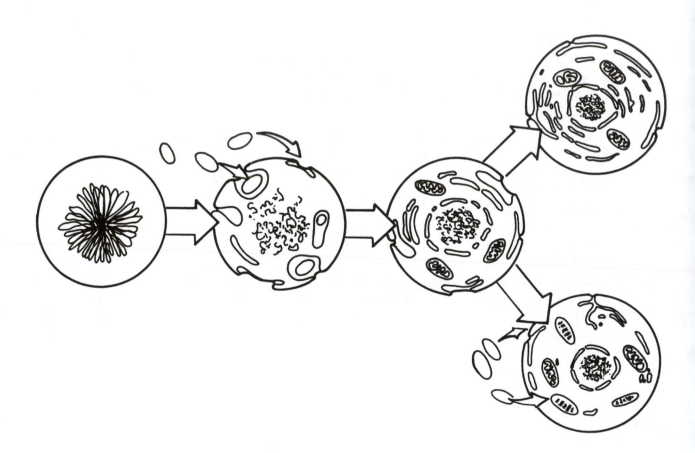

CHAPTER 21

The Evolution of Primates

Primates evolved from small, tree-dwelling, shrewlike mammals during the "Age of Reptiles." The first primates to evolve were the prosimians. Then, during the Oligocene epoch, the prosimians gave rise to the anthropoids. The early anthropoids branched into two groups, the New World Monkeys and the Old World Monkeys. The latter gave rise to apes that, in turn, gave rise to the hominids. Early hominids evolved a two-footed posture that has resulted in distinct characteristics of the skeleton and skull. Hominid evolution began in Africa. The earliest hominids belong to the genus *Australopithecus*. The first hominid to have enough human features to be placed in the same genus as modern humans is *Homo habilis*. Although it may have been a contemporary of *Australopithecus*, it had a larger brain, and it made tools from stone. *Homo erectus* was taller than *H. habilis* and fully erect. Also, its brain size was larger, it made more sophisticated tools, and it discovered how to use fire. *Homo sapiens* appeared approximately 200,000 years ago. Since then its cranial capacity has increased by over 60 percent. This evolutionary increase in human brain size made cultural evolution possible. Two of the most significant stages in cultural evolution have been the development of agriculture and the industrial revolution.

CHAPTER OUTLINE AND CONCEPT REVIEW
Fill in the blanks.

PRIMATES ARE ADAPTED FOR AN ARBOREAL EXISTENCE
1 There are two subgroups in the order Primates: the (a)_____ include lemurs, lorises, and tarsiers, and the (b)_____ include monkeys, apes, and humans.

Prosimians are primitive, arboreal primates
Anthropoids include monkeys, apes, and humans
2 Anthropoids evolved from (a)_____ ancestors during the (b)_____ epoch about 38 million years ago.

Hominids include apes and humans
3 Apes and humans are members of the superfamily _____.

4 The common names for the four genera of apes are _____
_____.

THE FOSSIL RECORD PROVIDES CLUES TO HOMINID EVOLUTION
The earliest hominids belong to the genus *Australopithecus*
5 Earliest hominids appeared about _____ years ago.

6 "Lucy" was one of the most ancient hominids, a member of the species _____.

Homo habilis **is the oldest member of the genus** *Homo*
Homo erectus **apparently evolved from** *Homo habilis*

7 *Homo habilis* appeared about _____ years ago. *H. habilis* was the first primate to consciously design useful tools.

8 *Homo erectus* is about _____ years old. It had a larger brain than *H. habilis*, made more sophisticated tools, and discovered how to use fire.

Homo sapiens **appeared approximately 200,000 years ago**

9 Modern *Homo sapiens* probably arose in _____ and migrated to Eurasia fairly recently.

HUMANS UNDERGO CULTURAL EVOLUTION

10 Cultural evolution is the progressive addition of _____ to the human experience. It is made possible by an evolutionary increase in brain size in humans.

11 Three of the most significant advances in cultural evolution were the _____
_____
_____.

Development of agriculture resulted in a more dependable food supply
Cultural evolution has had a profound impact on the ecosphere

BUILDING WORDS
Use combinations of prefixes and suffixes to build words for the definitions that follow.

Prefixes	The Meaning	Suffixes	The Meaning
anthrop(o)-	human	-oid	resembling, like
bi-	twice, two	-ped(al)	foot
pro-	"before"		
quadr(u)-	four		
supra-	above		

Prefix	Suffix	Definition
_____	_____	1. Member of the Suborder Anthropoidea; animals that "resemble" humans.
_____	_____	2. Pertains to an animal that walks on four feet.
_____	_____	3. Pertains to an animal that walks on two feet.
_____	-simian	4. Belonging or pertaining to the Suborder Prosimii; the primates that evolved before the simians (apes and monkeys).
homin-	_____	5. Member of the Superfamily Hominoidea; a gibbon, orangutan, gorilla, chimpanzee, or human.
_____	-orbital	6. Situated above the eye socket.

MATCHING
For each of these definitions, select the correct matching term from the list that follows.

____ 1. Any of a group of extinct and living humans.

____ 2. One of the earliest groups of *Homo sapiens.*

____ 3. Tree-dwelling.

____ 4. The earliest hominid genus.

____ 5. Adapted for grasping by wrapping around an object.

____ 6. Bearing living young.

____ 7. Scientists that study human evolution.

____ 8. The first hominid to have enough human features to be placed in the same genus as modern humans.

____ 9. To swing arm to arm, from one branch to another.

____10. The opening in the vertebrate skull through which the spinal cord passes.

Terms:

a. Arboreal
b. *Australopithecus*
c. Brachiate
d. Foramen magnum
e. Hominid

f. *Home erectus*
g. *Homo habilis*
h. *Homo sapiens*
i. Neanderthals
j. Paleoanthropologist

k. Prehensile
l. Viviparous
m. Oviparous

TABLE
Fill in the blanks.

Hominid	Time of Origin	Characteristic (s)
Australopithecus ramidus	4.4 million years ago	Most primitive of known hominids
#1	#2	Small bipedal hominids, upright posture, relatively small brain. No evidence of tool use
Homo habilis	#3	#4
Homo erectus	#5	#6
#7	#8	Thoroughly modern features, established a culture in what is now Spain and France
#9	200,000 - 27,000 years ago	Disappeared mysteriously

THOUGHT QUESTIONS
Write your responses to these questions.

1. What arboreal adaptations do primates have? Which of these do humans possess?
2. Create a table that shows the major similarities and differences among primates, prosimians, anthropoids, hominoids, and hominids.
3. Offer arguments and opposing views about the origin(s) of humans.
4. Describe some of the characteristics that distinguish the two major groups of primates.

MULTIPLE CHOICE
Place your answer(s) in the space provided. Some questions may have more that one correct answer.

_____ 1. The Peking man and Java man are classified as
    a. *H. habilis.*   d. apes.
    b. *H. erectus.*   e. hominids.
    c. *H. sapiens.*

_____ 2. The "Neanderthal man" is classified as
    a. *H. habilis.*   d. an ape.
    b. *H. erectus.*   e. a hominid.
    c. *H. sapiens.*

_____ 3. *Homo sapiens* appeared about how many years ago?
    a. 2,000   d. 200,000
    b. 10,000   e. one million
    c. 20,000

_____ 4. The first primates
    a. were hominids.   d. were anthropoids.
    b. had opposable thumbs.   e. were prosimians.
    c. had an ancestor in common with insectivores.

_____ 5. Anthropoids evolved directly from
    a. hominids.   d. apes.
    b. primates.   e. prosimians.
    c. insectivores.

_____ 6. Based on molecular similarities and other characteristics, it is thought that the nearest living relative of humans is the
    a. gorilla.   d. chimpanzee.
    b. monkey.   e. ape.
    c. gibbon.

_____ 7. Both apes and humans
    a. are hominoids.   d. lack tails.
    b. are hominids.   e. have opposable thumbs.
    c. can brachiate.

_____ 8. The first hominid that migrated to Europe and Asia is
    a. *H. habilis.*   d. australopithecines.
    b. *H. erectus.*   e. "Lucy" and her descendants.
    c. *H. sapiens.*

_____ 9. The earliest hominids
    a. had short canines.   d. appeared about 4.4 million years ago.
    b. are in the genus *Homo.*   e. evolved in Africa.
    c. are in the genus *Australopithecus.*

_____10. The earliest hominid to be placed in the genus *Homo* is
    a. *H. habilis.*   d. australopithecines.
    b. *H. erectus.*   e. "Lucy" and her descendants.
    c. *H. sapiens.*

VISUAL FOUNDATIONS
Color the parts of the illustration below as indicated.

RED ☐ cerebrum

GREEN ☐ jaw (mandible bone)

BLUE ☐ supraorbital ridge

YELLOW ☐ canine teeth

ORANGE ☐ other teeth

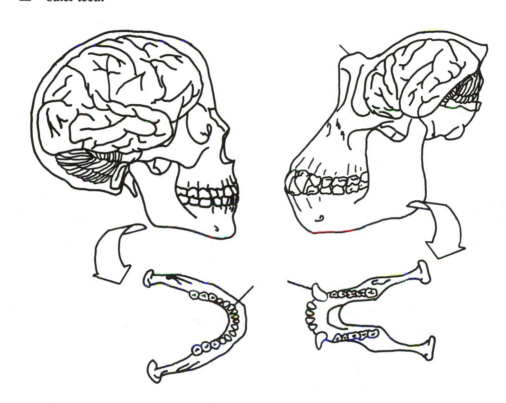

Color the parts of the illustration below as indicated.

RED	❑	pelvis
GREEN	❑	skull
YELLOW	❑	big toe
BLUE	❑	spine
ORANGE	❑	arm
BROWN	❑	leg

In each box, label the indicated anatomical difference.

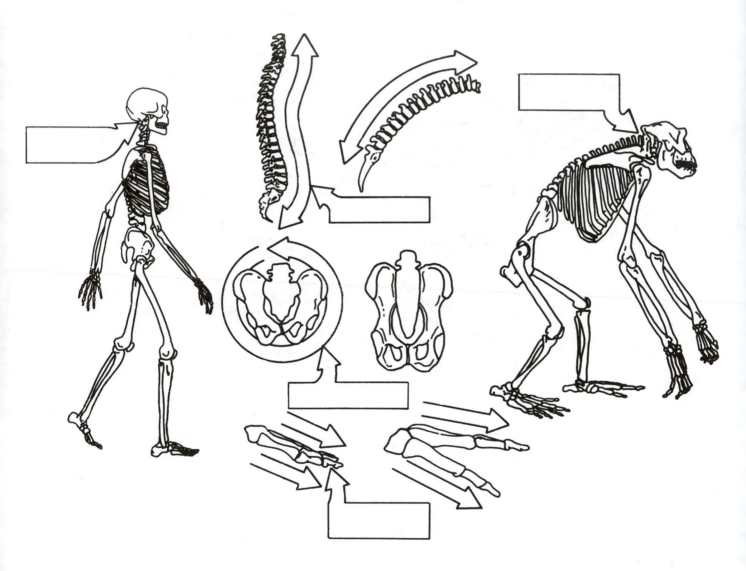

PART V

❑

The Diversity of Life

The Classification of Organisms

Organisms are classified according to a system developed in the mid-18th century. In this system, each organism is given a two-part name; the first name is the genus and the second is the species. The basic unit of classification is the species. Classification is hierarchical, with one or more related genera constituting a family, and one or more related families constituting an order, and so on. Currently, five major categories of organisms are recognized: Prokaryotae, Protista, Fungi, Plantae, and Animalia. Classification is based on evolutionary relationships. In some groupings, all of the organisms have a common ancestor. In others, all of the organisms are derived from different ancestors. Two or more different organisms are grouped together when evidence suggests that they have evolved from a common ancestor. The evidence might be shared homologous structures, or similarities in DNA and protein structure. There are three principal approaches to the classification of organisms. In one, organisms are classified according to the number of characteristics they have in common, whether homologous or analogous. In the second, organisms in a grouping all have a common ancestor and are classified based on how long ago one group of organisms branched off from another. In the third and most widely accepted and used system of classification, organisms are classified based on their evolutionary relationships as well as the extent of their divergence.

CHAPTER OUTLINE AND CONCEPT REVIEW
Fill in the blanks.

ORGANISMS ARE NAMED USING THE BINOMIAL SYSTEM OF NOMENCLATURE

1 _____ is the science of classifying and naming organisms.

2 The system developed by Carolus Linnaeus that assigns a two part name to each kind of organism is logically referred to as the (a)_____ of nomenclature. The name consists first of the (b)_____, which is capitalized, followed by the decapitalized (c)_____.

Each taxonomic level is more general than the one below it

3 The classification system consists of several levels of organization. For example, a group of closely related species is placed in the same genus. In similar fashion, closely related genera are grouped together in the same (a)_____, which in turn are grouped and placed in orders. The next level in the hierarchy above orders is the (b)_____, followed in succession by the (c)_____.

4 In the system for classifying organisms, each grouping, or level, such as a particular species, or genus, or order, and so forth is called a _____.

Subspecies may become species
Many biologists recognize five kingdoms

5 The five kingdoms proposed by R. H. Whitaker in 1969 are: _____
_____.

SYSTEMATICS IS CONCERNED WITH RECONSTRUCTING PHYLOGENY

6 Modern classification is based on the reconstruction of _____
_____, or phylogeny, as it is called.

An ideal taxon is monophyletic

7 The organisms in a (a)_____ taxon have a common ancestor, and the taxon containing the common ancestor and all of its descendant species is a (b)_____.

8 When the organisms in a given taxon have evolved from different ancestors, the taxon is said to be _____.

Homologous structures are imortant criteria for classification
Homologous structures are important criteria for classification
Derived characters have evolved more recently than ancestral characters

9 Shared *primitive* characters suggest (ancient or recent?) (a)_____ ancestry, whereas, shared *derived* characters indicate more (ancient or recent?) (b)_____ common ancestry.

Molecular biology provides taxonomic tools

10 Comparisons of the structure of the two molecules _____ are used to confirm evolutionary relationships.

TAXONOMISTS USE THREE MAIN APPROACHES

11 The three main approaches to taxonomy are _____
_____.

Phenetics is based on phenotypic similarities

12 The phenetic system is a numerical taxonomy in which organisms are grouped according to the number of _____ they share.

Cladistics emphasizes phylogeny

13 The cladistic approach emphasizes phylogeny, focusing on how long ago one group branched off from another. Each taxon contains a common ancestor and its descendants, and is therefore said to be _____.

Classical evolutionary taxonomy uses phylogenetic trees

BUILDING WORDS
Use combinations of prefixes and suffixes to build words for the definitions that follow.

Prefixes	The Meaning	Suffixes	The Meaning
eu-	good, well, "true"	-karyo(te)	nucleus
mono-	alone, single, one		
poly-	much, many		
sub-	under, below		

Prefix	Suffix	Definition
_____	-species	1. A unit of classification below species.
_____	-phyletic	2. Refers to a taxon in which all of the subgroups have a common (single) ancestry.
_____	-phyletic	3. Refers to a taxon consisting of several evolutionary lines and not including a common ancestor.

pro- _____   4. An organism that lacks a nuclear membrane.

_____ _____   5. An organism with a distinct nucleus surrounded by nuclear membranes.

MATCHING
For each of these definitions, select the correct matching term from the list that follows.

_____ 1. A group of interbreeding individuals that live within the same population.

_____ 2. The systematic study of organisms based on measurable morphological characters.

_____ 3. The science of naming, describing and classifying organisms.

_____ 4. A characteristic that has remained unchanged in a species.

_____ 5. A branching diagram based on shared derived characters to show the evolutionary relationships between organisms.

_____ 6. A taxonomic group of any rank.

_____ 7. The science of classifying organisms based on their evolutionary relationships.

_____ 8. A characteristic that is not present in the ancestral lineage.

_____ 9. A taxon that comprises related genera.

_____10. The evolutionary history of organisms.

Terms:

a. Ancestral character
b. Cladogram
c. Class
d. Derived character
e. Family

f. Order
g. Phenetics
h. Phylogeny
i. Species
j. Systematics

k. Taxon
l. Taxonomy

TABLE
Fill in the blanks.

Taxonomic Approach	Basis for Classification	Rational for Grouping Organisms	Emphasis
#1	Similarities of measurable characters	Number of characteristics that the organisms share	Quantitative comparison of characteristics
#2	Common ancestry	Presence of shared derived characters	Phylogeny
#3	Shared ancestral characteristics	Shared characteristics derived from a common ancestor and adaptations in related organisms	Evolutionary branching and extent of divergence

THOUGHT QUESTIONS
Write your responses to these questions.

1. What is the least number of characteristics one must use to distinguish each of the five kingdoms from the others?
2. Explain the rationale for each of the three major approaches to the classification of organisms.
3. What are some of the problems that arise when taxonomists attempt to delineate definitive characteristics that can be used to classify organisms?
4. What does "shared derived characteristics" mean? What does it have to do with classification of organisms?

MULTIPLE CHOICE
Place your answer(s) in the space provided. Some questions may have more that one correct answer.

_____ 1. The systematist would likely be most interested in
   a. lumping taxa.   d. ontogeny.
   b. splitting taxa.   e. ancestries.
   c. evolutionary relationships.

_____ 2. Humans are placed in the phylum Chordata because they have
   a. hair.   d. an opposable digit.
   b. an embryonic notochord.   e. a backbone.
   c. mammary glands.

_____ 3. Which of the following would the pheneticist emphasize when classifying organisms?
   a. phylogenetic trees.   d. number of shared characteristics.
   b. amino acid sequences.   e. antigen-antibody relationships.
   c. the point in time that branching occurred.

_____ 4. Which one of the following differs most?
   a. Animals and plants.   d. Prokaryotes and eukaryotes.
   b. Fungi and animals.   e. Plants and fungi.
   c. Algae and animals.

_____ 5. The fundamental unit of classification is the
   a. epithet.   d. species.
   b. binomial.   e. phylum or division.
   c. variety or subspecies.

_____ 6. Which of the following would the cladist emphasize when classifying organisms?
   a. phylogenetic trees.   d. number of shared characteristics.
   b. amino acid sequences.   e. antigen-antibody relationships.
   c. the point in time that branching occurred.

_____ 7. Which of the following is/are used extensively by the molecular biologist to determine relationships?
   a. carbohydrates.   d. proteins.
   b. nucleic acids.   e. hormones.
   c. lipids.

_____ 8. Fungi are generally not classified as plants because they do not
   a. photosynthesize.   d. have mitochondria.
   b. have a true nucleus.   e. reproduce sexually.
   c. have cell walls.

_____ 9. Which of the following is/are listed in order from taxa of greatest diversity to taxa of least diversity?
- a. Phylum, order, genus
- b. Family, class, order
- c. Class, family, genus
- d. Family, genus, species
- e. Class, order, division

VISUAL FOUNDATIONS

Color the parts of the illustration below as indicated. Inside each large box, write the specific name of each category used in classifying the depicted organisms.

RED ☐ specific epithet
GREEN ☐ phylum
YELLOW ☐ class
BLUE ☐ kingdom
ORANGE ☐ order
VIOLET ☐ family
PINK ☐ genus

CHAPTER 23

Viruses and Bacteria

Viruses are not true organisms. They consist of a core of DNA or RNA surrounded by a protein coat. Some are surrounded by an outer envelope too. Most can only be seen with an electron microscope. Whereas some viruses may or may not kill their hosts, others invariably do. Some integrate their DNA into the host DNA, conferring new properties on the host. Bacterial viruses released from host cells may contain some host DNA, which can then become incorporated into the genome of a new host. Some viruses infect humans and other animals, causing many diseases and cancers. Others infect plants, causing serious agricultural losses. Receptor molecules on the surface of a virus determine what type of cell it can infect. Viruses enter and exit cells in a variety of ways. Viruses are believed to have their origin in bits of nucleic acid that "escaped" from cellular organisms. Bacteria are in the kingdom Prokaryotae. Most are unicellular, but some form colonies or filaments. They have ribosomes, but lack membrane-bounded organelles. Their genetic material is a circular DNA molecule, not surrounded by a nuclear membrane. Some have flagella. Whereas most bacteria get their nourishment either from dead organic matter or symbiotically from other organisms, some manufacture their own organic molecules. Bacteria respire in a variety of ways. Although most reproduce asexually by transverse binary fission, some exchange genetic material. There are two major groups of bacteria; the Archaebacteria and Eubacteria. The former are thought to be the original bacteria from which all cellular life descended. Although most Eubacteria are harmless, a few are notorious for the diseases they cause. The mycoplasmas are the only bacteria without cell walls. Bacteria are classified on the basis of their ability to absorb and retain crystal violet stain. Those that do are said to be gram-positive; those that don't are gram-negative. Bacteria have important medical and industrial uses.

CHAPTER OUTLINE AND CONCEPT REVIEW
Fill in the blanks.

VIRUSES ARE TINY, INFECTIOUS AGENTS THAT ARE NOT ASSIGNED TO ANY OF THE FIVE KINGDOMS

1 Viruses are tiny particles consisting of a (a)_____ core surrounded by a protein coat called a (b)_____.

2 Viruses are thought to have had a multiple evolutionary origin; that is, some originated from plant cells, others from animal cells, and still others from bacteria. In all cases, regardless of their ancestral origins, they are thought to be _____
_____.

Bacteriophages are viruses that attack bacteria
Lytic viruses destroy the host cell

3 Bacteriophages that break apart host cells are said to be (a)_____,
while those that do not destroy the host cell are referred to as (b)_____.

4 The five steps that almost all lytic viral infections follow are _____
_____.

Temperate viruses can integrate their DNA into the host DNA

5   Temperate (lysogenic) viruses may or may not destroy their host cells; some integrate their nucleic acid into the host's DNA and multiply whenever the host does. Viral DNA that is integrated into host DNA is called a (a)_____, and host cells carrying integrated viral DNA are said to be (b)_____.

6   _____ occurs when bacteria exhibit new properties as a result of integration of temperate viral DNA.

7   Phages released from lysogenic cells may carry a portion of host DNA to a new host. This form of recombination in called _____.

Some viruses infect animal cells

8   RNA viruses called _____ transcribe their RNA genome into a DNA strand that is then used as a template to produce more viral RN HIV is one such virus.

Some viruses infect plants

9   _____ are disease-causing RNA particles that are smaller and simpler than viruses and infect plants.

BACTERIA MAKE UP KINGDOM PROKARYOTAE

Bacteria have three main shapes

10   Eubacteria cells are spherical, rod-shaped, or spirals. Spherical bacteria are called (a)_____. They may occur singly, in pairs called (b)_____, or in chains called (c)_____, or in clumps or bunches called (d)_____. Rod-shaped bacteria are known as (e)_____, and spiral-shaped bacteria are called (f)_____.

Prokaryotic cells lack membrane-bound organelles

The cell surface is generally covered by a cell wall

11   The bacterial cell wall owes its rigidity to _____, a molecule found only in prokaryotes.

12   Gram-positive bacteria have a thick-layered cell wall of (a)_____ that retains (b)_____ stain. Gram-negative bacteria have a thick outer layer of lipid compounds surrounding the peptidoglycan layer in the cell wall.

Many types of bacteria are motile

Bacteria have a single DNA molecule

13   The bacterial chromosome is a single, long, circular strand of DNA. Additional genetic information may also be found on small DNA loops called _____.

Bacteria reproduce by binary fission

14   Bacteria usually reproduce asexually by simple transverse binary fission, but sometimes genetic material is exchanged in one of several ways. In (a)_____, fragments of DNA from a cell are taken in by another cell. In (b)_____, bacterial genes are carried from one cell to another by phages, and, finally, bacterial "mating types" may exchange genetic material during (c)_____.

Some bacteria form endospores

Metabolic diversity is evident among bacteria

15   Most bacteria are heterotrophic (a)_____, organisms that obtain nutrients from dead organic matter. Others are (b)_____ _____ that produce organic molecules by oxidizing simple inorganic molecules.

TWO FUNDAMENTALLY DIFFERENT GROUPS OF PROKARYOTES ARE THE ARCHAEBACTERIA AND THE EUBACTERIA

The Archaebacteria include methanogens, halobacteria, and thermoacidophiles

16 Methanogens are anaerobes that produce methane from _____

_____.

Eubacteria are the most familiar prokaryotes

17 Spirochetes are spiral-shaped bacteria with flexible cell walls; they move by means of

_____.

18 Most of the gram-negative (a)_____ are photosynthetic autotrophs that
contain chlorophyll *a* and accessory pigments on internal membranes called
(b)_____.

19 Rickettsias are obligate intracellular parasites. Most of them parasitize members of the taxon

_____.

Bacteria are ecologically important
Some bacteria cause disease

20 Representatives of the group of gram-positive bacteria known as _____ can cause
scarlet fever, "strep throat," and other infections.

21 _____ are opportunistic pathogens that may cause boils, skin infections,
food poisoning, and possibly toxic shock syndrome.

22 Certain species of the anaerobic _____ cause tetanus, gas gangrene, and botulism.

23 (a)_____ form branching filaments and spores called
(b)_____. Some of them cause tuberculosis and leprosy.

Humans harness bacteria for medical and industrial purposes

BUILDING WORDS
Use combinations of prefixes and suffixes to build words for the definitions that follow.

Prefixes	The Meaning	Suffixes	The Meaning
an-	without, lacking	-gen	production of
arch(ae)-	primitive	-phage	eat, devour
bacterio-	bacteria		
endo-	within		
eu-	good, well, "true"		
exo-	outside, outer, external		
halo-	salt		
patho-	suffering, disease, feeling		

Prefix	Suffix	Definition
_____	_____	1. A virus that infects and destroys bacteria.
_____	-toxin	2. A poison that is secreted by bacterial cells into the external environment.
_____	-toxin	3. A poison that is a component "within" the cell wall of most gram-negative bacteria.

_____ -bacteria 4. Bacteria that inhabit extreme environments and do not have peptidoglycan in their cell walls; primitive bacteria.

_____ -bacteria 5. Bacteria that live only in extremely salty environments.

methano- _____ 6. A bacterium that produces methane from carbon dioxide and water.

_____ -spore 7. A thick-walled spore that forms within a bacterium in response to adverse conditions.

_____ -bacteria 8. The so-called "true bacteria."

_____ _____ 9. Any disease-producing organism.

_____ -aerobe 10. An organism that metabolizes only in the absence of (without) molecular oxygen.

MATCHING

For each of these definitions, select the correct matching term from the list that follows.

____ 1. A genetic transfer mechanism that produces new DNA in bacteria when DNA from a new organism is combined with the DNA of the host cell.

____ 2. A rod-shaped bacterium.

____ 3. A wall-less bacterium that is one of the smallest living organisms.

____ 4. Whip-like structure used for locomotion in some protists.

____ 5. Equal division of a cell or organisms into two cells or organisms.

____ 6. The protein covering of a virus.

____ 7. A modified protein or peptide possessing an attached carbohydrate.

____ 8. A method of reproduction in single-celled organisms in which two cells link and exchange nuclear material.

____ 9. A type of genetic recombination resulting from transfer of genes from one organism to another by a virus.

____10. A hair-like structure associated with prokaryotes.

Terms:

a. Bacillus
b. Binary fission
c. Capsid
d. Conjugation
e. Flagellum

f. Lysogenic conversion
g. Mycoplasma
h. Peptidoglycan
i. Pilus
j. Saprobe

k. Transduction
l. Transformation

TABLE
Fill in the blanks.

Microorganism	Genetic Material	Rigid Outer Covering	Disease(s)
Poxvirus	Double-stranded DNA	Capsid	Smallpox
Rhabdovirus	#1	#2	#3
Retrovirus	#4	#5	AIDS
#6	Single-stranded DNA	Capsid	Gastroenteritis
Treponema pallidum	#7	#8	Syphilis
Spiroplasma sp.	DNA	#9	Plant and insect diseases
#10	DNA	Gram-positive cell wall	Tuberculosis, leprosy
#11	DNA	#12	Botulism

THOUGHT QUESTIONS
Write your responses to these questions.

1. Present the pros and cons for considering viruses to be living organisms. How are viruses like cells? How do viruses and cells differ?
2. Describe how a virus enters an animal cell and replicates itself? How does this process differ in plant cells? In bacterial cells?
3. Using actual examples of each, explain the basic metabolic principles that distinguish autotrophic bacteria from heterotrophic bacteria. How do the terms aerobic, facultative anaerobic, and obligate anaerobic apply to autotrophy and heterotrophy?
4. Explain how transformation, conjugation, or transduction can improve a species' chances for survival by increasing the genetic variability in that species.
5. Describe some of the benefits that humans derive from bacteria. Describe diseases that are caused by some of the pathogenic prokaryotes.

MULTIPLE CHOICE
Place your answer(s) in the space provided. Some questions may have more that one correct answer.

_____ 1. Bacteria that retain crystal violet stain differ from those that do not retain the stain in having a
   a. thick lipoprotein layer.
   b. thick lipopolysaccharide layer.
   c. thicker peptidoglycan layer.
   d. cell wall.
   e. mesosome.

____ 2. The coat surrounding the nucleic acid core of a virus is
    a. a capsid.
    b. a protein.
    c. a virion.
    d. a viral envelope.
    e. composed of capsomeres.

____ 3. The simplest form of life capable of independent growth and metabolism is probably the
    a. virus.
    b. spirochete.
    c. azotobacterium.
    d. myxobacterium.
    e. mycoplasma.

____ 4. Bacteria that are adapted to extreme conditions and are perhaps most like the first organisms on the primitive Earth are the
    a. cyanobacteria.
    b. monera.
    c. eubacteria.
    d. archaebacteria.
    e. saprobes.

____ 5. An important difference between virulent and temperate viruses is that only the temperate virus
    a. destroys the host cell.
    b. lyses the host cell.
    c. does not lyse host cells in the lysogenic cycle.
    d. contains DNA.
    e. infects eukaryotic cells.

____ 6. Prokaryotic cells are distinguished from eukaryotic cells by
    a. absence of a nuclear membrane.
    b. absence of mitochondria.
    c. absence of DNA in the genetic material.
    d. absence of a plasma membrane.
    e. bacteriorhodopsin in cell walls.

____ 7. Asexual reproduction in bacteria typically involves
    a. DNA replication.
    b. cytokinesis.
    c. exchange of genetic material (DNA) between donors and recipients.
    d. inward growth of the cell wall.
    e. expansion of a central plate.

____ 8. When a host bacterium exhibits new properties because of a prophage the phenomenon is called
    a. lysogenic conversion.
    b. assembly.
    c. transduction.
    d. viroid induction.
    e. reverse transcription.

____ 9. Viruses are usually grouped (classified) according to
    a. size.
    b. shape.
    c. evolutionary relationships.
    d. phylogeny.
    e. type of nucleic acid.

____ 10. The majority of bacteria are
    a. autotrophs.
    b. heterotrophs.
    c. pathogens.
    d. saprobes.
    e. photosynthesizers.

____ 11. The phenomenon by which a virus introduces DNA from a bacterium it previously infected into a new host bacterium is called
    a. lysogenic conversion.
    b. assembly.
    c. transduction.
    d. viroid induction.
    e. reverse transcription.

____ 12. A prion
    a. is a subviral pathogen.
    b. contains DNA.
    c. is composed of glycoprotein.
    d. contains RNA.
    e. is larger than a typical virus.

_____13. The group of bacteria that have molecular traits so different from other prokaryotes as to be considered a separate kingdom by some biologists is the
   a. cyanobacteria.   d. archaebacteria.
   b. monera.   e. saprobes.
   c. eubacteria.

_____14. A layer in some viruses that contains lipid, carbohydrate, and protein is a
   a. capsid.   d. viral envelope.
   b. viral core.   e. capsomere.
   c. virion.

_____15. The characteristic(s) of life that viruses do not exhibit is/are
   a. presence of nucleic acids.   d. a cellular structure.
   b. independent movement.   e. independent metabolism.
   c. reproduction.

_____16. Exchange of genetic material between bacteria sometimes occurs by
   a. fusion of gametes.   d. transfer of genes in bacteriophages.
   b. conjugation.   e. transduction.
   c. incorporation by a bacterium of DNA fragments from another bacterium.

_____17. A small piece of DNA that is separate from the main chromosome is
   a. called a plasmid.   d. found in archaebacteria and cyanobacteria, but not eubacteria.
   b. a mesosome.   e. found in gram-negative, but not gram-positive bacteria.
   c. responsible for photosynthesis.

VISUAL FOUNDATIONS
Color the parts of the illustration below as indicated.

RED   ❑   genetic material
GREEN   ❑   outer membrane
YELLOW   ❑   inner (plasma) membrane
BLUE   ❑   peptidoglycan layer
ORANGE   ❑   capsule

Also label cell wall and nuclear region.

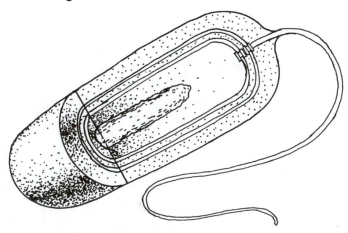

Color the parts of the illustration below as indicated.

RED	❑ DNA
GREEN	❑ RNA
YELLOW	❑ capsid
BLUE	❑ spikes
ORANGE	❑ fibers
BROWN	❑ tail

Color the parts of the illustration below as indicated.

RED	❑ transport protein
GREEN	❑ lipoprotein
YELLOW	❑ outer membrane
BLUE	❑ polysaccharides
ORANGE	❑ peptidoglycan layer
BROWN	❑ plasma membrane

CHAPTER 24

❑

The Protist Kingdom

The protist kingdom, an unnatural assemblage of mostly aquatic organisms, includes three major groupings — protozoa, slime and water molds, and algae. Although eukaryotic, the cell structure of protists is more complex than that of either animal or plant cells. Protists range in size from microscopic single cells to meters-long multicellular organisms. They do not have specialized tissues. Whereas some protists are free-living, others form symbiotic associations ranging from mutualism to parasitism. Reproductive strategies vary considerably among the different groups of protists; most reproduce both sexually and asexually, others only asexually. Whereas some are nonmotile, most move by either flagella, pseudopods, or cilia. The protozoa are the animal-like protists, so considered because they ingest their food like animals. Included in this group are the amoebas, foraminiferans, actinopods, flagellates, ciliates, and sporozoans. The slime and water molds are the fungus-like protists. There are two groups of slime molds — the plasmodial slime molds and the cellular slime molds. The algae are the plant-like protists, so considered because most are photosynthetic. Algae include the dinoflagellates, diatoms, euglenoids, and the green, red and brown algae. Protists are believed to be the first eukaryotes. Flagellates appear to be the most primitive protists, and are thought to have given rise to animals. Green algae are believed to have given rise to plants, and red algae to fungi. Multicellular organisms may have evolved from colonial forms.

CHAPTER OUTLINE AND CONCEPT REVIEW
Fill in the blanks.

INTRODUCTION
1 Organisms are placed in the kingdom Protista for convenience. They do not comprise a naturally occurring phylogenetic group, but they are all eukaryotes and most of them are unicellular. Others are colonial, or multicellular, or single multinucleated cells, the latter a condition known as

_____.

PROTISTS EXHIBIT REMARKABLE VARIATION
2 Most protists are aquatic, and many of them are floating microscopic forms called _____.

PROTOZOA ARE ANIMAL-LIKE PROTISTS
Amoebas move by forming pseudopodia
3 Amoebas glide along surfaces by flowing into cytoplasmic extensions called _____, meaning "false feet."

4 The parasitic species _____ causes amebic dysentery.

Foraminiferans extend cytoplasmic projections through tests
Actinopods project slender axopods
5 Actinopods utilize long, filamentous cytoplasmic projections called _____ to entrap prey.

Flagellates move by means of flagella

6 The parasitic flagellate _____ causes African sleeping sickness.

7 The _____, or "collared flagellates," as their name implies, resemble sponge cells.

Ciliates use cilia for locomotion

8 Many ciliates possess _____, which are small organelles that eject filaments to trap their prey.

9 Ciliates are distinct in having one or more small (a)_____ that function during the sexual process and a larger (b)_____ that regulates other cell functions.

10 "Cross fertilization" occurs among many ciliates during the sexual process known as _____.

Sporozoans are spore-forming parasites of animals

11 A sporozoan in the genus _____ can enter human RBCs and cause malaria.

ALGAE ARE PLANTLIKE PROTISTS

Most dinoflagellates are a part of marine plankton

12 The dinoflagellates are mostly unicellular, biflagellate, photosynthetic organisms with a cellulose shell-like covering. The _____, however, are photosynthetic endosymbionts that lack cellulose "shells" and flagella.

Diatoms have shells composed of two parts

13 Diatoms are mostly single-celled, with silica impregnated in their cell walls. When they die, their shells accumulate, forming a sediment called _____.

Euglenoids are unique freshwater unicellular flagellates

14 Euglenoids are single-celled, flagellated protists containing (a)_____, the same three pigments as those found in green algae and higher plants. They store food in the form of the polysaccharide (b)_____.

Green algae share many similarities with plants

15 Green algae exhibit a wide diversity in size, complexity, and reproduction. They resemble plants, but some are endosymbionts, and others are symbiotic with fungi in a mutually obligatory "biorganismic" entity known as a _____.

16 Green algae exhibit three kinds of sexual reproduction. In (a)_____ reproduction, the two gametes that fuse are apparently identical, whereas (b)_____ reproduction involves gametes of different sizes. Fusion of a nonmotile "egg" and a motile sperm-like gamete is called (c)_____ reproduction.

Red algae do not produce motile cells

17 Most red algae are attached to a surface by means of a root-like structure called the (a)_____. Their chloroplasts contain chlorophyll *a* and carotenoids, and in addition, the red pigment (b)_____ and the blue pigment (c)_____.

Brown algae are multicellular seaweeds

18 All brown algae are multicellular and produce flagellated cells during reproduction. Kelp are the largest brown algae; in addition to the holdfast, they possess differentiated leaf-like (a)_____ and stem-like (b)_____.

SLIME MOLDS AND WATER MOLDS ARE FUNGUS-LIKE PROTISTS

Plasmodial slime molds feed as multinucleate plasmodia

19 Plasmodial slime molds form intricate stalked reproductive structures called _____, within which meiosis occurs. The resulting haploid nuclei are then surrounded by a layer of chitin or cellulose, forming an extremely resistant spore.

20 When environmental conditions are suitable, spores open and one-celled haploid gametes of two types, one a flagellated form called a (a)_____, the other an ameboid (b)_____, emerge and fuse.

Cellular slime molds feed as single ameboid cells

21 The cellular slime molds live vegetatively as individual, single-celled organisms, except during reproduction they form a multicellular aggregate called a _____, which then differentiates and forms spores.

Water molds produce flagellated reproductive cells

22 The water molds, like the fungi, have a body called the (a)_____. They reproduce asexually by biflagellate (b)_____ and sexually by means of (c)_____.

THE EARLIEST EUKARYOTES WERE PROTISTS
Multicellularity evolved in the protist kingdom

BUILDING WORDS
Use combinations of prefixes and suffixes to build words for the definitions that follow.

Prefixes	The Meaning	Suffixes	The Meaning
cyto-	cell	-gamy	united
iso-	equal, "same"	-pod	foot, footed
macro-	large, long, great, excessive	-zoa	animal
micro-	small		
proto-	first, earliest form of		
pseudo-	false		
syn-	with, together		
tricho-	hair		

Prefix	Suffix	Definition
_____	_____	1. Sexual reproduction; the coming together and the uniting of the gametes.
_____	_____	2. Single-celled, animal-like protists including amoebae, ciliates, flagellates, and sporozoans; members of this group are believed to have given rise to the earliest form of animal life.
_____	-pharynx	3. The gullet of a protozoan, a single-celled organism.
_____	_____	4. A temporary protrusion of the cytoplasm of an ameboid cell that the cell uses for feeding and locomotion; a "false foot."
_____	-cyst	5. A cellular organelle of some ciliates that can discharge a hair-like filament that may aid in trapping and holding prey.
_____	-nucleus	6. A small nucleus found in ciliates.
_____	-nucleus	7. A large nucleus found in ciliates.

_____ -plasmodium
8. The aggregation of cells for reproduction in cellular slime molds; not a true plasmodium (multinucleate ameboid mass) as occurs in the plasmodial slime molds.

_____ _____
9. Reproduction resulting from the union of two gametes that are identical ("the same") in size and structure.

MATCHING
For each of these definitions, select the correct matching term from the list that follows.

____ 1. One of the long, filamentous cytoplasmic projections that protrude through pores in the skeletons of actinopods.

____ 2. A spore case, found in certain protists, fungi, and plants.

____ 3. The vegetative body of fungi; consists of a mass of hyphae.

____ 4. Free-floating, mainly microscopic aquatic organisms found in the upper layers of the water.

____ 5. Marine protozoans that produce chalky, many-chambered shells or tests.

____ 6. A sexual process in which two individuals come together and exchange genetic material.

____ 7. One of the filaments composing the mycelium of a fungus.

____ 8. The basal structure in multicellular algae that attaches to solid surfaces.

____ 9. Multinucleate, ameboid mass of living matter that constitutes the vegetative phase of the life cycle of slime molds.

____10. A reproductive cell that gives rise to individual offspring in plants, algae, fungi and some protozoans.

Terms:

a. Actinopod
b. Axopod
c. Conjugation
d. Foraminiferan
e. Holdfast
f. Hypha
g. Mycelium
h. Oospore
i. Plankton
j. Plasmodium
k. Sporangium
l. Spore

TABLE
Fill in the blanks.

Organism	Life-Style	Mode of Nutrition	Locomotion
Amebae	Free-living, some parasitic	Heterotrophic (ingest food)	Pseudopodia
#1	Free-living, some symbiotic	#2	Flagella
Sporozoans	#3	Heterotrophic (absorb food)	#4
Dinoflagellates	#5	#6	Flagella
Green algae	#7	#8	#9
Cellular slime molds	#10	#11	Pseudopodia
Water molds	Free-living	#12	Flagellated reproductive cells

THOUGHT QUESTIONS
Write your responses to these questions.

1. What are some of the principal morphological characteristics, methods of reproduction, and means of obtaining nutrients exemplified among the protistans?
2. Create a table or several tables that describe some of the major characteristics for each of the following:
   - Protozoans, including amoebae, foraminiferans, actinopods, flagellates, ciliates, sporozoans
   - Algae, including dinoflagellates, diatoms, euglenoids, green algae, red algae, brown algae
   - Protists, including plasmodial slime molds, cellular slime molds, water molds
3. Describe features of *Chlamydomonas* that might help us understand how multicellularity originated.

MULTIPLE CHOICE
Place your answer(s) in the space provided. Some questions may have more that one correct answer.

_____ 1. The parasite that causes malaria is in the phylum
    a. Ciliophora.   d. Dinoflagellata.
    b. Apicomplexa.   e. Euglenophyta.
    c. Sarcomastigophora.

_____ 2. Which of the following does the text include among protists?
    a. fungi.   d. protozoa.
    b. slime molds.   e. algae.
    c. azobacteria.

_____ 3. A protozoan whose cells bear a striking resemblance to cells in sponges is the
    a. foraminiferan.   d. ameba.
    b. diatom.   e. radiolarian.
    c. choanoflagellate.

_____ 4. The ameboid protozoans that form the grey mud on the bottom of oceans and that are used as indicators of geophysical changes in the environment are the
    a. foraminiferans.   d. amoebae.
    b. dinoflagellates.   e. radiolarians.
    c. choanoflagellates.

_____ 5. A protist rich in mineral nutrients and containing an agent used to thicken ice cream is a
    a. foraminiferan.   d. red alga.
    b. Phaeophyta.   e. brown alga.
    c. Rhodophyta.

_____ 6. A protist covering a large area of a piece of decaying tree bark would most likely be a
    a. Myxomycota.   d. alga.
    b. fungus.   e. slime mold.
    c. plasmodium.

_____ 7. One of the reasons that protists are placed in a separate kingdom is that many of them possess both plantlike and animal-like characteristics, which is particularly well-illustrated in the genus
    a. *Paramecium.*   d. *Ameba.*
    b. *Euglena.*   e. *Didinium.*
    c. *Plasmodium.*

_____ 8. Ciliates are
    a. all motile.
    b. all sessile.
    c. sometimes motile, sometimes sessile.
    d. protozoa.
    e. binucleated or multinucleated.

_____ 9. Which of the following is/are characteristic of protists and clearly distinguish(es) them from monerans?
    a. Membrane-bound nucleus.
    b. Plasma membrane.
    c. Cell walls.
    d. Mitosis and meiosis.
    e. Eukaryotic cell structure.

_____10. Red algae are placed in the phylum
    a. Phaeophyta.
    b. Rhodophyta.
    c. Acrasiomycota.
    d. Oomycota.
    e. Apicomplexa.

_____11. A protozoan with two nuclei and projectile filaments is classified as
    a. Ciliophora.
    b. Apicomplexa.
    c. Sarcomastigophora.
    d. Dinoflagellata.
    e. Euglenophyta.

VISUAL FOUNDATIONS
Color the parts of the illustration below as indicated.

RED ☐ micronucleus
GREEN ☐ trichocyst
YELLOW ☐ oral groove
BLUE ☐ contractile vacuole
ORANGE ☐ cilia
BROWN ☐ food vacuole
TAN ☐ anal pore
VIOLET ☐ macronucleus

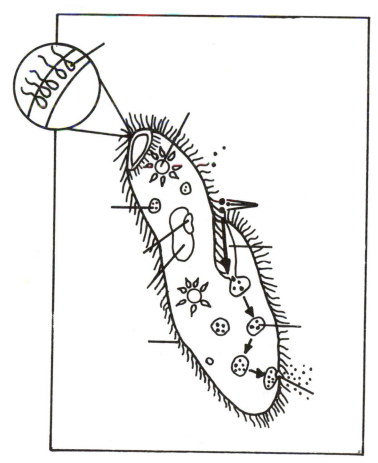

Color the parts of the illustration below as indicated.

RED	☐	nucleus
GREEN	☐	chloroplast
YELLOW	☐	eye spot
BLUE	☐	contractile vacuole
ORANGE	☐	locomotory flagellum
PINK	☐	mitochondria
VIOLET	☐	paramylon body

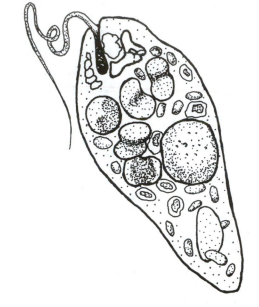

Color the parts of the illustration below as indicated. Label the portions of the illustration depicting sexual reproduction, and the portions depicting asexual reproduction.

RED	☐	positive strain
GREEN	☐	negative strain
YELLOW	☐	zygote

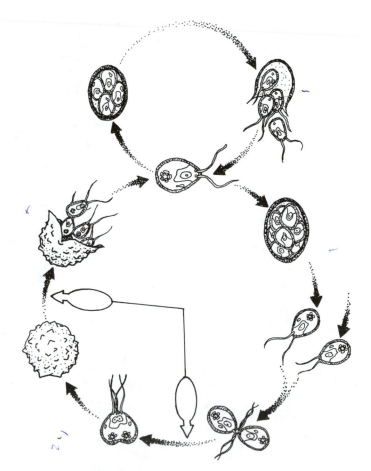

CHAPTER 25

Kingdom Fungi

Fungi are eukaryotes with cell walls. They digest their food outside their body and absorb the nutrients through the cell wall and cell membrane. Most fungi are decomposers, and as such play a critical role in the cycle of life. Others however form symbiotic relationships with other organisms. Some, for example, are parasites, causing diseases in animals and plants. And some are involved in mutualistic relationships, such as that between fungi and the roots of plants known as fungus-roots or mycorrhizae, and that between fungi and certain algae and cyanobacteria known as lichens. Fungi are nonmotile and reproduce by means of spores that are formed either sexually or asexually. Some fungi are unicellular (the yeast form) and some are multicellular, having a filamentous body plan. Fungi are classified into three phyla: Zygomycota, Ascomycota, and Basidiomycota. In addition, fungi which share similarities but with no known sexual stage or common ancestry, are assigned, as a matter of convenience, to the Deuteromycota. Fungi are of both positive and negative economic importance. Some damage stored goods and building materials. Some provide food for humans. Some function in brewing beer and in baking. Some produce antibiotics and other drugs and chemicals of economic importance, and some cause serious diseases in economically important animals, humans, and plants.

CHAPTER OUTLINE AND CONCEPT REVIEW
Fill in the blanks.

INTRODUCTION
1 Cell walls of most fungi contain _____, which is highly resistant to microbial breakdown.

MOST FUNGI HAVE A FILAMENTOUS BODY PLAN
2 Body structures vary in complexity from single-celled (a)_____ to filamentous, sometimes elaborately complex multicellular (b)_____.

MOST FUNGI REPRODUCE BY SPORES
3 Some form of sexual conjugation often occurs, sometimes producing hyphal cells with two distinctly different unfused nuclei. Such cells are said to be (a)_____, whereas cells containing only one haploid nucleus are (b)_____.

FUNGI ARE CLASSIFIED INTO THREE PHYLA
4 Based mainly on the characteristics of the (a)_____, fungi are classified into three divisions: (b)_____.

Zygomycetes reproduce sexually by forming zygospores
5 The zygomycete species (a)_____ is the well known black bread mold. It is (b)_____, meaning a fungal mycelium will only mate with a *different* mating type, not with its own type.

Ascomycetes (sac fungi) reproduce sexually by forming ascospores

6 Ascomycetes produce sexual spores in sacs called (a)_____, and asexual spores called (b)_____ at the tips of (c)_____.

7 Sexual spore sacs (asci) develop on specialized dikaryotic (n + n) fruiting bodies known as _____.

Basidiomycetes (club fungi) reproduce sexually by forming basidiospores

8 Basidiomycetes develop an enlarged, club-shaped hyphal cell called a (a)_____, on the surface of which four spores called (b)_____ develop.

9 The mushroom we eat is actually a compact mass of hyphae that form a stalk and cap called the (a)_____. The mushroom is technically a sporocarp, or (b)_____.

Imperfect fungi are fungi with no known sexual stage

10 Deuteromycetes are called "imperfect fungi" simply because no one has observed them to have a _____ in their life cycles. If one is discovered in a species, it is moved to another division.

LICHENS ARE DUAL "ORGANISMS"

11 A lichen is a symbiotic combination of a fungus and a phototroph in which the fungus benefits from the photosynthetic activity of the phototroph. Lichens typically assume one of three forms: (a)_____, a low, flat growth, or (b)_____, a moderately flat growth with leaf-like lobes, or (c)_____, an erect and branching shrub-like growth.

12 The phototroph portion of the lichen is usually a (a)_____ _____. The fungus is sometimes a basidiomycete, but usually an (b)_____.

13 Lichens reproduce asexually by fragmentation, wherein a piece of dried lichen called a _____ breaks off and begins growing when it lands on a suitable substrate.

FUNGI ARE ECOLOGICALLY IMPORTANT

14 Most fungi are heterotrophic decomposers, more specifically they are (a)_____ that absorb nutrients from externally digested nonliving organic matter. In the process, they release (b)_____ to the atmosphere and return (c)_____ to the soil.

15 Some fungi are parasites. Others are mutualistic symbionts, notable among which are the _____ that live on the roots of most plants.

FUNGI ARE ECONOMICALLY IMPORTANT

Fungi provide beverages and food

16 (a)_____ are the fungi used to make wine and beer, and to produce baked goods. Wine is produced when the fungus ferments (b)_____, beer results from fermentation of (c)_____, and (d)_____ bubbles cause bread to rise.

17 Roquefort and Camembert cheeses are produced using the genus (a)_____, and the imperfect fungus (b)_____ is used to make soy sauce from soybeans.

18 Some of the most toxic mushrooms, such as the "destroying angel" and "death angel," belong to the genus _____.

19 The chemical _____ found in some mushrooms causes hallucinations and intoxication.

Fungi produce useful drugs and chemicals

20 Alexander Fleming noticed that bacterial growth is inhibited by the mold
(a)_____, and this discovery eventually led to the development
of the most widely used of all antibiotics, (b)_____.

21 An ascomycete that infects cereal plant flowers produces a structure called an (a)_____ where
seeds would normally form. When livestock or humans eat contaminated grain or grain products,
they may develop the disease (b)_____, a condition characterized by convulsions,
delusions, and sometimes gangrene and death.

Fungi cause many important plant diseases

22 All plants are susceptible to fungal diseases. Fungi enter plants through stomata, or wounds, or by
dissolving a portion of cuticle with the enzyme (a)_____. Parasitic fungi often extend
specialized hyphae called (b)_____ into host cells in order to obtain nutrients from
host cell cytoplasm.

Fungi cause certain animal diseases

BUILDING WORDS

Use combinations of prefixes and suffixes to build words for the definitions that follow.

Prefixes	The Meaning	Suffixes	The Meaning
coeno-	common	-cyt(ic)	cell
hetero-	different	-karyo(tic)	nucleus
homo-	same	-phore	bearer
mono-	alone, single, one	-troph	nutrition, growth, "eat"
sapro-	rotten		

Prefix	Suffix	Definition
_____	-troph	1. An organism that absorbs ("eats") its nutrients through its cell membrane following extracellular digestion of dead and decaying (rotten) life forms.
_____	_____	2. Pertains to an organism made up of a multinucleate, continuous mass of cytoplasm enclosed by one cell wall (all nuclei share a common cell).
_____	_____	3. Pertains to hyphae that contain only one nucleus per cell.
_____	-thallic	4. Pertains to an organism that has two different mating types.
conidio-	_____	5. Hyphae in the ascomycetes that bear conidia.
_____	-thallic	6. Pertains to an organism that can mate with itself.

MATCHING

For each of these definitions, select the correct matching term from the list that follows.

____ 1. Compound organism composed of symbiotic algae and fungi.

____ 2. The vegetative body of fungi; consists of a mass of hyphae.

____ 3. The sexual spores produced by an ascomycete.

____ 4. The fruiting body of a basidiomycete.

_____ 5. Asexual reproductive body produced by lichens.

_____ 6. Mutualistic associations of fungi and plant roots that aid in the absorption of materials.

_____ 7. Asexual reproduction in which a small part of the parent's body separates from the rest and develops into a new individual.

_____ 8. The club-shaped spore-producing organ of certain fungi.

_____ 9. Cells having two nuclei.

_____10. An insoluble, horny protein-polysaccharide that forms the cell wall in fungi.

Terms:

a. Ascocarp
b. Ascospore
c. Basidiocarp
d. Basidium
e. Budding

f. Chitin
g. Conidium
h. Dikaryotic
i. Haustorium
j. Lichen

k. Mycelium
l. Mycorrhizae
m. Soredium
n. Zygospore

TABLE
Fill in the blanks.

Phyla	Mode of Sexual Reproduction	Representative (s) of the Group
Zygomycota	Zygospores	Black bread mold
Ascomycota	#1	#2
#3	Basidiospores	Mushrooms, puffballs, rusts, and smuts
#4	#5	*Aspergillus tamarii*

THOUGHT QUESTIONS
Write your responses to these questions.

1. What is the smallest number of characteristics that can be used to distinguish fungi from all members of other kingdoms?
2. Create a table that includes distinguishing characteristics, a typical life cycle, and a specific example of zygomycetes, ascomycetes, basidiomycetes, and imperfect fungi.
3. How do fungi function as decomposers? What is the significance of this role in nature?

MULTIPLE CHOICE
Place your answer(s) in the space provided. Some questions may have more that one correct answer.

_____ 1. Yeast participates in the brewing of beer by
   a. adding vital amino acids.
   b. fermenting grain.
   c. fermenting fruit sugars.
   d. producing ethyl alcohol.
   e. converting barley to hops.

_____ 2. The course of history was changed when Peter the Great's cavalry was killed by the fungus-caused disease known as

a. autoimmune response.
b. St. Anthony's fire.
c. histoplasmosis.
d. ergotism.
e. candidiasis.

_____ 3. A fungus infection throughout the body obtained by exposure to bird droppings is likely

a. an autoimmune response.
b. St. Anthony's fire.
c. histoplasmosis.
d. ergotism.
e. candidiasis.

_____ 4. The fungus that makes minerals available to a tree through mutualism with the tree's roots is a

a. water mold.
b. mycorrhizae.
c. lichen.
d. saprophyte.
e. yeast.

_____ 5. A collection of filamentous hyphae is called a

a. hypha.
b. mycelium.
c. conidium.
d. thallus.
e. ascocarp.

_____ 6. A hypha containing two genetically distinct nuclei can result from

a. conjugation.
b. meiosis.
c. fusion of hyphae from two genetically different mating types.
d. polynucleate mitosis.
e. dikaryosis.

_____ 7. If you eat just any mushroom that you find in the wild, there's a chance that you will

a. die.
b. become intoxicated.
c. see colors that aren't really there.
d. not become nauseated or die.
e. ingest the hallucinogenic drug psilocybin.

_____ 8. A common fungal infection of the mucous membranes of the mouth or vagina is

a. an autoimmune response.
b. St. Anthony's fire.
c. histoplasmosis.
d. ergotism.
e. candidiasis.

_____ 9. The black fungus growing on a piece of bread

a. is heterothallic.
b. has male and female strains.
c. has coenocytic hyphae.
d. has sporangia on the tips of stolons.
e. is in the division Zygomycota.

_____10. When you eat a mushroom, you are eating

a. a club fungus.
b. a sporocarp.
c. the bulk of the vegetative body.
d. a Basidiomycota.
e. an ascocarp.

_____11. A lichen can be composed of a/an

a. alga and fungus.
b. phototroph and fungus.
c. cyanobacterium and ascomycete.
d. alga and basiodiomycete.
e. alga and ascomycete.

_____12. The genus *Penicillium*

a. is an ascomycete.
b. is a sac fungus.
c. produces the flavor in Roquefort cheese.
d. produces the antibiotic penicillin.
e. is an imperfect fungus.

____13. Lichens can be used as indicators of air pollution because they
   a. cannot excrete absorbed elements.   d. do not grow well in polluted areas.
   b. tolerate sulfur dioxide.   e. overgrow polluted areas.
   c. can endure large quantities of toxins.

____14. Fungi can reproduce
   a. sexually.   d. by fission.
   b. asexually.   e. by budding.
   c. by spore formation.

VISUAL FOUNDATIONS
Color the parts of the illustration below as indicated.

RED	☐	haustorium
GREEN	☐	hypha
YELLOW	☐	epidermal cell
BLUE	☐	cuticle

Color the parts of the illustration below as indicated.

RED	☐	released basidiospore	ORANGE	☐	base
GREEN	☐	mycelium	BROWN	☐	stalk
YELLOW	☐	gills	TAN	☐	cap
BLUE	☐	basidium			

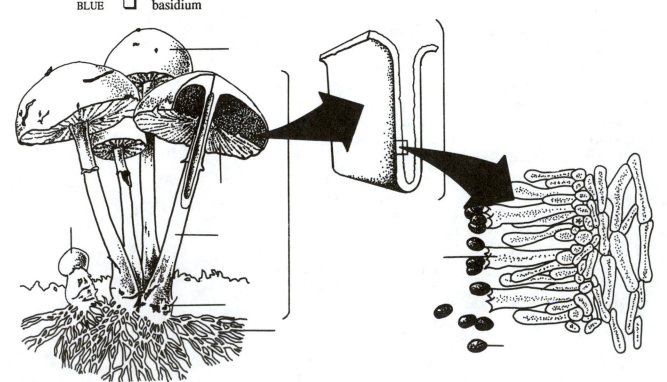

CHAPTER 26

The Plant Kingdom: Seedless Plants

Plants, believed to have evolved from green algae, are multicellular and exhibit great diversity in size, habitat, and form. They photosynthesize to obtain their energy. To survive on land, plants evolved 1) a waxy surface layer that protects against water loss, 2) tiny openings in the surface layer that allow for the gas exchange that is essential for photosynthesis, and 3) a strengthening polymer in the cell walls that enables them to grow tall and dominate the landscape. Plants spend part of their life cycle in a haploid stage that produces gametes (egg and sperm) by mitosis. The gametes fuse in a process known as fertilization, forming the diploid stage of the life cycle. To complete the life cycle, the diploid stage produces haploid spores by meiosis that then divide mitotically producing the haploid stage. The seedless plants include the bryophytes and the ferns and their allies. The bryophytes, i.e., the mosses, liverworts, and hornworts, are the only land plants without vascular tissues, and are therefore restricted in size. Most of the bryophytes require a moist environment. Ferns and their allies, i.e., the whisk ferns, club mosses, and horsetails, have a vascular system and are therefore more advanced than the bryophytes. A vascular system allows plants to achieve larger sizes because water and nutrients can be transported over great distances to all parts of the plant. The ferns and their allies also have a larger and more dominant diploid stage, which is a trend in the evolution of land plants. Most of the seedless plants produce only one kind of spore as a result of meiosis. A few however produce two types of spores, an evolutionary development that led to the evolution of seeds.

CHAPTER OUTLINE AND CONCEPT REVIEW
Fill in the blanks.

INTRODUCTION

1 Because of the numerous characteristics that they share, plants are thought to have evolved from ancient forms of the (a)_____. For example, both groups have the same photosynthetic pigments, which are (b)_____
_____; they both store excess carbohydrates as (c)_____;
(d)_____ is the major component of their cell walls; and a (e)_____
forms during cytokinesis in both groups.

2 There are four major groups of plants. Of these, only the (a)_____
_____ lack a vascular (conducting) system. Of the remaining three groups, only
the (b)_____ reproduce entirely by spores. The
(c)_____ have "naked seeds," often in cones, while the
(d)_____ produce seeds within a fruit.

PLANTS HAVE ADAPTED TO LIFE ON LAND

3 The adaptation of plants to land environments involved a number of important adaptations. Among them was the (a)_____ that covers aerial parts, protecting the plant against water loss. Openings in this covering also developed. These (b)_____ facilitate the gas exchange that is necessary for photosynthesis.

4 The sex organs of plants, or (a)_____, as they are called, are multicellular structures containing gametes. Both the male sex organ, the (b)_____, and the female sex organ, the (c)_____, contain gametes, sperm and egg respectively, that are protected from desiccation by an outer layer or jacket of "sterile" cells.

5 _____ is a polymer in the cell walls of large, vascular plants that strengthens and supports the plant and its conducting tissues.

THE PLANT LIFE CYCLE ALTERNATES BETWEEN TWO DIFFERENT GENERATIONS

6 Plants have alternation of generations, spending part of their life cycle in the haploid (gametophyte) stage and part in the diploid (sporophyte) stage. The first stage in the sporophyte generation is the (a)_____, and the haploid (b)_____ are the first stage of the gametophyte generation.

MOSSES AND OTHER BRYOPHYTES ARE NONVASCULAR PLANTS

7 Bryophytes can be separated into three classes, known in the vernacular as the _____. All of them are nonvascular and have similar life cycles.

Mosses have a dominant gametophyte generation

8 Mosses are land plants without _____. They are therefore restricted in size and most of them require a moist environment for growth and reproduction.

9 Among the most important mosses are those with large water storage vacuoles in their cells. These mosses are in the genus (a)_____, and are known by their common name, the (b)_____.

Liverworts are either thalloid or leafy

10 Liverworts reproduce sexually and asexually. In sexual reproduction, the thallus may simply branch and grow. Another form of asexual reproduction involves the formation of small masses of cells called _____ on the thallus, which can grow into a new liverwort thallus.

Hornworts are inconspicuous thalloid plants

Seedless vascular plants did not descend from the bryophytes

SEEDLESS VASCULAR PLANTS INCLUDE FERNS AND THEIR ALLIES

11 Ferns and fern allies have several advancements over the bryophytes, including the possession of vascular tissue and a dominant sporophyte generation. They also have two basic forms of leaves: the small (a)_____ with its single vascular strand and the larger (b)_____ with multiple vascular strands.

Ferns have a dominant sporophyte generation

12 In ferns, the dominant (conspicuous), diploid sporophyte generation is composed mainly of an underground stem called the (a)_____, from which extend roots and leaves called (b)_____.

13 (a)_____, the spore cases on the leaves of ferns, often occur in clusters called (b)_____. The haploid spore develops into a tiny, heart-shaped mature gametophyte plant called a (c)_____.

Whisk ferns are the simplest vascular plants

14 The main organ of photosynthesis in sporophyte whisk ferns is the (a)_____, which exhibits (b)_____ branching (dividing into two *equal* halves). The small haploid gametophyte is a nonphotosynthetic subterranean plant that apparently obtains nourishment through a symbiotic relationship with a (c)_____.

Horsetails have hollow, jointed stems

15 The few surviving species of horsetails are all in the genus _____.

Club mosses are small plants with rhizomes and short, erect branches

More advanced plants are less dependent on water as a transport medium for reproductive cells

Some ferns and club mosses are heterosporous

16 The bryophytes, horsetails, whisk ferns, and most ferns and club mosses produce only one type of spore, a condition known as (a)_____. Some ferns and club mosses exhibit (b)_____, the production of two different types of spores.

17 The strobilus of the club moss *Selaginella* has two kinds of sporangia. The small (a)_____ produce small "mother cells" or (b)_____ that undergo meiosis to form tiny, haploid (c)_____ that develop into sperm-producing male gametophytes. Egg-producing female gametophytes develop from (d)_____ that in turn result from the meiosis of (e)_____.

BUILDING WORDS

Use combinations of prefixes and suffixes to build words for the definitions that follow.

Prefixes	The Meaning		Suffixes	The Meaning
arch(e)-	primitive		-angi(o)(um)	vessel, container
bryo-	moss		-gon(o)(ium)	sexual, reproductive
gamet(o)-	sex cells, eggs and sperm		-phyte	plant
hetero-	different, other		-spor(e)(y)	spore
homo-	same			
mega-	large great			
micro-	small			
spor(o)-	spore			
xantho-	yellow			

Prefix	Suffix	Definition
_____	-phyll	1. Yellow plant pigment.
_____	_____	2. Special structure (container) of plants, protists, and fungi in which gametes are formed.
_____	_____	3. The female reproductive organ in primitive land plants.
_____	_____	4. The gamete-producing stage in the life cycle of a plant.
_____	_____	5. The spore-producing stage in the life cycle of a plant.
_____	-phyll	6. A small leaf that contains one vascular strand.
_____	-phyll	7. A large leaf that contains multiple vascular strands.
_____	_____	8. Special structure (container) of certain plants and protists in which spores and sporelike bodies are produced.
_____	_____	9. Production of one type of spore in plants (all spores are the same type).

_____   _____   10. Production of two different types of spores in plants, microspores and megaspores.

_____   _____   11. Large spore formed in a megasporangium.

_____   _____   12. Small spore formed in a microsporangium.

_____   _____   13. Mosses, liverworts, and their relatives.

MATCHING
For each of these definitions, select the correct matching term from the list that follows.

____ 1. Vascular tissue that conducts food in plants.

____ 2. The substance responsible for the hard, woody nature of plant stems and roots.

____ 3. Colorless, hair-like absorptive filaments analogous to roots that extend from the base of the stem of mosses, liverworts and fern prothallia.

____ 4. A cluster of sporangia (in the ferns).

____ 5. A type of branching in which the branches or veins always branch into two more or less equal parts.

____ 6. The heart-shaped, haploid gametophyte plant found in ferns, whisk ferns, club mosses, and horsetails.

____ 7. The male gametangium in certain plants.

____ 8. A conelike structure that bears sporangia.

____ 9. Vascular tissue that conducts water and dissolved minerals in plants.

____10. A horizontal underground stem.

Terms:

a. Antheridium   f. Protonema   k. Strobilus
b. Dichotomous   g. Rhizoid   l. Thallus
c. Lignin   h. Rhizome   m. Xylem
d. Phloem   i. Sorus
e. Prothallus   j. Stoma

TABLE
Fill in the blanks.

Plant	Nonvascular or Vascular	Method of Reproduction	Dominant Generation
Ferns	Vascular	Seedless, reproduce by spores	Sporophyte
Mosses	#1	#2	#3
Horsetails	Vascular	#4	Sporophyte
Angiosperms	#5	Seeds enclosed within a fruit	#6
Hornworts	#7	#8	Gametophyte
Whisk ferns	Vascular	#9	#10
Club mosses	#11	Seedless, reproduce by spores	#12
Gymnosperms	#13	#14	Sporophyte

THOUGHT QUESTIONS
Write your responses to these questions.

1. What are some of the structural and physiological developments that enabled plants to survive on land?
2. Using words and an illustration, explain alternation of generations.
3. What are the distinguishing characteristics of bryophytes? Of ferns?
4. Compare and contrast homosporous and heterosporous plants.

MULTIPLE CHOICE
Place your answer(s) in the space provided. Some questions may have more that one correct answer.

_____ 1. A strobilus is
   a. on a diploid plant.
   b. on a haploid plant.
   c. on a vascular plant.
   d. found on horsetails.
   e. found on plants that do not have true leaves.

_____ 2. Mosses
   a. are in the class Bryopsida.
   b. are in the division Bryophyta.
   c. may have antheridia and archegonia on the same plant.
   d. contributed significantly to coal deposits.
   e. are vascular plants.

_____ 3. The sporophyte generation of a plant
   a. is diploid.
   b. is haploid.
   c. produces haploid gametes.
   d. produces haploid spores by mitosis.
   e. produces haploid spores by meiosis.

_____ 4. The spore cases on a fern are
   a. usually on the fronds.
   b. formed by the haploid generation.
   c. called sporangia.
   d. often arranged in a sorus.
   e. precursors to the fiddlehead.

_____ 5. Land plants are thought to have evolved from
   a. fungi.
   b. green algae.
   c. bryophytes.
   d. Euglena-like autotrophs.
   e. mosses.

_____ 6. Plants all have
   a. chlorophyll a.
   b. chlorophyll b.
   c. xanthophyll.
   d. carotene.
   e. yellow pigments.

_____ 7. The gametophyte generation of a plant
   a. is diploid.
   b. is haploid.
   c. produces haploid spores.
   d. produces haploid gametes by mitosis.
   e. produces haploid gametes by meiosis.

_____ 8. Plant sperm form in
   a. diploid gametophyte plants.
   b. haploid sporophyte plants.
   c. haploid gametophyte plants.
   d. antheridia.
   e. archegonia.

____ 9. The leaves of vascular plants that evolved from several branches
     a. are megaphylls.   d. contain one vascular strand.
     b. are microphylls.   e. contain more than one vascular strand.
     c. evolutionarily derived from stem tissue.

____10. Plants in the class Bryopsida
     a. are mosses.   d. have true leaves and stems.
     b. have rhizoids.   e. have true roots.
     c. do not have vascular systems.

____11. The leafy green part of a moss is the
     a. sporophyte generation.   d. product of buds from a protonema.
     b. gametophyte generation.   e. thallus.
     c. rhizoid.

____12. Liverworts
     a. are vascular plants.   d. can produce archegonia and antheridia on a haploid thallus.
     b. are bryophytes.   e. are in the same class as hornworts.
     c. contain a medicine that cures liver disease.

____13. When a gamete produced by an archegonium fuses with a gamete produced by a male gametophyte plant, the result is
     a. an anomaly.   d. the first stage in the gametophyte generation.
     b. a diploid zygote.   e. called fertilization.
     c. the first stage in the sporophyte generation.

____14. The conspicuous fern plant we often see in yards and homes is a
     a. Psilophyta.   d. haploid sporophyte.
     b. Pterophyta.   e. diploid sporophyte.
     c. gamete producer.

____15. Spores grow
     a. into gametophyte plants.   d. to form a plant body by meiosis.
     b. into sporophyte plants.   e. to form a plant body by mitosis.
     c. into a haploid plant.

VISUAL FOUNDATIONS
Color the parts of the illustration below as indicated.

RED ☐ zygote

GREEN ☐ archegonium

YELLOW ☐ egg

BLUE ☐ sperm

ORANGE ☐ antheridium

BROWN ☐ capsule

TAN ☐ gametophyte

PINK ☐ sporophyte

VIOLET ☐ spore

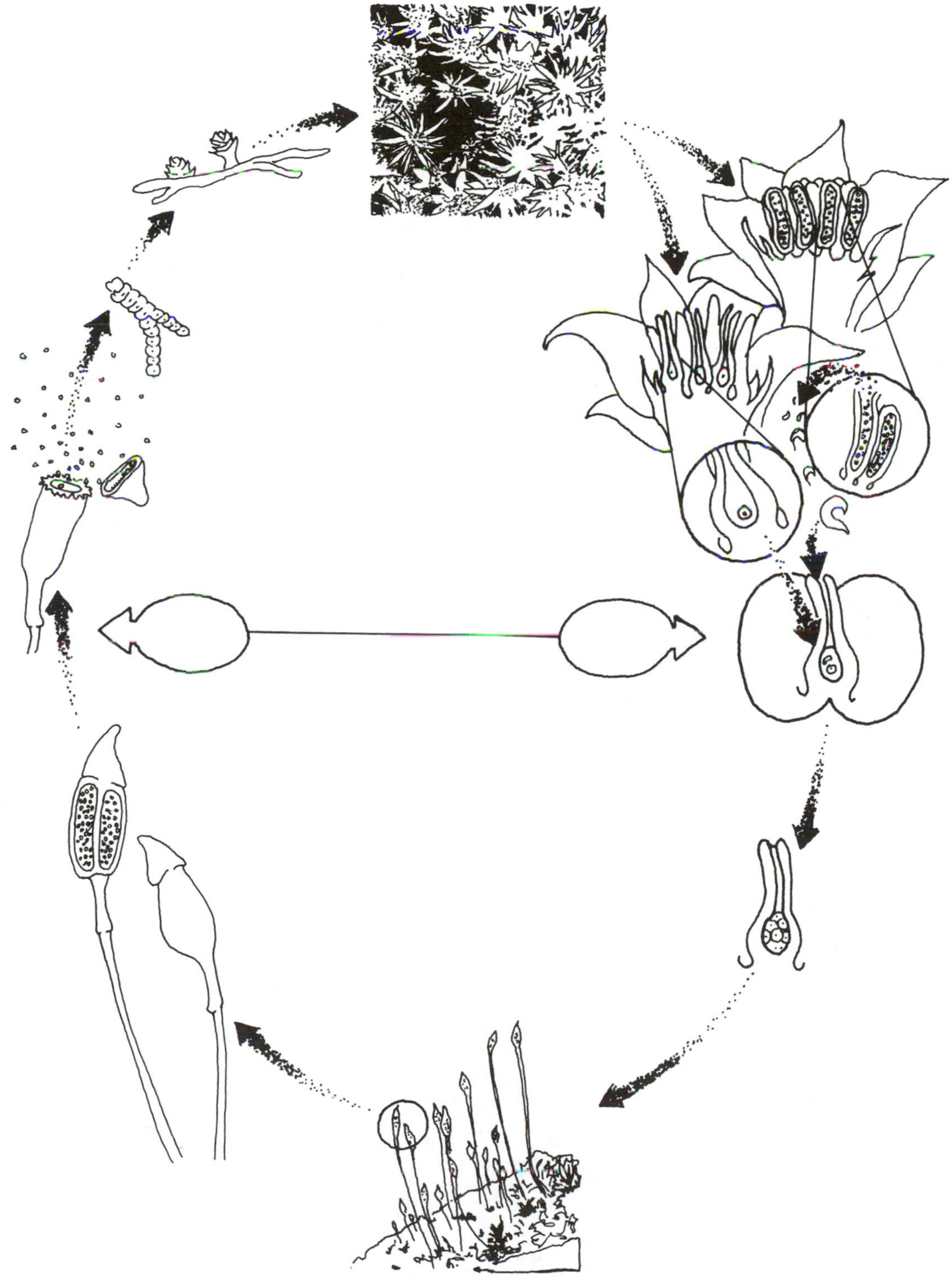

Color the parts of the illustration below as indicated.

RED	☐	zygote	BROWN	☐	sporangium
GREEN	☐	archegonium	TAN	☐	gametophyte
YELLOW	☐	egg	PINK	☐	mature sporophyte
BLUE	☐	sperm	VIOLET	☐	spores
ORANGE	☐	antheridium			

The Plant Kingdom: Seed Plants

The most successful plants on earth are the seed plants, also known as the gymnosperms and angiosperms (flowering plants). They are the dominant plants in most habitats. Whereas gymnosperm seeds are totally exposed or borne on the scales of cones, those of angiosperms are encased within a fruit. All seeds are multicellular, consisting of an embryonic root, stem, and leaves. They are protected by a resistant seed coat that allows them to survive until conditions are favorable for germination, and they contain a food supply that nourishes the seed until it becomes self-sufficient. Seed plants have vascular tissue, and they exhibit alternation of generations in which the gametophyte generation is significantly reduced. They are heterosporous, producing both microspores and megaspores. There are four phyla of gymnosperms — the Coniferophyta, the Ginkgophyta, the Cycadophyta, and the Gnetophyta. Although there is only one phylum of angiosperms — the Magnoliophyta — it is a huge one and is further divided into two major groups, the monocots and the dicots. Seeds and seed plants have been intimately connected with the development of human civilization. Human survival, in fact, is dependent on angiosperms, for they are a major source of human food. Angiosperm products play a major role in world economics. The organ of sexual reproduction in angiosperms is the flower. Flowering plants may be pollinated by wind or animals. Flowering plants have several evolutionary advancements that account for their success. The fossil record indicates that seed plants evolved from seedless vascular plants.

CHAPTER OUTLINE AND CONCEPT REVIEW
Fill in the blanks.

INTRODUCTION

1 The two groups of seed plants are the (a)_____, from the Greek "naked seed," and the (b)_____, meaning "seed in a case."

2 Seed plants all have two types of vascular tissues: (a)_____ for conducting water and (b)_____ to conduct food.

GYMNOSPERMS ARE THE "NAKED SEED" PLANTS

3 The four divisions of plants with "naked seeds" are: _____.

Conifers are woody plants that bear their seeds in cones

4 The conifers are the largest group of gymnosperms. They are woody plants that bear needle leaves and produce their seeds in cones. Most of them are _____, which means that the male and female reproductive organs are in different places in the same plant.

5 A few examples of conifers include _____.

6 The leaflike structures that bear sporangia on male cones of conifers are (a)_____, at the base of which are two microsporangia containing many "microspore mother cells" called (b)_____. These cells develop into the (c)_____
_____.

7   Female cones have (a)_____ at the base of bracts, within which are megasporocytes. Each of these cells produces four (b)_____, three of which disintegrate, while the fourth develops into the egg producing (c)_____.

Cycads are gymnosperms with compound leaves and simple seed cones

8   The cycads are palmlike or fernlike gymnosperms that reproduce in a manner similar to pines, except cycads are _____, meaning that male cones and female cones are on *separate* plants. There are only a few extant members of this once large division.

Ginkgo is the only living species in its phylum

Gnetophytes include three unusual gymnosperms

9   The gnetophytes share a number of advances over the rest of the gymnosperms, one of which is the presence of distinct _____ in their xylem tissues.

FLOWERING PLANTS PRODUCE FLOWERS, FRUITS, AND SEEDS

Monocots and dicots are the two classes of flowering plants

10 The monocots have floral parts in multiples of (a)_____, and their seeds contain (b)_____(#?) cotyledons. The nutritive tissue in their mature seeds is the (c)_____.

11 The dicots have floral parts in multiples of (a)_____, and their seeds contain (b)_____(#?) cotyledons. The nutritive tissue in their mature seeds is usually in the (c)_____.

12 A few examples of monocots include (a)_____ _____. A few examples of dicots include (b)_____ _____.

Flowers are involved in sexual reproduction

13 The four main organs in flowers are the (a)_____ _____, each of them arranged in whorls. The "male" organs are the (b)_____ and the "female" organs are the (c)_____. Flowers with both "male" and "female" parts are said to be (d)_____.

14 Pollen forms in the (a)_____, a saclike structure on the tip of a stamen, and ovules form within the (b)_____ at the base of a carpel.

The life cycle of flowering plants includes double fertilization

15 The megasporocyte in an ovule produces four haploid (a)_____, three of which disintegrate while the remaining one develops into the female gametophyte or (b)_____, as it is called.

16 Microsporocytes in the anther produce four haploid (a)_____, each of which develops into a (b)_____. A cell in this structure forms two male gametes called (c)_____.

17 Double fertilization is a phenomenon that is unique to flowering plants. It results in the formation of two structures, namely, the _____.

Flowering plants are the most successful plant group

THE FOSSIL RECORD PROVIDES VALUABLE CLUES ABOUT THE EVOLUTION OF SEED PLANTS

18 (a)_____ is an extinct group of plants with features intermediate to ancient seedless plants and modern gymnosperms. This group gave rise to conifers and to another group of extinct plants called the (b)_____, which in turn are thought to be the ancestors of cycads and ginkgoes. Flowering plants evolved from (c)_____.

BUILDING WORDS
Use combinations of prefixes and suffixes to build words for the definitions that follow.

Prefixes	The Meaning
angio-	vessel, container
di-	two, twice, double
gymno-	naked, bare, exposed
mon(o)-	alone, single, one

Prefix	Suffix	Definition
_____	-sperm	1. A plant having its seeds enclosed in an ovary (a "vessel" or "container").
_____	-sperm	2. A plant having its seeds exposed or naked.
_____	-oecious	3. Pertains to plant species in which the male and female cones or flowers are on one plant.
_____	-oecious	4. Pertains to plant species in which the male and female flowers are on two different plants.

MATCHING
For each of these definitions, select the correct matching term from the list that follows.

_____ 1. The part of the stamen in flowers that produces microspores and, ultimately, pollen.

_____ 2. The nutritive tissue that is found at some point in all flowering plant seeds.

_____ 3. The union of a male gamete and a female gamete to produce a zygote.

_____ 4. The colored cluster of modified leaves that constitute the next-to-outermost portion of a flower.

_____ 5. In seed plants, the transfer of pollen from the male to the female part of the plant.

_____ 6. The neck connecting the stigma to the ovary of a carpel.

_____ 7. The part of a plant that develops into seed after fertilization.

_____ 8. The outermost parts of a flower, usually leaflike in appearance, that protect the flower as a bud.

_____ 9. The collective term for the sepals of a flower.

_____10. The female reproductive unit of a flower that bears the ovules.

Terms:

a.	Anther	f.	Fertilization	k.	Pollination
b.	Calyx	g.	Ovary	l.	Sepals
c.	Carpel	h.	Ovule	m.	Style
d.	Corolla	i.	Petals		
e.	Endosperm	j.	Pistil		

TABLE
Fill in the blanks.

Plant Structure	Monocot	Dicot
Flowers	Floral parts in multiples of 3	Floral parts in multiples of 4 or 5
Nutritive material in mature seeds	#1	#2
Seeds	#3	2 cotyledons
Stems	Scattered vascular bundles	#4
Leaf venation	#5	#6
Leaf shape	#7	Broad to narrow leaves
Growth tissue type	Herbaceous	#8

THOUGHT QUESTIONS
Write your responses to these questions.

1. What is a plant?
2. What is an angiosperm? What is a gymnosperm?
3. Illustrate, describe, and compare the life cycles of a typical gymnosperm and a typical angiosperm.
4. List the characteristics that differentiate dicots and monocots.
5. Outline the evolutionary development of gymnosperms and angiosperms.

MULTIPLE CHOICE
Place your answer(s) in the space provided. Some questions may have more that one correct answer.

_____ 1. Lilies are/have
    a. monocots.
    b. angiosperms.
    c. in the division Magnoliophyta.
    d. in the class Liliopsida.
    e. flower parts in multiples of three.

_____ 2. Dioecious plants with naked seeds, motile sperm, and the ability to affect fertilization without water are
    a. ginkgoes.
    b. cycads.
    c. extinct.
    d. gnetophytes.
    e. gymnosperms.

_____ 3. The sepals of a flower are collectively known as the
    a. corolla.
    b. calyx.
    c. gynoecium.
    d. androecium.
    e. florus perfecti.

_____ 4. In angiosperms, of the four haploid megaspores that result from meiosis of the megaspore mother cell, one of them becomes the
    a. male gametophyte generation.
    b. female gametophyte generation.
    c. embryo sac.
    d. archegonium.
    e. antheridium.

_____ 5. A traditional Christmas tree would likely be

a. a gymnosperm.
b. a conifer.
c. the gametophyte generation.
d. in the division Coniferophyta.
e. heterosporous.

_____ 6. Which of the following is/are correct about a pine tree?

a. Sporophyte generation is dominant.
b. Male gametophyte produces an antheridium.
c. Gametophyte is dependent on sporophyte for nourishment.
d. Nutritive tissue in seed is gametophyte tissue.
e. Pollen grain is an immature male gametophyte.

_____ 7. A pine tree has

a. sporophylls.
b. megasporangia on the female cones.
c. separate male and female parts on the same tree.
d. two sizes of spores in separate cones.
e. female cones that are larger than male cones.

_____ 8. In gymnosperms, the pollen grain develops from

a. microspore cells.
b. spores.
c. the male gametophyte.
d. the gametophyte generation.
e. meiosis of cells in the microsporangium.

_____ 9. The triploid endosperm of angiosperms develops from fusion of

a. two sperm and one polar nucleus.
b. two polar nuclei and one sperm.
c. three polar nuclei.
d. one diploid ovule and one haploid sperm.
e. one haploid egg and one diploid sperm.

_____10. A perfect flower has

a. stamens only.
b. carpels only.
c. both stamens and carpels.
d. anthers.
e. ovules.

Visual Foundations on next page ➤

VISUAL FOUNDATIONS
Color the parts of the illustration below as indicated.

RED	☐	zygote
GREEN	☐	female cone
YELLOW	☐	megasporangium, megaspore,
BLUE	☐	pollen grain
ORANGE	☐	microsporangium, microspore

BROWN	☐	male cone
TAN	☐	sporophyte
PINK	☐	gametophyte
VIOLET	☐	embryo

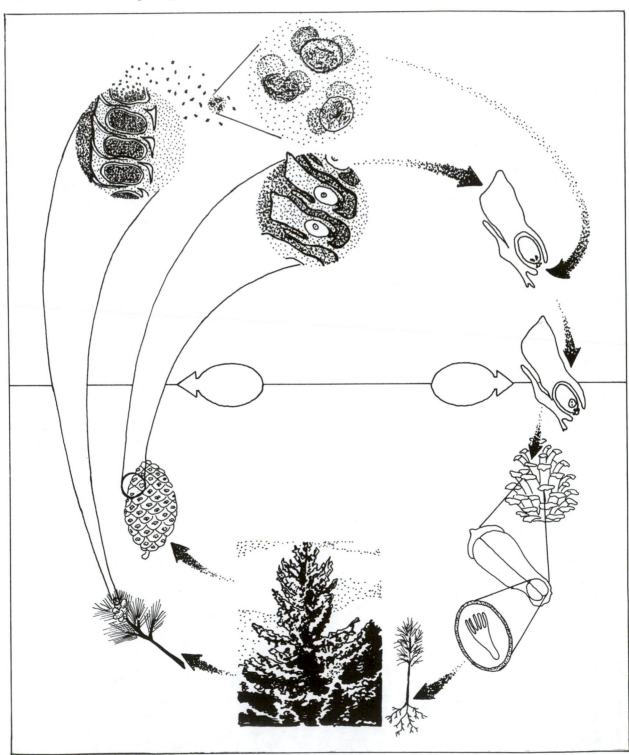

Color the parts of the illustration below as indicated.

RED ❏ zygote	BROWN ❏ male floral part	
GREEN ❏ female floral part	TAN ❏ sporophyte	
YELLOW ❏ megasporangium, megaspore	PINK ❏ gametophyte	
BLUE ❏ pollen tube	VIOLET ❏ embryo	
ORANGE ❏ microsporangium, microspore		

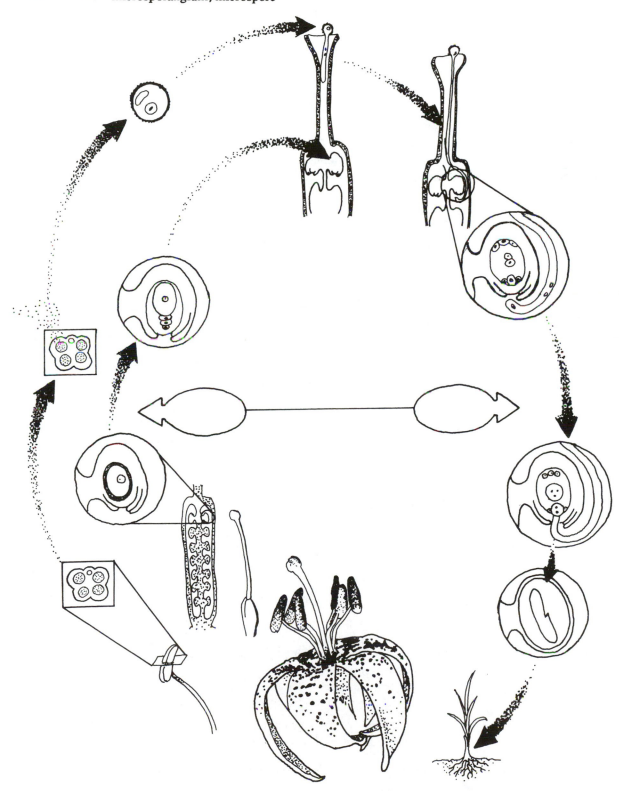

CHAPTER 28

The Animal Kingdom: Animals Without a Coelom

Animals are eukaryotic, multicellular, heterotrophic organisms. They are made up of localized groupings of cells with specialized functions. One grouping, for example, might function in locomotion, another might respond to external stimuli, and another might function in sexual reproduction. Animals inhabit virtually every environment on earth. They may be classified in a variety of ways — as radially symmetrical or bilaterally symmetrical; as those with no body cavity (the acoelomates), a body cavity between mesoderm and endoderm (the pseudocoelomates), or a body cavity within mesoderm (the coelomates); or those in which the first opening that forms in the embryonic gut becomes either the mouth (the protostomes) or the anus (the deuterostomes). This chapter briefly discusses the acoelomates and the pseudocoelomates. The acoelomates include the sponges, the cnidarians (hydras, jelly fish, and corals), the ctenophores (comb jellies), the platyhelminthes (flatworms), and the nemerteans (proboscis worms). The pseudocoelomates include the nematodes (roundworms), and the rotifers (wheel animals).

CHAPTER OUTLINE AND CONCEPT REVIEW
Fill in the blanks.

ANIMALS INHABIT MOST ENVIRONMENTS OF THE ECOSPHERE

1 Of the three major environments, the most hospitable to animals is (a)_____.
   (b)_____ environments are hypotonic to tissue fluids and
   (c)_____ environments tend to dehydrate animals.

ANIMALS CAN BE CLASSIFIED ACCORDING TO BODY STRUCTURE OR PATTERN OF DEVELOPMENT
Animals can be classified according to body symmetry

2 There are many anatomical terms that are used to describe the locations of body parts. Among them
   are: (a)_____ for front and (b)_____ for the rear; (c)_____ for
   a back surface and (d)_____ for the under or "belly" side; (e)_____ for parts
   closer to the midline and (f)_____ for parts farther from the midline, towards the sides;
   (g)_____ towards the head end and (h)_____ when away from the head, toward
   the tail.

Animals can be grouped according to type of body cavity

3 Animals can also be grouped according to the presence or absence and, when present, the type of
   body cavity. For example, the "lowly" cnidarians and flatworms are called (a)_____
   because they do not have a body cavity. The rest of the animals are either
   (b)_____, possessing a body cavity of sorts between the mesoderm and
   endoderm, or (c)_____ with a true body cavity within the mesoderm.

Animals can be classified as protostomes or deuterostomes

4 "Higher" animals can also be grouped by shared developmental characteristics. In mollusks, annelids, and arthropods — all protostomes — the blastopore develops into the (a)_____, whereas in echinoderms and chordates — both deuterostomes — the blastopore becomes the (b)_____. Early cell divisions in deuterostomes are either parallel to or at right angles to the polar axis, which is called (c)_____, while protostomes early embryonic divisions are at oblique angles to the polar axis, a type of division referred to as (d)_____.

PHYLUM PORIFERA CONSISTS OF THE SPONGES

5 Because they have collar cells, sponges are thought to have evolved from the protozoan _____. They appear to be an evolutionary "dead end."

6 Sponges are divided into three main classes on the basis of the type of skeleton they secrete. The
(a)_____ secrete spicules of calcium carbonate, the spicules of
(b)_____ (the glass sponges) contain silicon, and the
(c)_____ which have protein fibers called (d)_____, or silicon spicules, or a combination of both.

7 The sponge is a cellular sac perforated by numerous "pores." Water enters the internal cavity or
(a)_____ through these openings, then exits through the (b)_____, which is the only open end in the body.

HYDROZOANS, JELLYFISH, AND CORALS BELONG TO PHYLUM CNIDARIA

8 Cnidarians are characterized by radial symmetry; stinging cells called (a)_____; two definite tissue layers, the outer (b)_____ and the inner (c)_____ separated by a gelatinous (d)_____; and a nerve net.

9 The phylum is divided into three classes: Hydras, hydroids, and the Portuguese man-of-war are in the class (a)_____; jellyfish are in (b)_____; and sea anemones, corals, and sea whips and sea fans are in (c)_____.

Class Hydrozoa includes solitary and colonial forms

10 The epidermis of hydra contains specialized "stinging cells" called (a)_____, within which are "thread capsules" called (b)_____. These capsules release a long thread that entraps and sometimes also paralyzes their prey.

The jellyfish belong to class Scyphozoa

11 The familiar jellyfish is the (a)_____ stage of this scyphozoan. The (b)_____ stage is a small, relatively inconspicuous larva.

The corals belong to class Anthozoa

PHYLUM CTENOPHORA INCLUDES THE COMB JELLIES

12 Phylum ctenophora consists of the comb jellies, which are fragile, luminescent,
(a)_____ symmetrical marine animals with (b)_____(#?) rows of cilia resembling a comb. They have two tentacles with adhesive (c)_____ that trap prey.

FLATWORMS BELONG TO PHYLUM PLATYHELMINTHES

13 Flatworms are characterized by (a)_____ symmetry, cephalization, (b)_____(#?) definite tissue layers, well-developed organs, a simple brain and nervous system, and specialized excretory structures called (c)_____.

14 The four classes of Platyhelminthes are (a)_____, the free-living flatworms;
(b)_____, the two classes of flukes; and (c)_____, the tapeworms.

Class Turbellaria includes planarians

15 Planarians trap prey in a mucous secretion, then extend the _____ outward to "vacuum" them into their digestive tube.

The flukes belong to classes Trematoda and Monogenea

16 Flukes are similar to free-living flatworms, except they are parasitic and they have _____ for clinging to their hosts.

The tapeworms belong to class Cestoda

17 Most tapeworms have suckers and/or hooks on the (a)_____ for attachment to their hosts. Each segment, or (b)_____, contains both male and female reproductive organs. They absorb nutrients through the body wall.

PHYLUM NEMERTEA COMPRISES THE PROBOSCIS WORMS

18 Members of the phylum nemertea have a tube-within-a-tube body plan, a complete digestive tract with mouth and anus, and a separate circulatory system. Their most distinctive feature is the _____, which is a long, hollow, muscular tube that can be everted from the anterior end of the body for use in seizing food or in defense.

ROUNDWORMS BELONG TO PHYLUM NEMATODA

19 Nematodes include species of great ecological importance and species that are parasitic in plants and animals. They are characterized by (a)_____(#?) definite tissue layers, a body cavity called a (b)_____, (c)_____ symmetry, and a complete digestive tract.

Ascaris is a parasitic roundworm
Several other parasitic roundworms infect humans

20 The _____ larva enters the body through the skin, then migrates to the intestine where it matures.

21 (a)_____ enter the intestines of humans who eat undercooked, infected pork. Their larvae migrate to skeletal muscles where they form (b)_____.

22 Female _____ in the intestine migrate to the host's anal region to deposit eggs.

PHYLUM ROTIFERA ARE WHEEL ANIMALS

23 "Wheel animals" are aquatic, pseudocoelomate, microscopic worms that exhibit (a)_____ _____, which means that each member of a given species is composed of exactly the same number of cells. Their nickname comes from the circular aggregate of (b)_____ that seems to be spinning like a wheel.

BUILDING WORDS
Use combinations of prefixes and suffixes to build words for the definitions that follow.

Prefixes	The Meaning	Suffixes	The Meaning
deutero-	second	-coel(om)(y)	cavity
ecto-	outer, outside, external	-cyte	cell
entero-	intestine	-stome	mouth
gastro-	stomach		
meso-	middle		
proto-	first, earliest form of		
pseudo-	false		
schizo-	split		

Prefix	Suffix	Definition
_____	_____	1. A body cavity between the mesoderm and endoderm; not a true coelom.
_____	-derm	2. The outermost germ layer.
_____	-derm	3. The middle layer of the three basic germ layers.
_____	_____	4. Major division of the animal kingdom in which the mouth forms from the first opening in the embryonic gut (the blastopore); the anus forms secondarily.
_____	_____	5. Major division of the animal kingdom in which the mouth forms from the second opening in the embryonic gut; the first opening (the blastopore) forms the anus.
_____	_____	6. The process of coelom formation in which the mesoderm splits into two layers.
_____	_____	7. The process of coelom formation in which the mesoderm forms as "outpocketings" of the developing intestine, eventually separating and forming pouches which become the coelom.
spongo-	_____	8. The central cavity in the body of a sponge.
amoebo-	_____	9. Ameba-like cell found in many invertebrates.
_____	-dermis	10. The tissue lining the gut cavity ("stomach") that is responsible for digestion and absorption in certain phyla.

MATCHING

For each of these definitions, select the correct matching term from the list that follows.

_____ 1. The outer covering of some animals.

_____ 2. The flame-cell excretory organs of lower invertebrates and of some larva of higher animals.

_____ 3. Possessing sex organs of both the male and the female.

_____ 4. Without a body cavity (coelom).

_____ 5. The main body cavity of most animals.

_____ 6. A unique cell having a flagellum surrounded by a thin cytoplasmic collar.

_____ 7. The clustering of neural tissues at the anterior (leading) end of an animal.

_____ 8. Referring to the belly aspect of an animal's body.

_____ 9. A stinging structure found in cnidarians used for anchorage, defense and capturing prey.

_____10. Permanently attached to one location.

Terms:

a.	Acoelomate	f.	Cuticle	k.	Protonephridia
b.	Auricle	g.	Dorsal	l.	Sessile
c.	Cephalization	h.	Hermaphroditic	m.	Ventral
d.	Coelom	i.	Invertebrate		
e.	Collar cells	j.	Nematocyst		

TABLE
Fill in the blanks.

Category	Phylum Porifera	Phylum Cnidaria	Phylum Platyhelminthes	Phylum Nematoda
Representative animals	Sponges	Hydras, jellyfish, corals	Flatworms (planarians, flukes, tapeworms)	Roundworms (ascarids, hookworms, nematodes)
Level of organization	#1	Tissues	#2	#3
Type of body cavity, if present	Not present	#4	#5	Pseudocoelom
Nervous system	#6	Nerve net	#7	Cephalization, dorsal and ventral nerve cord, sense organs
Circulation	#8	Diffusion	#9	Diffusion
Digestion	#10	Intra- and extracellular digestion	#11	Complete digestive tract
Reproduction	#12	Asexual: budding Sexual: separate sexes	Asexual: fission Sexual: hermaphroditic	#13
Life style	#14	Capture food with cnidocytes, tentacles	#15	Carnivores, scavengers, parasites

THOUGHT QUESTIONS
Write your responses to these questions.

1. What is an animal?
2. How are body symmetry, type of body cavity, and patterns of development used to determine phylogenetic relationships? Does one of these characteristics seem more relevant than the others? Why? Does one seem least relevant? Why?
3. Create a table that illustrates the salient differences between the "acoelomic" phyla of animals. Include a specific example from each phylum.
4. What are the adaptive advantages, if any, of bilateral symmetry, a complete digestive system, hermaphroditism, and cephalization?

MULTIPLE CHOICE

Place your answer(s) in the space provided. Some questions may have more that one correct answer.

_____ 1. The parasitic roundworms
   a. are pseudocoelomates.
   b. have a mastax.
   c. are in the phylum Platyhelminthes.
   d. include hookworms.
   e. are nematodes.

_____ 2. The taxon that has "thread capsules" in epidermal stinging cells is
   a. Scyphozoa.
   b. Hydrozoa.
   c. Porifera.
   d. Cnidaria.
   e. Ctenophora.

_____ 3. Deuterostomes characteristically have
   a. radial cleavage.
   b. spiral cleavage.
   c. indeterminate cleavage.
   d. schizocoely.
   e. enterocoely.

_____ 4. Sessile, marine animals with no medusa stage and a partitioned gastrovascular cavity include
   a. hydrozoans.
   b. scyphozoans.
   c. anthozoans.
   d. jellyfish.
   e. corals, sea anemones, and their close relatives.

_____ 5. Examples of acoelomate eumetazoans include
   a. sponges.
   b. insects.
   c. flatworms.
   d. those groups whose coelom developed from a blastocoel.
   e. cnidarians.

_____ 6. The simplest bilaterians are members of
   a. Ctenophora.
   b. Porifera.
   c. Platyhelminthes.
   d. the phylum that includes round worms.
   e. deuterostomes.

_____ 7. The phyla of animals that are placed in a separate branch of the Eumetazoa because they are radially symmetrical are the
   a. Parazoa.
   b. Radiata.
   c. sponges.
   d. cnidarians.
   e. ctenophores.

_____ 8. With the possible exception of sponges, all animals
   a. are heterotrophs.
   b. are eukaryotes.
   c. are multicellular.
   d. reproduce sexually.
   e. are capable of locomotion.

_____ 9. The group of animals in which every individual has the same number of cells is
   a. Ctenophora.
   b. Nematoda.
   c. Porifera.
   d. Rotifera.
   e. Trematoda.

_____ 10. When a group of cells moves inward to form a pore in early embryonic development, and that pore ultimately develops into a mouth, the group of animals is considered to be a
   a. pseudocoelomate.
   b. deuterostome.
   c. protostome.
   d. parazoan.
   e. schizocoelomate.

____11. Animals that are parasitic and have a scolex and proglottids belong in the taxon/taxa
    a. Platyhelminthes.
    b. Nematoda.
    c. Turbellaria.
    d. Cestoda.
    e. Trichina.

____12. The taxon that has biradial symmetry, a mesoglea, and eight rows of cilia that move the animal is
    a. Scyphozoa.
    b. Hydrozoa.
    c. Porifera.
    d. Cnidaria.
    e. Ctenophora.

____13. The phylum/phyla that does/do not have specialized nerve cells is/are
    a. Porifera.
    b. Cnidaria.
    c. Platyhelminthes.
    d. Nemertea.
    e. Echinodermata.

____14. The one "lower" invertebrate phylum with radial symmetry and distinctly different tissues is
    a. Porifera.
    b. Cnidaria.
    c. Platyhelminthes.
    d. Nemertea.
    e. Echinodermata.

VISUAL FOUNDATIONS
Color the parts of the illustration below as indicated.

RED ☐ mesoderm
GREEN ☐ endoderm
YELLOW ☐ body cavity
BLUE ☐ ectoderm
ORANGE ☐ mesenchyme

Label the animals depicted as acoelomate, pseudocoelomate, or coelomate.

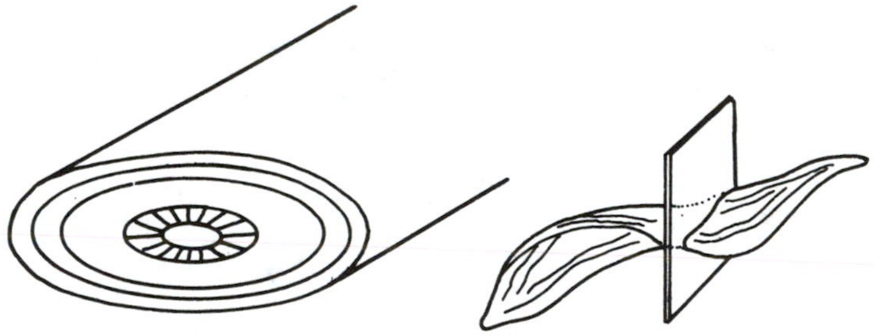

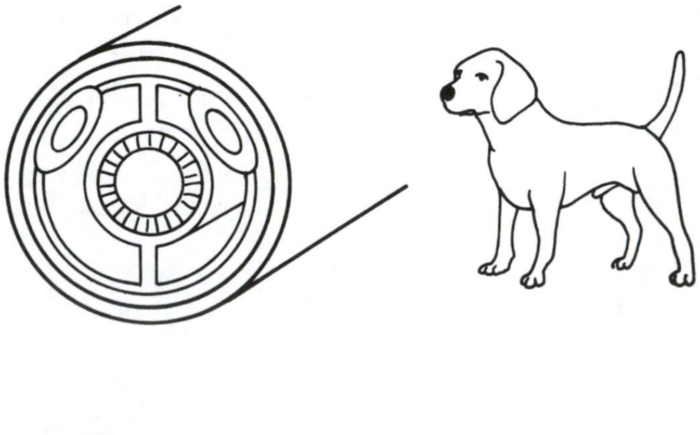

CHAPTER 29

The Animal Kingdom: The Coelomate Protostomes

The coelomate protostomes are animals with true coeloms in which the first opening that forms in the embryonic gut gives rise to the mouth. They have a complete digestive tract, and most have well-developed circulatory, excretory, and nervous systems. The coelom confers advantages to these animals over those without a coelom. For one thing, it separates muscles of the body wall from those of the digestive tract, permitting movement of food independent of body movements. Additionally, it serves as a space in which organs develop and function, and it can be used as a hydrostatic skeleton. Also, coelomic fluid helps transport food, oxygen, and wastes. Life on land requires many adaptations. To minimize water loss, terrestrial organisms have developed special body coverings as well as respiratory surfaces located deep inside the animal. To support the body against the pull of gravity, land animals have developed a supporting skeleton, located either within the body or externally covering the body. To meet the requirements of reproduction in a dry environment, some land animals return to water where their eggs and sperm are shed directly into the water. Most, however, copulate, depositing sperm directly into the female body. To prevent the developing embryo from drying out, it is either surrounded by a tough protective shell, or it develops within the moist body of the mother. The coelomate protostomes include the mollusks (chitons, snails, clams, oysters, squids, and octopuses), annelids (earthworms, leeches, and marine worms), onycophorans, and arthropods (spiders, scorpions, ticks, mites, lobsters, crabs, shrimp, insects, centipedes, and millipedes).

CHAPTER OUTLINE AND CONCEPT REVIEW
Fill in the blanks.

INTRODUCTION

1 The coelomate protostomes include the mollusks, annelids, and arthropods, as well as some minor phyla. The blastopore develops into the (a)_____, and the coelom is formed within and is lined by the (b)_____ between the complete digestive tube and body wall.

2 Among the many advantages derived from having a coelom are: _____
_____
_____.

LIFE ON LAND REQUIRES MANY ADAPTATIONS

3 Most invertebrate coelomates are still aquatic, however there are successful terrestrial groups as well. The first terrestrial, air-breathing animals were probably scorpion-like representatives of the phylum
_____.

4 Terrestrial animals must overcome serious problems inherent in a land-based "life style." Three that are critical to their continued existence are: They must develop a body covering that prevents (a)_____, a support system (skeleton) that can withstand (b)_____, and reproductive adaptations that compensate for the lack of a supportive watery environment.

MOLLUSKS HAVE A FOOT, VISCERAL MASS, AND MANTLE

5 Mollusks are soft-bodied animals usually covered by a shell; they possess a ventral foot for locomotion and a (a)_____ that covers the visceral mass. With about 60,000 species, the phylum Mollusca is second only to (b)_____ in number of species.

6 Most mollusks have an open circulatory system, meaning the blood flows through a network of open sinuses called the blood cavity or _____.

7 Most marine mollusks have larval stages, the first a free-swimming, ciliated form called a (a)_____ larva, and the second, when present, a (b)_____ larva with a shell and foot.

Class Polyplacophora includes the chitons

8 A distinctive feature of chitons is their shell, which is composed of _____(#?) overlapping plates.

Gastropods are the largest group of mollusks

9 Class gastropoda, the largest and most successful group of mollusks, includes the snails, slugs, and whelks. It is a class second in size only to the _____.

10 The gastropod shell (when present) is coiled, and the visceral mass is twisted, a phenomenon known as _____.

Bivalves typically burrow in the mud

11 Foreign matter lodged between the shell and (a)_____ of bivalves may cause deposition of the compound (b)_____, forming a pearl.

12 The _____ is the part of a scallop that we humans eat.

Cephalopods are active, predatory animals

13 Class cephalopoda includes the squids and octopods, which are active predatory animals. The foot is divided into (a)_____ that surround the mouth, (b)_____(#?) of them in squids, (c)_____(#?) of them in octopods.

ANNELIDS HAVE SEGMENTED BODIES

14 Phylum annelida, the segmented worms, includes many aquatic worms, earthworms, and leeches. Their body segments are separated from one another by partitions called _____.

15 A pair of bristle-like structures called _____ project from the segments in annelids. These structures anchor portions of the worms' body during locomotion.

16 _____ are pairs of excretory organs in each annelid segment.

Polychaetes have parapodia

17 Class polychaeta consists of marine worms characterized by bristled _____, used for locomotion and gas exchange.

Earthworms are familiar annelids

18 The earthworm is in the phylum (a)_____, the class (b)_____, and the genus and species (c)_____.

19 The two parts of the earthworm's stomach are the (a)_____ where food is stored and the (b)_____ where ingested materials are ground into a mulch.

20 The respiratory pigment in earthworms is (a)_____; the excretory system consists of paired (b)_____ in most segments; and the gas exchange surface is the (c)_____.

The leeches belong to class Hirudinea

ARTHROPODS HAVE JOINTED APPENDAGES AND AN EXOSKELETON OF CHITIN

21 The word "Arthropoda" means (a)_____, which is one of the phylum's distinguishing characteristics. Another is the (b)_____, an armor-like body covering composed of (c)_____.

22 The three body segments in arthropods are the _____.

23 Gas exchange in terrestrial arthropods occurs in a system of thin, branching tubes called _____.

Living arthropods can be assigned to three subphyla

24 The three subphyla of contemporary arthropods are (a)_____, the arthropods such as horseshoe crabs and arachnids that lack antennae; the (b)_____, which includes lobsters, crabs, shrimp, and barnacles that all have biramous appendages and two pairs of antennae; and (c)_____, which have unbranched appendages and one pair of antennae.

Arthropods are closely related to annelids

25 The most primitive arthropods are the (a)_____, which co-existed in Paleozoic waters with (b)_____, the subphylum that is thought to have arisen from these ancient arthropods.

Subphylum Chelicerata includes the horseshoe crabs and arachnids

26 Subphylum Chelicerata includes class (a)_____ (the horseshoe crabs) and class (b)_____ (spiders, mites, and their relatives).

27 The arachnid body consists of a (a)_____ and abdomen; there are (b)_____(#?) pairs of jointed appendages, of which (c)_____(#?) pairs serve as legs.

28 Gas exchange in arachnids occurs in tracheae and/or across thin, vascularized plates called _____.

29 Silk glands in spiders secrete an elastic protein that is spun into web fibers by organs called _____.

Subphylum Crustacea includes the lobsters, crabs, shrimp, and their relatives

30 Crustaceans are characterized by the (a)_____, the third pair of appendages used for biting food; (b)_____ appendages; and (c)_____(#?) pairs of antennae. In most crustaceans, (d)_____(#?) pairs of appendages are adapted for walking.

31 _____ are the only sessile crustaceans.

32 Lobsters, crayfish, crabs, and shrimp are members of the largest crustacean order, the _____.

33 Among the many highly specialized appendages in decapods are the (a)_____, two pairs of feeding appendages just behind the mandibles; the (b)_____ that are used to chop food and pass it to the mouth; the (c)_____ or pinching claws on the fourth thoracic segments; and the pairs of (d)_____ on the last four thoracic segments.

34 Appendages on the first abdominal segment of decapods are part of the (a)_____ system. (b)_____ on the next four abdominal segments are paddle-like structures adapted for swimming and holding eggs.

Subphylum Uniramia includes the insects, centipedes, and millipedes

35 Insects are described as articulated, meaning (a)_____, tracheated, meaning (b)_____, hexapods, which means (c)_____.

36 Adult insects typically have (a)_____(#?) pairs of legs, (b)_____(#?) pairs of wings, and (c)_____(#?) pair(s) of antennae. The excretory organs are called (d)_____.

Classes Chilopoda and Diplopoda include centipedes and millipedes

37 The members of the class (a)_____ or "centipedes" have (b)_____ pair(s) of legs per body segment, whereas the members of the class (c)_____ or millipedes have (d)_____(#?) pair(s) of legs per body segment.

BUILDING WORDS

Use combinations of prefixes and suffixes to build words for the definitions that follow.

Prefixes	The Meaning	Suffixes	The Meaning
arthro-	joint, jointed	-pod	foot, footed
bi-	twice, two		
cephalo-	head		
endo-	within		
exo-	outside, outer, external		
hexa-	six		
tri-	three		
uni-	one		

Prefix	Suffix	Definition
_____	-skeleton	1. Bony and cartilaginous supporting structures within the body that provide support from within.
_____	-skeleton	2. An external skeleton, such as the shell of arthropods.
_____	-valve	3. A mollusk that has two shells (valves) hinged together, as the oyster, clam, or mussel; a member of the class Bivalvia.
_____	-ramous	4. Consisting of or divided into two branches.
_____	-ramous	5. Unbranched; consisting of only one branch.
_____	-lobite	6. An extinct marine arthropod characterized by two dorsal grooves that divide the body into three longitudinal parts (or lobes).
_____	-thorax	7. Anterior part of the body in certain arthropods consisting of the fused head and thorax.
_____	_____	8. Having six feet; an insect.
_____	_____	9. An animal with paired, jointed legs.

MATCHING

For each of these definitions, select the correct matching term from the list that follows.

____ 1. Network of large spaces where body tissues are directly bathed in blood.

____ 2. A rasplike structure in the digestive track of certain mollusks.

____ 3. The first pair of appendages in arthropods.

____ 4. The shedding and replacement of the arthropod exoskeleton.

____ 5. Transition from one developmental stage to another, such as from a larva to an adult.

____ 6. The main body cavity of most animals.

____ 7. One of the microscopic air ducts branching throughout the body of most terrestrial arthropods and some terrestrial mollusks.

____ 8. A larval form found in mollusks and many polychaetes.

____ 9. The division of the animal body into a series of similar segments.

____10. An external mouth part in certain arthropods.

Terms:

a. Chelicerae
b. Coelom
c. Hemocoel
d. Malpighian tubule
e. Mandible

f. Metamerism
g. Metamorphosis
h. Molting
i. Radula
j. Tracheae

k. Trochophore

TABLE
Fill in the blanks.

Category	Mollusks	Annelids	Arthropods
Representative animals	Clams, snails, squid	Earthworms, marine worms	Crustaceans, insects
Level of organization	#1	#2	Organ system
Type of body cavity	Coelomate protostomes	#3	#4
Nervous system	#5	Cephalization, ventral nerve cords, sense organs	#6
Circulation	Open system, closed in cephalopods	#7	#8
Digestion	#9	Complete digestive tract	#10
Reproduction	Sexual: separate sexes	#11	Sexual: separate sexes
Life style	Herbivores, carnivores, scavengers, suspension feeders	#12	Herbivores, carnivores, scavengers

THOUGHT QUESTIONS
Write your responses to these questions.

1. What are some of the structural and physiological developments that enabled animals to survive on land?
2. Create a table that illustrates the salient differences between the phyla of animals that possess a coelom. Include a specific example from each phylum.
3. List distinguishing characteristics for each class in the phyla Mollusca, Annelida, and Arthropoda.

MULTIPLE CHOICE
Place your answer(s) in the space provided. Some questions may have more that one correct answer.

_____ 1. An octopus is a
   a. vertebrate.
   b. mollusk.
   c. chilopod.
   d. protostome.
   e. cephalopod.

_____ 2. Butterflies, potato bugs, termites, and black widows all
   a. are insects.
   b. are arthropods.
   c. have paired, jointed appendages.
   d. have closed circulatory systems.
   e. have an abdomen.

_____ 3. The most primitive arthropods are
   a. crustaceans.
   b. trilobites.
   c. in the class Chilopoda.
   d. spiders.
   e. insects.

_____ 4. Animals with most of the organs in a visceral mass that is covered by a heavy fold of tissue are
   a. annelids.
   b. oligochaets.
   c. onychophorans.
   d. mollusks.
   e. cheliceratans.

_____ 5. The animal that has a shell consisting of eight separate, overlapping transverse plates is
   a. an insect.
   b. a chiton.
   c. in the same class as squids.
   d. in the class Polyplacophora.
   e. a mollusk.

_____ 6. Earthworms are
   a. annelids.
   b. oligochaets.
   c. in the phylum Polychaeta.
   d. protostomes.
   e. acoelomate animals.

_____ 7. Garden snails and abalone, considered by many to be gourmet delicacies, are
   a. in different phyla.
   b. in the same phylum.
   c. in different classes.
   d. in the same class.
   e. nudibranchs.

_____ 8. A marine animal with parapodia and a trochophore larva might very well be
   a. a tubeworm.
   b. an oligochaet.
   c. bilaterally symmetrical.
   d. a leech.
   e. a polychaet.

_____ 9. Animals thought by many zoologists to be a link between annelids and arthropods are
  a. wormlike.
  b. acoelomates.
  c. found in humid tropical rain forests.
  d. onychophorans.
  e. in the class Hirudinea.

_____10. Earthworms are held together in copulation by mucous secretions from the
  a. clitellum.
  b. typhlosole.
  c. seminal receptacles.
  d. epidermis.
  e. prostomium.

VISUAL FOUNDATIONS
Color the parts of the illustration below as indicated.

RED ❑ digestive tract
GREEN ❑ shell
YELLOW ❑ foot

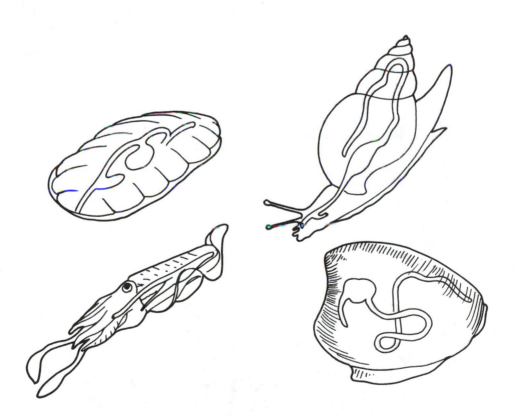

Color the parts of the illustration below as indicated.

RED   ❑   dorsal blood vessel
GREEN   ❑   nephridium
YELLOW   ❑   ventral nerve cord
BLUE   ❑   cerebral ganglia
ORANGE   ❑   septa
BROWN   ❑   pharynx
TAN   ❑   coelom

Color the parts of the illustration below as indicated.

RED   ❑   heart
GREEN   ❑   Malpighian tubules
YELLOW   ❑   brain, nerve cord
BLUE   ❑   ovary
ORANGE   ❑   digestive gland
BROWN   ❑   intestine

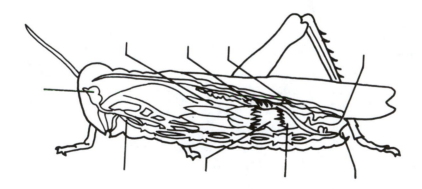

CHAPTER 30

❑

The Animal Kingdom: The Deuterostomes

Deuterostomes all have a true coelom. They owe their name to the fact that the second opening that develops in the embryo becomes the mouth, and the first becomes the anus. The deuterostomes include the echinoderms, chaetognaths, lophophorates, hemichordates, and the chordates. The echinoderms and the chordates are by far the more ubiquitous and dominant deuterostomes. The echinoderms all live in the sea. They have spiny skins, a water vascular system, and tube feet. There are five groups of echinoderms: the feather stars and sea lilies, the starfish, the basket stars and brittle stars, the sea urchins and sand dollars, and the sea cucumbers. The chordates include animals that live both in the sea and on land. They have a notochord, a dorsal tubular nerve cord, and pharyngeal gill slits, at least at some time in their life cycle. There are three groups of chordates: the tunicates, the lancelets, and the vertebrates. The vertebrates are characterized by a vertebral column, a cranium, pronounced cephalization, a differentiated brain, muscles attached to an endoskeleton, and two pairs of appendages. There are seven major groups of vertebrates: jawless fishes, cartilaginous fishes, bony fishes, amphibians, reptiles, birds, and mammals. Mammals are classified first on the basis of the site of embryonic development. Those that complete embryonic development in an egg are the monotremes; the duck-billed platypus is a monotreme. Those that complete embryonic development in a maternal pouch are the marsupials; the kangaroo is a marsupial. Those that complete embryonic development within an organ of exchange, i.e., a placenta, are the placental mammals; the human is an example of a placental mammal.

CHAPTER OUTLINE AND CONCEPT REVIEW
Fill in the blanks.

INTRODUCTION

1 Echinoderms and chordates are thought to be related because they are both deuterostomes and therefore share many developmental characteristics. Among other things, this means that the (a)_____ forms from the first embryonic opening, and the (b)_____ forms from a second opening.

ECHINODERMS ARE "SPINY-SKINNED" MARINE ANIMALS

2 Phylum echinodermata includes marine animals with (a)_____ skins, a water vascular system, and tube feet; larvae have (b)_____ symmetry; most adults have (c)_____ symmetry. Echinoderms also have a true (d)_____ that houses the internal organs.

Class Crinoidea includes the feather stars and sea lilies

3 Class Crinoidea includes sea lilies and feather stars. Unlike other echinoderms, the oral surface of crinoids is located on the _____ surface.

Class Asteroidea includes the sea stars

4 Sea stars have a (a)_____ from which radiate five or more arms or rays. Each ray contains hundreds of small (b)_____, each of which contains a canal that leads to a central canal.

Class Ophiuroidea includes basket stars and brittle stars

Class Echinoidea includes sea urchins and sand dollars

5 Class echinoidea includes the sea urchins and sand dollars, animals that lack arms; they have a solid shell called a _____, and their body is covered with spines.

Class Holothuroidea includes sea cucumbers

6 Class Holothuroidea consists of sea cucumbers, animals with flexible bodies. The mouth is surrounded by a circle of modified _____ that serve as tentacles.

CHORDATES HAVE A NOTOCHORD, A DORSAL TUBULAR NERVE CORD, AND PHARYNGEAL GILL SLITS DURING SOME TIME IN THEIR LIFE CYCLE

7 At some time in its life cycle, a chordate has a notochord, a dorsal tubular nerve cord, and pharyngeal gill slits. Chordates are all coelomates with _____ symmetry.

8 The three characteristics that distinguish chordates from all other groups are the dorsal, flexible rod called the (a)_____; the single, dorsal, hollow (b)_____; and the embryonic (c)_____ that develop into functional respiratory structures in aquatic chordates and entirely different structures in terrestrial chordates.

Subphylum Urochordata includes the tunicates

9 Subphylum Urochordata comprises the tunicates, which are sessile, filter-feeding marine animals that have tunics made of _____.

Subphylum Cephalochordata includes the lancelets

10 Subphylum Cephalochordata consists of the lancelets, small segmented fishlike animals that exhibit all three chordate characteristics. A common genus, *Branchiostoma*, also known as _____, is often selected as a representative, typical chordate.

11 The oral hood is a vestibule of the digestive tract just anterior to the mouth. Food particles swept into the mouth are trapped in mucus in the _____, then passed back to the intestine.

The success of the vertebrates is linked to the evolution of key adaptations

12 Subphylum Vertebrata includes animals with a backbone or (a)_____, a braincase or (b)_____, and concentration of nerve cells and sense organs in a definite head, a phenomenon known as (c)_____.

13 The classes in the subphylum Vertebrata include the jawless fish or (a)_____; the cartilaginous fish or (b)_____; the bony fish or (c)_____; tetrapod frogs, toads, and salamanders in the class (d)_____; the lizards, snakes, turtles, and alligators in the class (e)_____; birds or (f)_____; and animals with hair and mammary glands in the class (g)_____.

The jawless fishes are the most primitive vertebrates

14 It is thought that the now extinct _____, that existed some 500 million years ago, gave rise to the present day lampreys and hagfishes.

The earliest jawed fishes are now extinct

Class Chondrichthyes includes the sharks, rays, and skates

15 The skin of cartilaginous fish contains numerous _____, toothlike structures composed of layers of enamel and dentine.

16 (a)_____ are sensory grooves along the sides of fish that contain cells capable of perceiving water movements. Similarly, the (b)_____ on the head can sense very weak electrical currents, like those generated by another animal's muscle contractions.

17 Sharks may be (a)_____, that is, lay eggs; or (b)_____, which means their eggs hatch internally; or (c)_____, giving live birth to their young.

Bony fish belong to class Osteichthyes

18 The lobe-finned (a)_____, generally considered to be the ancestors of land vertebrates, are survived by the genus *Latimeria* in the suborder of (b)_____.

19 In the Devonian period, bony fishes diverged into two major groups: the (a)_____ fish or actinopterygians and the (b)_____ with their fleshy, lobed fins and lungs. The actinopterygians gave rise to the (c)_____, the modern bony fish in which lungs became modified as (d)_____.

Class Amphibia includes frogs, toads, and salamanders

20 The three orders of modern amphibians are: (a)_____, the amphibians with long tails, including salamanders, mud puppies, and newts; the (b)_____, tailless frogs and toads; and (c)_____, the wormlike caecilians.

21 Most amphibians return to water to reproduce; frog embryos develop into tadpoles, which undergo metamorphosis to become adults. Some salamanders, such as the mud puppy, retain several larval characteristics in adulthood, a process known as _____.

22 Amphibians use lungs and their (a)_____ for gas exchange. They have a (b)_____(#?)-chambered heart with systemic and pulmonary circulations, and they have (c)_____ in the skin that help keep the body surface moist.

Class Reptilia includes turtles, lizards, snakes, and alligators

23 Reptiles are true terrestrial animals; fertilization is internal; most secrete a protective shell around the egg; the embryo develops an amnion and other extraembryonic membranes. Most reptiles have a (a)_____(#?)-chambered heart, excrete nitrogenous wastes in the form of (b)_____, and are (c)_____, which means they cannot regulate body temperature.

24 Living reptiles are in three orders: the (a)_____ includes turtles and tortoises; the lizards, snakes, iguanas, and geckos in the order (b)_____; and (c)_____ with its crocodiles, alligators, and caimans.

25 Reptiles dominated the earth during the _____ era; then, during the Cretaceous period, most of them, including all of the dinosaurs, became extinct.

The birds belong to class Aves

26 Adaptations for flight in birds include feathers, wings, and light hollow bones. Birds also have a (a)_____(#?)-chambered heart, very efficient lungs, a high metabolic rate, and they are (b)_____, meaning they can maintain a constant body temperature. They excrete wastes as semi-solid (c)_____. Birds also have a well-developed nervous system and excellent vision and hearing. Of all these characteristics, birds are the only animals that have (d)_____.

27 The saclike portion of a bird's digestive system that temporarily stores food is the (a)_____, whereas the (b)_____ portion of the stomach secretes powerful gastric juices, and the (c)_____ grinds the food.

28 The earliest know bird, _____, was a medium sized bird with rather feeble wings. With its teeth, long tail, and clawed wing tips, it looked somewhat like a reptile.

Mammals have hair and mammary glands

29 The distinguishing characteristics of mammals are _____
_____. They also maintain a constant body temperature, and they have a highly developed nervous system and a muscular diaphragm.

30 Mammals probably evolved from the reptilian group (a)_____, about 200 million years ago in the (b)_____ period.

31 _____ are mammals that lay eggs. They include the duck-billed platypus and the spiny anteater.

32 (a)_____ are pouched mammals. The young are born immature and complete development in the (b)_____, where they are nourished from the mammary glands.

33 The placental mammals are characterized by an organ of exchange (placenta) that develops between the embryo and the mother; this organ supplies _____ to the fetus, enabling it to complete development within the uterus.

BUILDING WORDS

Use combinations of prefixes and suffixes to build words for the definitions that follow.

Prefixes	The Meaning	Suffixes	The Meaning
a-	without	-derm	skin
Chondr(o)-	cartilage	-gnath(an)	jaw
echino-	sea urchin, "spiny"	-ichthy(es)	fish
Oste(o)-	bone	-pod(al)	foot, footed
tetra-	four	-ur(o)(an)	tail
Uro-	tail		

Prefix	Suffix	Definition
_____	_____	1. A fish without jaws (jawless fish); a member of the class of vertebrates including lampreys and hagfishes.
an-	_____	2. An amphibian with legs but no tail; a tailless frog or toad.
_____	_____	3. Pertains to those amphibians with no feet or legs, i.e., the wormlike caecilians.
_____	_____	4. The class comprising the cartilaginous fishes.
_____	_____	5. The class comprising the bony fishes.
_____	_____	6. A vertebrate with four legs; a member of the superclass Tetrapoda.
_____	-dela	7. The order which comprises the amphibians with long tails, i.e., the salamanders, mudpuppies, and newts.
_____	_____	8. A spiny-skinned animal.

MATCHING

For each of these definitions, select the correct matching term from the list that follows.

____ 1. Sexual maturity in a larval or otherwise immature stage.

____ 2. The flexible, longitudinal rod in the anteroposterior axis that serves as an internal skeleton in the embryos of all chordates and in the adults of some.

____ 3. The pouch in which marsupial young develop.

____ 4. Animals that are egg-layers.

____ 5. An extra-embryonic membrane that forms a fluid-filled sac for the protection of the developing embryo.

____ 6. An extinct jawed fish.

____ 7. A group of mammal-like reptiles of the Permian period that gave rise to the mammals.

____ 8. The partly fetal and partly maternal organ whereby materials are exchanged between fetus and mother in the uterus of eutherian mammals.

____ 9. A chordate possessing a bony vertebral column.

____10. Bearing living young that develop within the body of the mother.

Terms:

a.	Amnion	f.	Marsupium	k.	Placenta
b.	Amphibian	g.	Neoteny	l.	Placoderm
c.	Coelacanth	h.	Notochord	m.	Therapsid
d.	Cotylosaur	i.	Oviparous	n.	Vertebrate
e.	Labyrinthodont	j.	Ovoviviparous	o.	Viviparous

TABLE
Fill in the blanks.

Vertebrate Class	Representative Animals	Heart	Skeletal Material	Other Characteristics
Agnatha	Hagfish	Two-chambered heart	Cartilage	Jawless, most primitive vertebrates, gills,
Chondrichthyes	#1	Two-chambered heart	#2	#3
#4	Salmon, tuna	#5	#6	Gills, swim bladder
Amphibia	#7	#8	bone	Aquatic larva metamorphoses into terrestrial adult, moist skin and lungs for gas exchange
#9	#10	Three-chambered heart	#11	Tetrapods with hard scales and dry skin, reproduction adapted for land
#12	Birds	#13	#14	Tetrapods with feathers, many adaptations for flight, constant body temperature, high metabolic rate
#15	Monotremes, marsupials, placentals	Four-chambered heart	#16	#17

Write your responses to these questions.

1. What is a deuterostome? Give examples of deuterostomes.
2. Describe the classes of echinoderms and name an example of each.
3. Compare and contrast characteristics possessed by each of the chordate subphyla.
4. Describe characteristics possessed by each of the vertebrate classes and name two examples of each.
5. Describe characteristics possessed by the orders of placental mammals and name an example of each.
6. Sketch a phylogenetic tree for the major vertebrate taxa.

MULTIPLE CHOICE

Place your answer(s) in the space provided. Some questions may have more that one correct answer.

_____ 1. The group(s) of animals with a part of the stomach that secretes gastric juices and a separate part of the stomach that grinds food is/are
    a. Reptilia.   d. Amphibia.
    b. Mammalia.   e. Pisces.
    c. Aves.

_____ 2. Dolphins and porpoises are members of the taxon(a)
    a. Lagomorpha.   d. that includes placental mammals.
    b. Rodentia.   e. Cetacea.
    c. that has chisel-like incisors that grow continually.

_____ 3. The first successful land vertebrates were
    a. tetrapods.   d. amphibians.
    b. reptiles.   e. in the superclass Pisces.
    c. labyrinthodonts.

_____ 4. A frog is a member of the taxon(a)
    a. Reptilia.   d. Amphibia.
    b. Chelonia.   e. Anura.
    c. Urodela.

_____ 5. A turtle is a member of the taxon(a)
    a. Reptilia.   d. Amphibia.
    b. Chelonia.   e. Anura.
    c. Urodela.

_____ 6. Unique features of the echinoderms include
    a. radial larvae.   d. gas exchange by diffusion.
    b. ciliated larvae.   e. water vascular system.
    c. endoskeleton composed of $CaCO_3$ plates.

_____ 7. Mammals probably evolved from
    a. amphibians.   d. monotremes.
    b. a type of fish.   e. therapsids.
    c. reptiles.

_____ 8. The group(s) of animals that maintain a constant internal body temperature is/are
    a. Reptilia.   d. Amphibia.
    b. Mammalia.   e. Pisces.
    c. Aves.

_____ 9. The lateral line organ is found
    a. only in sharks.   d. in all fish.
    b. only in bony fish.   e. in all fish except placoderms.
    c. only in agnathans.

_____10. The acorn worm
    a. has a notochord.   d. is in the phylum Chordata.
    b. is a deuterostome.   e. is a hemichordate.
    c. is in the same subphylum as *Amphioxus*.

_____11. A sting ray is
    a. a bony fish.   d. in the same superclass as a rainbow trout.
    b. a jawless fish.   e. in the class Chondrichthyes.
    c. in the same class as sharks.

_____12. The lungs of the ancestors of most modern fish became modified as
    a. gills.   d. a hydrostatic organ.
    b. a vascular pump.   e. a swim bladder.
    c. an organ that stores oxygen.

_____13. The group(s) of animals with both four-chambered hearts and double circuit of blood flow is/are the
    a. reptiles.   d. class Aves.
    b. birds.   e. amphibians.
    c. mammals.

_____14. Members of the phylum Chordata all have
    a. a notochord.   d. well-developed germ layers.
    b. gill slits.   e. bilateral symmetry.
    c. a dorsal, tubular nerve cord.

_____15. The phylum(a) that many biologists believe had a common ancestry with our own phylum is/are
    a. mollusca.   d. echinodermata.
    b. annelida.   e. chordata.
    c. arthropoda.

_____16. Humans are members of the taxon(a)
    a. Lagomorpha.   d. that includes placental mammals.
    b. Rodentia.   e. Cetacea.
    c. that has chisel-like incisors that grow continually.

_____17. The "spiny-skinned" animals include
    a. sea stars.   d. lancets.
    b. sand dollars.   e. marine worms.
    c. snails.

_____18. A rabbit is a member of the taxon(a)
    a. Lagomorpha.   d. that includes placental mammals.
    b. Rodentia.   e. Cetacea.
    c. that has chisel-like incisors that grow continually.

_____19. A hermaphroditic animal that traps food in mucus secreted by cells of the endostyle is a/an
    a. holothuroidean.   d. sea cucumber.
    b. enchinoderm.   e. tunicate.
    c. urochordate.

_____20. Guinea pigs are members of the taxon(a)
    a. Lagomorpha.   d. that includes placental mammals.
    b. Rodentia.   e. Cetacea.
    c. that has chisel-like incisors that grow continually.

_____21. Cats and dogs are members of the taxon(a)
    a. Lagomorpha.   d. that includes placental mammals.
    b. Rodentia.   e. Cetacea.
    c. that has chisel-like incisors that grow continually.

VISUAL FOUNDATIONS
Color the parts of the illustration below as indicated.

      RED   ❑   heart
    GREEN   ❑   mouth
  YELLOW   ❑   brain and dorsal, hollow nerve tube
    BLUE   ❑   notochord
 ORANGE   ❑   pharyngeal gill slits
  BROWN   ❑   pharynx, intestine
     TAN   ❑   muscular segments

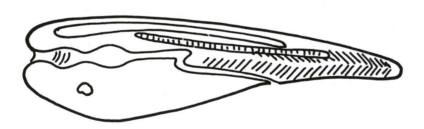

PART VI

□

Structure and Life Process in Plants

CHAPTER 31

Plant Structure, Growth, and Differentiation

The plant body is organized into a root system and a shoot system. The latter system consists of the stems and leaves. Plants are composed of cells that are organized into tissues, and tissues that are organized into organs. Each organ performs a single function or group of functions, and is dependent on other organs for its survival. Leaves, for example, photosynthesize, providing sugar molecules to the rest of the plant, and roots absorb water and minerals, making them available to the entire plant body. All vascular plants have three tissue systems: the dermal tissue system provides a covering for the plant body; the vascular tissue system conducts various substances throughout the plant body; and the ground tissue system has a variety of functions. The tissue systems are composed of various simple and complex tissues. The major plant tissues are parenchyma, collenchyma, sclerenchyma, xylem, phloem, epidermis, and periderm. Plants grow by increasing both in girth and length. Unlike growth in animals, plant growth is localized in areas of unspecialized cells, and involves cell proliferation, elongation, and differentiation.

CHAPTER OUTLINE AND CONCEPT REVIEW
Fill in the blanks.

INTRODUCTION

1 Plants with secondary growth that live year-after-year are (a)_____, while those that survive for only one season and have only primary growth are (b)_____. Plants that complete their life cycles in two years are (c)_____.

ROOTS, STEMS, LEAVES, FLOWERS, AND FRUITS MAKE UP THE PLANT BODY

THE PLANT BODY IS COMPOSED OF CELLS AND TISSUES

2 As is the case in animals, plant cells are organized in aggregations that perform specific functions. These groups of cells are (a)_____, which in turn are organized to form (b)_____ such as roots, stems, and leaves.

The ground tissue system is composed of three simple tissues

3 Parenchyma tissue is composed of relatively unspecialized, living cells with very thin cell walls. Their three primary functions include _____

_____.

4 Collenchyma tissue is composed of living cells that function primarily to _____ the plant.

The vascular tissue system consists of two complex tissues

5 The tracheid and vessel elements in xylem conduct (a) _____ to stems and leaves. Xylem also contains (b)_____ for storage and (c)_____ for support.

6 Phloem is a complex tissue that functions to conduct _____ throughout the plant.

7 Highly specialized cells in phloem called (a)_____ conduct food in solution. They are stacked to form tubes with continuous cytoplasm running through holes, the (b)_____, between the cells.

The dermal tissue system consists of two complex tissues

8 The epidermis covers the plant body and functions primarily for (a)_____. They secrete a layer called the (b)_____ that restricts water loss. (c)_____ are openings in this layer through which gases diffuse.

9 The periderm covers the plant body in plants with secondary growth. Its primary function is for _____.

PLANTS EXHIBIT LOCALIZED GROWTH AT MERISTEMS

10 Plant growth is localized in regions called (a)_____, and involves three cell activities, namely, (b)_____.

11 Plants have two kinds of growth: (a)_____ growth, an increase in length; and (b)_____ growth, an increase in width of the plant.

Primary growth takes place at apical meristems

12 The root apical meristem consists of three zones, which, in order from the root cap inward, are the (a)_____, (b)_____, and (c)_____.

13 The root apical meristem contains embryonic leaves called (a)_____ and (b)_____ or embryonic buds.

Secondary growth takes place at lateral meristems

14 Two lateral meristems are responsible for secondary growth (increase in girth), the (a)_____ _____, which forms a cylinder of cells around the stem, and the (b)_____ _____, consisting of patches of cells in the outer bark.

BUILDING WORDS
Use combinations of prefixes and suffixes to build words for the definitions that follow.

Prefixes	The Meaning
bi-	two
epi-	on, upon
stom-	mouth
trich-	hair

Prefix	Suffix	Definition
_____	-ome	1. A hair or other appendage growing out from the epidermis of plants.
_____	-ennial	2. A plant that takes two years to complete its life cycle.
_____	-dermis	3. Along with the periderm, provides a protective covering on the surface of plants.
_____	-a	4. A small pore ("mouth") in the epidermis of plants.

MATCHING
For each of these definitions, select the correct matching term from the list that follows.

_____ 1. Vascular tissue that conducts water and dissolved minerals through the plant.

_____ 2. The chief type of water-conducting cell in the xylem of gymnosperms.

_____ 3. An area of dividing tissue located at the tips of roots and shoots in plants.

_____ 4. Secondary meristem that produces the secondary xylem and secondary phloem.

_____ 5. The type of tissue system that gives rise to the epidermis.

_____ 6. Plant cell that has thick secondary walls, is dead at maturity and functions in support.

_____ 7. A nucleated cell in the phloem responsible for loading and unloading sugar into the sieve tube member.

_____ 8. A plant that lives longer than two years.

_____ 9. General term for all localized areas of mitosis and growth in the plant body.

_____10. Plant cells that are relatively unspecialized, are thin walled, and function in photosynthesis and in the storage of nutrients.

Terms:

a. Apical meristem
b. Companion cell
c. Dermal tissue
d. Ground tissue
e. Lateral meristem

f. Meristem
g. Parenchyma
h. Perennial
i. Periderm
j. Root system

k. Sclerenchyma
l. Shoot system
m. Tracheid
n. Vascular cambium
o. Xylem

TABLE
Fill in the blanks.

Tissue	Tissue System	Function	Location in Plant Body
Parenchyma	Ground tissue	Photosynthesis, storage, secretion	Throughout plant body
Collenchyma	#1	Flexible structural support	#2
#3	Ground tissue	#4	Throughout plant body; common in stems and certain leaves, some nuts and pits of stone fruit
#5	#6	Conducts water, dissolved minerals	Extends throughout plant body
Phloem	Vascular tissue	#7	#8
Epidermis	#9	#10	#11
#12	Dermal tissue	#13	Covers body of woody plants

THOUGHT QUESTIONS
Write your responses to these questions.

1. Create a table that delineates the major structural and functional features of each of the following plant organs and tissues: roots, shoots, parenchyma, collenchyma, sclerenchyma, xylem, phloem, epidermis, and periderm.
2. Considering that many plants get along just fine without one, why is a vascular system considered to be a very important development in plants?
3. What is growth? How does growth differ in plants and animals?

MULTIPLE CHOICE
Place your answer(s) in the space provided. Some questions may have more that one correct answer.

_____ 1. Localized areas of plant growth are called
  a. primordia.
  b. mitotic zones.
  c. meristems.
  d. primary growth centers.
  e. secondary growth centers.

_____ 2. The types of cells in phloem include
  a. tracheids.
  b. vessel elements.
  c. parenchyma.
  d. sieve tube members.
  e. fiber cells.

_____ 3. The types of cells in xylem include
  a. tracheids.
  b. vessel elements.
  c. parenchyma.
  d. sieve tube members.
  e. fiber cells.

_____ 4. The kind of growth that results in an increase in the width of the plant is known as
  a. differentiation.
  b. elongation.
  c. apical meristem growth.
  d. primary growth.
  e. secondary growth.

_____ 5. Hairlike outgrowths of plant epidermis are called
  a. periderm.
  b. fiber elements.
  c. lateral buds.
  d. trichomes.
  e. companion cells.

_____ 6. The cell type found throughout the plant body that often functions in photosynthesis and storage is called
  a. sclerenchyma.
  b. collenchyma.
  c. parenchyma.
  d. companion cells.
  e. a tracheid.

_____ 7. All plant cells have
  a. cell walls.
  b. secondary growth.
  c. the capacity to form a complete plant.
  d. primary cell walls.
  e. secondary cell walls.

_____ 8. Increase in the girth of a plant is due to growth of the
  a. vascular cambium.
  b. cork cambium.
  c. area of cell maturation.
  d. area of cell elongation.
  e. lateral meristems.

VISUAL FOUNDATIONS
Color the parts of the illustration below as indicated.

RED	❏	apical meristem
GREEN	❏	area of cell elongation
YELLOW	❏	root cap
BLUE	❏	developing lateral root
ORANGE	❏	emerging root hairs
BROWN	❏	dying root hairs
TAN	❏	root hair area

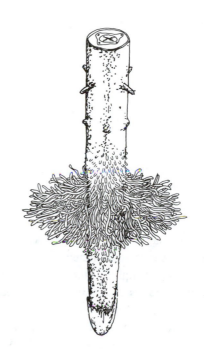

Color the parts of the illustration below as indicated.

RED	❏	flower
GREEN	❏	blade
YELLOW	❏	internode
BLUE	❏	petiole
ORANGE	❏	node
BROWN	❏	taproot
TAN	❏	branch root

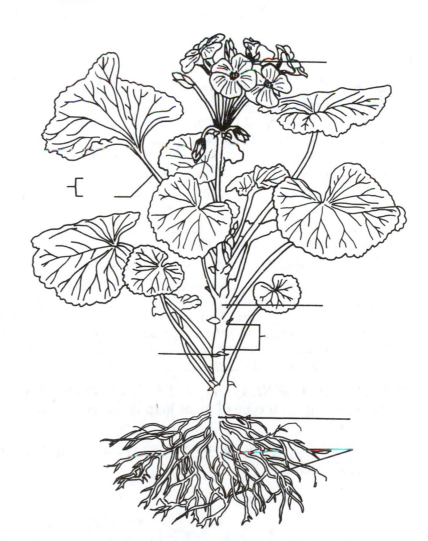

CHAPTER 32

Leaf Structure and Function

Leaves, as the principal photosynthetic organs in plants, are highly adapted to collect radiant energy, convert it into the chemical bonds of carbohydrates, and transport the carbohydrates to the rest of the plant. They are also adapted to permit gas exchange, and to receive water and minerals transported into them by the plant vascular system. Water loss from leaves is controlled in part by the cell walls of the epidermal cells, by presence of a surface waxy layer secreted by epidermal cells, and by tiny pores in the leaf epidermis that open and close in response to various environmental factors. Most of the water absorbed by land plants is lost, either in the form of water vapor or as liquid water. The rate of water loss is affected by various environmental factors. Survival at low temperatures is facilitated in some plants by the loss of leaves. The process is complex, involving both physiological and anatomical changes. The leaves of many plants are modified for functions other than photosynthesis. Some, for example, have modifications for protection, others for grasping and holding onto structures, others for storing water or food, and still others for trapping animals.

CHAPTER OUTLINE AND CONCEPT REVIEW
Fill in the blanks.

FOLIAGE LEAVES VARY GREATLY IN EXTERNAL FORM

EPIDERMIS, MESOPHYLL, XYLEM, AND PHLOEM ARE THE MAJOR TISSUES OF THE LEAF

1 Leaves are highly adapted for photosynthesis. The transparent epidermis allows light to penetrate into the photosynthetic tissue, the (a)_____. When this tissue is divided into two regions, the upper layer, nearest to the upper epidermis, is called the (b)_____ layer, and the lower portion is called the (c)_____ layer.

2 (a)_____ in veins of a leaf conduct water and essential minerals to the leaf, while (b)_____ in veins conducts sugar produced by photosynthesis to the rest of the plant. Veins may be surrounded by a (c)_____, consisting of one or more layers of parenchyma or sclerenchyma cells.

Leaf structure differs in dicots and monocots

3 Monocot and dicot leaves can be distinguished based on their external morphology and their internal anatomy. Dicot leaves usually have (a)_____ venation and monocots usually have (b)_____ veins. Some monocot leaves have bulliform cells along the midvein that may be involved in folding the leaf during drought. In dicots, epidermal cells surrounding guard cells are like other epidermal cells, whereas in some monocots the guard cells are associated with special epidermal cells called (c)_____ cells.

STRUCTURE IS RELATED TO FUNCTION IN LEAVES
The leaves of each type of plant help it survive in the environment to which it is adapted

STOMATAL OPENING AND CLOSING ARE DUE TO CHANGES IN GUARD CELL TURGIDITY

4 Stomates open during the day and close at night. Factors that affect opening and closing of stomates include _____

_____

The potassium ion mechanism explains stomatal opening and closing

5 Light triggers an influx of K^+ ions into the (a)_____ of (b)_____ cells, thereby increasing internal solute concentration. Water then moves into these cells changing their shape causing the pores to (c)_____(open or close?). In relatively low light, K^+ ions are pumped out of the cells, water leaves, the cells collapse, and the pores (d)_____(open or close?).

LEAVES LOSE WATER BY TRANSPIRATION AND GUTTATION
Some plants exude water as a liquid

6 Loss of liquid water from leaves is (a)_____ and water vapor loss is (b)_____. Environmental factors that affect the rate of water loss include (c)_____.

LEAF ABSCISSION ALLOWS PLANTS IN TEMPERATE CLIMATES TO SURVIVE WINTER

7 Leaf abscission involves complex physiological and anatomical changes, all of them regulated by _____.

In many leaves, abscission occurs at an abscission zone near the base of the petiole

8 The area where the petiole attaches to the stem is the (a)_____; it is a weak area because it contains relatively few strengthening (b)_____.

LEAVES WITH FUNCTIONS OTHER THAN PHOTOSYNTHESIS EXHIBIT MODIFICATIONS IN STRUCTURE

9 Leaves are variously modified for a plethora of specialized functions. For example, the hard, pointed (a)_____ on a cactus are leaves modified for protection, and (b)_____ are specialized leaves that anchor long, climbing vines to the supporting structures on which they are growing.

Some of the most remarkable examples of modified leaves are those of insectivorous plants

BUILDING WORDS
Use combinations of prefixes and suffixes to build words for the definitions that follow.

Prefixes	The Meaning
ab-	from, away, apart
circ-	around
meso-	middle
trans-	across, beyond

Prefix	Suffix	Definition
_____	-scission	1. The separation and falling away from the plant stem of leaves, fruit and flowers.
_____	-adian	2. Pertains to something that cycles at approximately 24-hour intervals.
_____	-phyll	3. The photosynthetic tissue of the leaf sandwiched between (in the middle of) the upper and lower epidermis.
_____	-piration	4. The loss of water vapor from the plant body across leaf surfaces.

MATCHING
For each of these definitions, select the correct matching term from the list that follows.

_____ 1. A ring of cells surrounding the vascular bundle in monocot and dicot leaves.

_____ 2. A leaf or stem that is modified for holding or attaching to objects.

_____ 3. Vascular tissue that transports sugars produced by photosynthesis.

_____ 4. A hair or other appendage growing out of the epidermis in plants.

_____ 5. A non-cellular waxy covering over the epidermis of the above-ground portion of plants.

_____ 6. A leaf that is modified for protection.

_____ 7. The part of a leaf that attaches to a stem.

_____ 8. One of two cells that collectively form a stoma.

_____ 9. In plants, vascular bundles in leaves.

_____10. The emergence of liquid water droplets on leaves, forced out through special water pores by root pressure.

Terms:

a.	Blade	f.	Petiole	k.	Trichome
b.	Bundle sheath	g.	Phloem	l.	Vascular bundle
c.	Cuticle	h.	Spine	m.	Vein
d.	Guard cell	i.	Stipule	n.	Xylem
e.	Guttation	j.	Tendril		

TABLE
Fill in the blanks.

Structure of Leaf	Function of Leaf Structure
Thin, flat shape	Maximizes light absorption, efficient gas diffusion
Ordered arrangement on stem	#1
Waxy cuticle	#2
#3	Allow gas exchange between plant and atmosphere
Relatively transparent epidermis	#4
#5	Photosynthetic tissue
#6	Allows for rapid diffusion of CO_2 to mesophyll cell surface
#7	Provide support to prevent leaf from collapsing
Xylem in veins	#8
#9	Carry sugar produced in photosynthesis to other plant parts

THOUGHT QUESTIONS
Write your responses to these questions.

1. Describe the structures and functions of the major leaf tissues. How do they differ in monocots and dicots?
2. Describe how leaf structures interact during photosynthesis.
3. Explain how and why stomata open and close.
4. How is transpiration accomplished? How does it benefit a plant?
5. Explain how and why abscission occurs?
6. Create a table that illustrates the basic structure and function(s) for various specialized leaves. Give an example of a plant for each leaf type.

MULTIPLE CHOICE
Place your answer(s) in the space provided. Some questions may have more that one correct answer.

_____ 1. The abscission zone of plants
    a. secretes suberin.
    b. is characteristic of desert plants.
    c. is composed primarily of thin-walled cells.
    d. is largely parenchymal cells.
    e. is an active area of photosynthesis.

_____ 2. Factors that tend to affect the amount of transpiration include
    a. the cuticle.
    b. wind velocity.
    c. ambient temperature.
    d. relative humidity.
    e. amount of light.

_____ 3. Trichomes are found on/in the
    a. palisade layer.
    b. spongy layer.
    c. entire mesophyll.
    d. epidermis.
    e. guard cells.

_____ 4. In general, leaf epidermal cells
    a. are living.
    b. lack chloroplasts.
    c. protect mesophyll cells from sunlight.
    d. are absent on the lower leaf surface.
    e. have a cuticle.

_____ 5. Guard cells generally
    a. have chloroplasts.
    b. form a pore.
    c. are found in the epidermis.
    d. are found only in monocots.
    e. are found only in dicots.

_____ 6. Active transport of potassium ions into guard cells
    a. requires ATP.
    b. closes stomates.
    c. causes guard cells to shrink and collapse.
    d. occurs more in daylight than at night.
    e. indirectly causes pores to open.

_____ 7. The portion of mesophyll that is usually composed of loosely and irregularly arranged cells is
    a. the palisade layer.
    b. the spongy layer.
    c. toward the leaf's underside.
    d. toward the leaf's upperside.
    e. an area of photosynthesis.

_____ 8. The process by which plants secrete water as a liquid is
    a. evaporation.
    b. transpiration.
    c. guttation.
    d. the potassium ion mechanism.
    e. found only in monocots.

_____ 9. The leaves of dicots usually have
    a. bulliform cells.   d. special subsidiary cells.
    b. netted venation.   e. bean-shaped guard cells.
    c. differentiated palisade and spongy tissues.

VISUAL FOUNDATIONS
Color the parts of the illustration below as indicated.

RED   ❑ xylem
GREEN   ❑ mesophyll
YELLOW   ❑ stoma
BLUE   ❑ phloem
ORANGE   ❑ bundle sheath
BROWN   ❑ cuticle
TAN   ❑ epidermis

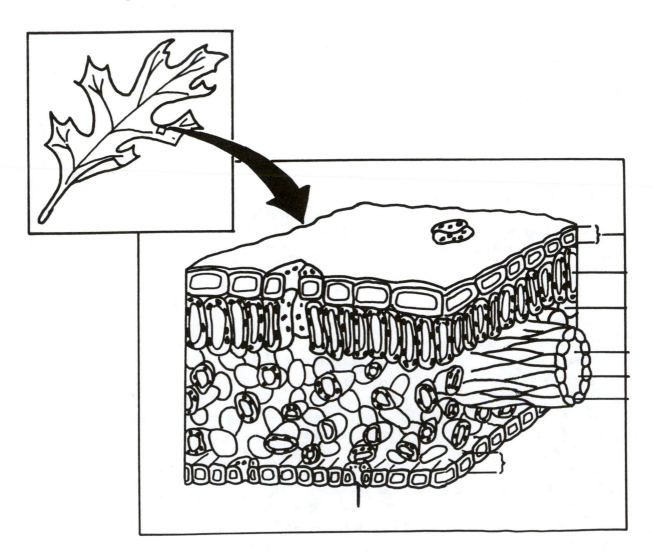

CHAPTER 33

❑

Stems and Plant Transport

Stems function to support leaves and reproductive structures, conduct materials absorbed by the roots and produced in the leaves to other parts of the plant, and produce new tissue throughout the life of the plant. Specialized stems have other functions as well. The stems of some plants increase only in length. These are the herbaceous, or nonwoody, plants. Although all herbaceous stems have the same basic tissues, the arrangement of the tissues in the monocot stem differs from that in the dicot stem. The stems of other plants increase in both length and girth. These include all gymnosperms and some dicots. Increases in the length of plants result from mitotic activity in apical meristems, and increases in the girth of plants result from mitotic activity in lateral meristems. Simple physical forces are responsible for the movement of food, water, and minerals in multicellular plants.

CHAPTER OUTLINE AND CONCEPT REVIEW
Fill in the blanks.

INTRODUCTION
1 The primary functions of stems are to _____
_____.

A WOODY TWIG DEMONSTRATES THE EXTERNAL STRUCTURE OF STEMS
STEMS ORIGINATE AND DEVELOP AT MERISTEMS
HERBACEOUS DICOT AND MONOCOT STEMS CAN BE DISTINGUISHED BY THE ARRANGEMENT OF VASCULAR TISSUES
Vascular bundles of herbaceous dicot stems are arranged in a circle in cross section

2 Patches of vascular tissues in herbaceous dicots are called (a)_____. Each patch contains two vascular tissues, the (b)_____, and a single layer of cells, the (c)_____, sandwiched between them. Vascular tissue patches in dicot stems are arranged in a circle between the central (d)_____ and the peripheral tissue layer called the (e)_____.

Vascular bundles are scattered throughout monocot stems

3 The parenchyma tissue in monocot stems that contains vascular bundles is called _____.

WOODY PLANTS HAVE STEMS WITH SECONDARY GROWTH
4 Secondary growth occurs in some dicots and all gymnosperms as a result of cell divisions in two lateral meristems, the (a)_____, which gives rise to secondary xylem and phloem, and the (b)_____, which gives rise to a periderm.

Cork cambium produces periderm

6   Cork cambium cells form tissues toward the inside and the outside. The layer toward the outside consists of (a)_____, heavily suberized cells that protect the plant. To the inside, cork cambium forms the (b)_____, which functions for storage.

Common terms associated with wood are based on plant structure

7   The older, brownish wood in the center of a tree is called (a)_____, and the newer, more peripheral wood is (b)_____.

8   Annual rings are composed of two types of cells arranged in alternating concentric circles, each layer appropriately named for the season in which it developed. The (a)_____ has thin-walled, large vessels and tracheids and few fibers, whereas (b)_____ has thicker-walled, narrower vessels and tracheids and numerous fibers.

TRANSPORT IN PLANTS OCCURS IN XYLEM AND PHLOEM
WATER AND MINERALS ARE TRANSPORTED IN XYLEM

Water movement can be explained by the concept of water potential

9   Water and dissolved minerals move upward in the xylem from the root to the stem to the leaves. One of the principal forces behind water movements through plants is a function of a cell's ability to absorb water by osmosis, also know as the "free energy of water" or the (a)_____. Water containing solutes has (b)_____(more or less?) free energy than pure water. Water moves from regions containing (c)_____(more or less?) free energy of water to regions containing (d)_____(more or less?) free energy of water.

10 Root pressure is caused by the differences in water potential between

    _____.

Tension-cohesion pulls water up a stem

11 The (a)_____ "pulls" water up the plant, and the evaporation-pull of (b)_____ causes a tension at the top of the plant due to a gradient in water potentials from soil up through the plant to the atmosphere.

Root pressure pushes water from the root up a stem

SUGAR IN SOLUTION IS TRANSLOCATED IN PHLOEM

The pressure-flow hypothesis explains translocation in phloem

12 Dissolved food, predominantly sucrose, is transported up or down in the phloem. Movement of materials in the phloem is explained by the (a)_____ hypothesis. Sugar is actively loaded into the sieve tubes at the (b)_____, causing water to move into sieve tubes by osmosis, then sugar is actively unloaded from the sieve tubes at the (c)_____, causing water to leave sieve tubes by osmosis.

BUILDING WORDS
Use combinations of prefixes and suffixes to build words for the definitions that follow.

Prefixes	The Meaning
trans-	across, beyond

Prefix	Suffix	Definition
_____	-location	1. The movement of materials (across distances) in the vascular tissues of a plant.

MATCHING
For each of these definitions, select the correct matching term from the list that follows.

_____ 1. An area of dividing tissue located at the tips of plant stems and roots.

_____ 2. Technical term for the outer bark of woody stems and roots.

_____ 3. Large, thin-walled parenchyma cells found in the innermost tissue in many plants.

_____ 4. A lateral meristem in plants that produces cork cells and cork parenchyma.

_____ 5. The cortex and the pith are parts of this tissue system.

_____ 6. Lateral meristem that gives rise to secondary vascular tissues.

_____ 7. The vascular tissue responsible for transporting water and dissolved minerals in all plants.

_____ 8. The positive pressure in the sap in the roots of plants.

_____ 9. A chain of parenchyma cells that functions for lateral transport of food, water, and minerals in woody plants.

_____10. Masses of cells in some plants that rupture the epidermis and form porous swellings in stems, facilitating the exchange of gases.

Terms:

a.	Apical meristem	f.	Lenticels	k.	Terminal bud
b.	Cork cambium	g.	Periderm	l.	Vascular cambium
c.	Ground tissue	h.	Pith	m.	Xylem
d.	Lateral bud	i.	Ray		
e.	Lateral meristem	j.	Root pressure		

TABLE
Fill in the blanks.

Tissue	Source	Location	Function
Secondary xylem	Produced by vascular cambium	Wood	Conducts water and dissolved minerals
#1	Produced by vascular cambium	Inner bark	#2
Cork parenchyma	#3	Periderm	#4
#5	Produced by cork cambium	#6	Replacement for epidermis
#7	A lateral meristem produced by procambium tissue	Between wood and inner bark in vascular bundles	#8
Cork cambium	#9	In epidermis or outer cortex	#10

THOUGHT QUESTIONS
Write your responses to these questions.

1. What is a stem? How are stems different from other major plant regions? What are their functions?
2. Differentiate between primary growth and secondary growth.
3. List the tissues that are derived from each type of lateral meristem.
4. Sketch cross sections of monocot and dicot stems and describe the similarities and differences.
5. What phenomena come into play to overcome the tremendous force of gravity as water moves upward from the roots of a plant to its leaves?
6. What is the pressure-flow hypothesis?

MULTIPLE CHOICE
Place your answer(s) in the space provided. Some questions may have more that one correct answer.

_____ 1. One daughter cell from a mother cell in the vascular cambium remains as part of the vascular cambium, the other divides to form
    a. wood or secondary phloem.   d. secondary tissue.
    b. wood or inner bark.   e. primary xylem and phloem.
    c. outer bark.

_____ 2. If soil water contains 0.1% dissolved materials and root water contains 0.2% dissolved materials, one would expect
    a. a negative water pressure in soil water.   d. water to flow from soil into root.
    b. a negative water pressure in root water.   e. water to flow from root into soil.
    c. less water pressure in roots than in soil.

_____ 3. Which of the following is/are true of dicot stems?
    a. stem covered with epidermis   d. vascular bundles scattered through the stem
    b. vascular tissues embedded in ground tissue   e. vascular bundles arranged in circles
    c. stem has distinct cortex and pith

_____ 4. Which of the following is/are true of monocot stems?
    a. stem covered with epidermis   d. vascular bundles scattered through the stem
    b. vascular tissues embedded in ground tissue   e. vascular bundles arranged in circles
    c. stem has distinct cortex and pith

_____ 5. When water is plentiful, wood formed by the vascular cambium is
    a. springwood.   d. composed of thin-walled vessels.
    b. summerwood.   e. composed of thick-walled vessels.
    c. late summerwood.

_____ 6. Newer wood closer to the bark is known as
    a. heartwood.   d. hardwood.
    b. sapwood.   e. softwood.
    c. tracheids.

_____ 7. If a plant is placed in a beaker of pure distilled water, then
    a. water will move out of the plant.   d. water pressure in the beaker is zero.
    b. water will move into the plant.   e. water pressure in the plant is less than zero.
    c. water pressure in the plant is positive relative to water in the beaker.

_____ 8. Secondary xylem and phloem are derived from
    a. cork parenchyma.   d. periderm.
    b. cork cambium.   e. epidermis.
    c. vascular cambium.

_____ 9. The outer portion of bark is formed primarily from
    a. cork parenchyma.   d. periderm.
    b. cork cambium.   e. epidermis.
    c. vascular cambium.

_____ 10. The pull of water up through a plant is due in part to
    a. cohesion.   d. transpiration.
    b. adhesion.   e. a water pressure gradient between soil and the atmosphere.
    c. the low water pressure in the atmosphere.

VISUAL FOUNDATIONS

Color the parts of the illustration below as indicated.

RED ❑ vascular cambium
GREEN ❑ secondary phloem
YELLOW ❑ pith
BLUE ❑ primary phloem
ORANGE ❑ secondary xylem
BROWN ❑ primary xylem
TAN ❑ periderm

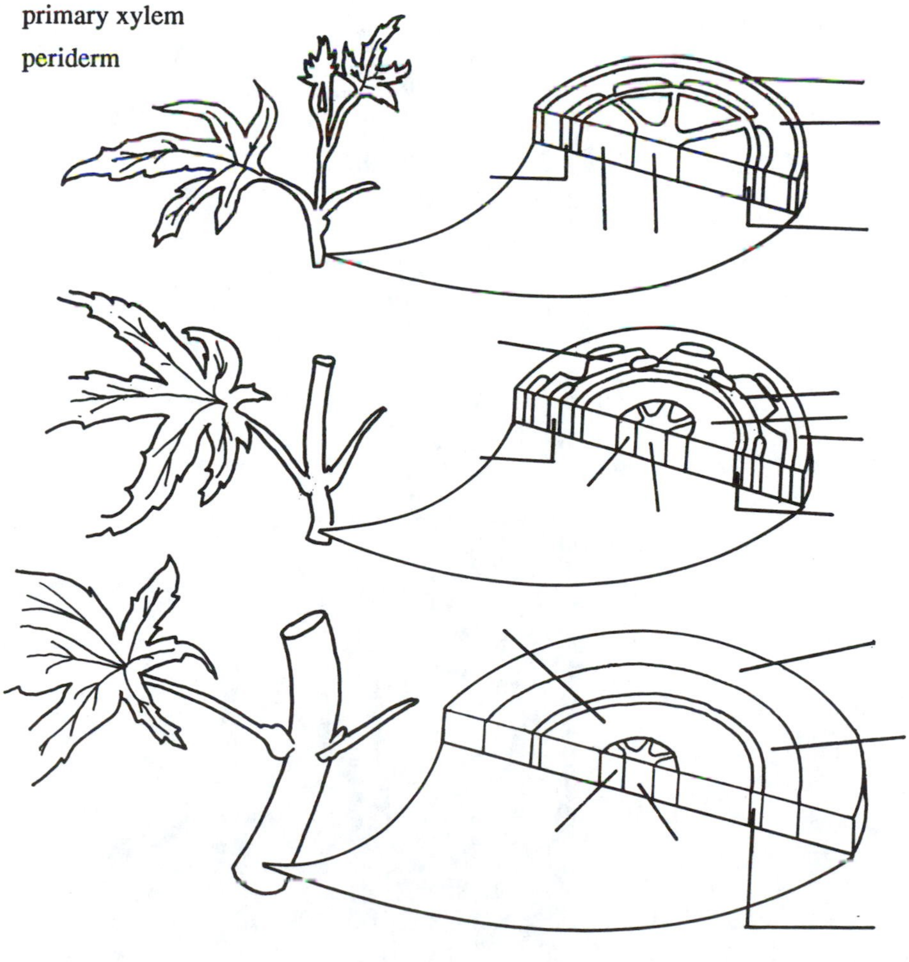

Color the parts of the illustration below as indicated.

RED	☐	sucrose
GREEN	☐	plasmodesmata
YELLOW	☐	companion cell
BLUE	☐	water
ORANGE	☐	sink
BROWN	☐	sieve tube member
PINK	☐	source

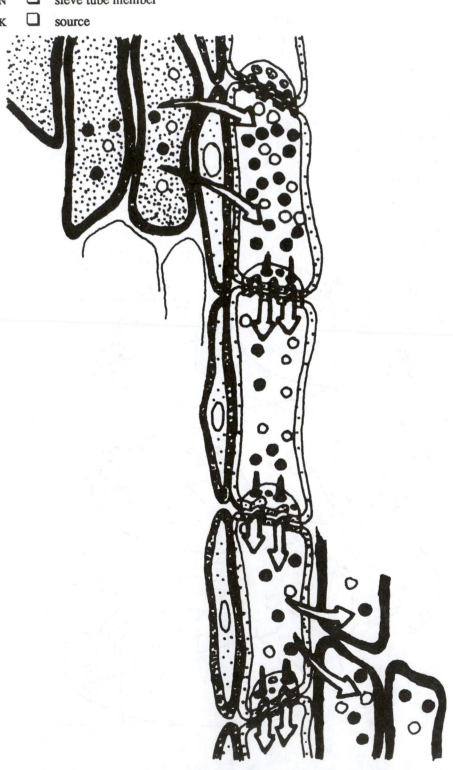

CHAPTER 34

Roots and Mineral Nutrition

The roots of most plants function to anchor plants to the ground, to absorb water and dissolved minerals from the soil, and, in some cases, to store food. Additionally, the roots of some plants are modified for support, aeration, and/or photosynthesis. Although herbaceous roots have the same tissues and structures found in stems, they also have other tissues and structures as well. There are structural differences between monocot and dicot roots. The gymnosperms and some dicots produce woody stems and roots that grow by increasing in both length and girth. Various factors influence soil formation. Soil is composed of inorganic minerals, organic matter, soil organisms, soil atmosphere, and soil water. Plants require at least 16 essential nutrients for normal growth, development, and reproduction. Some human practices, such as harvesting food crops, depletes the soil of certain essential elements, making it necessary to return these elements to the soil in the form of organic or inorganic fertilizer.

CHAPTER OUTLINE AND CONCEPT REVIEW
Fill in the blanks.

INTRODUCTION

1 Roots function in _____.

PLANTS HAVE ONE OF TWO BASIC TYPES OF ROOT SYSTEMS
UNLIKE STEMS, ROOTS POSSESS A ROOT CAP AND ROOT HAIRS
THE ARRANGEMENT OF VASCULAR TISSUES DISTINGUISHES HERBACEOUS DICOT ROOTS AND MONOCOT ROOTS
Xylem and phloem are not organized into bundles in herbaceous dicot roots

2 The inner layer of cortex in a dicot root is the (a)_____, the cells of which possess a bandlike region on their radial and transverse walls. This "band," or (b)_____ as it's called, contains suberin.

3 Just inside the endodermis is a layer of parenchyma cells with meristematic characteristics called the (a)_____. The central portion of dicot roots is occupied by (b)_____ tissue.

4 Most of the water that enters roots first moves along the (a)_____ rather than moving through intracellular spaces. In part, this is due to the powerful absorptive qualities of (b)_____. Later, when water reaches the waterproof walls of the (c)_____ cells, it does permeate cell membranes. Lateral movement of water through root tissues can be summarized as: root hair —> (d)_____ —> cortex —> (e)_____ —> pericycle —> (f)_____, where it is then transported upward.

5 The primary function of the root cortex is (a)_____. Multicellular branch roots originate in the (b)_____.

Xylem does not form the central tissue in some monocot roots
WOODY PLANTS HAVE ROOTS WITH SECONDARY GROWTH
SOME ROOTS ARE SPECIALIZED FOR UNUSUAL FUNCTIONS

6 Roots that grow from unusual places on plants, such as aerial roots growing laterally from a stem, are known as (a)_____ roots. Two other types of such roots are (b)_____ roots that support the plant, like those in corn, and (c)_____ roots that pull bulbs deeper into the ground.

ROOTS FORM RELATIONSHIPS WITH OTHER SPECIES
SOIL ANCHORS PLANTS AND PROVIDES WATER AND MINERALS

7 The five major components of soil are _____

_____.

8 Good, loamy agricultural soil contains about 40% of (a)_____ and about 20% of (b)_____. The organic portion of soil, or (c)_____, is also an important component. The spaces between soil particles are filled with (d)_____.

Soil pH affects soil characteristics and plant growth

SOIL PROVIDES MOST OF THE MINERALS FOUND IN PLANTS

9 In order to identify elements that are essential for plant growth, investigators must reduce the number of variables to a minimum. To do this, since soil is very complex, plants are grown in aerated water containing known quantities of elements, a process known as _____.

Sixteen elements are essential for plant growth

10 The 16 elements that plants require for normal growth, development, and reproduction are called (a)_____. Of these, nine are needed in relatively large amounts. These macronutrients include: (b)_____

_____. The remaining seven elements are needed in trace amounts; these micronutrients include: (c)_____.

SOIL CAN BE DAMAGED BY HUMAN MISMANAGEMENT

Mineral depletion occurs in soils that are farmed

11 Some essential elements may be added to soil as organic or inorganic fertilizer. Three essential factors for plant growth are water, light, and essential elements. The three elements that most often limit plant growth are _____.

Soil erosion is the loss of soil from the land
Salt accumulates in soil that is improperly irrigated

BUILDING WORDS
Use combinations of prefixes and suffixes to build words for the definitions that follow.

Prefixes	The Meaning
hydro-	water
macro-	large, long, great, excessive
micro-	small

Prefix	Suffix	Definition
_____	-ponics	1. Growing plants in water (not soil) containing dissolved inorganic minerals.
_____	-nutrient	2. An essential element that is required in fairly large amounts for normal plant growth.
_____	-nutrient	3. An essential element that is required in trace (small) amounts for normal plant growth.

MATCHING

For each of these definitions, select the correct matching term from the list that follows.

_____ 1. Soil fungi associated with the roots of vascular plants that aid in the absorption of materials.

_____ 2. Organic matter in various stages of decomposition in the soil.

_____ 3. The innermost layer of the cortex in the plant root.

_____ 4. A root that arises in an unusual position on a plant

_____ 5. A band of waterproof material around the radial and transverse walls of endodermal root cells.

_____ 6. An adventitious root that arises from the stem and provides additional support for plants.

_____ 7. Aerial "breathing roots" of some plants in swampy and tidal environments that assist in getting oxygen to submerged roots.

_____ 8. The type of root present in a plant that has several main roots without a dominant root.

_____ 9. An extension of an epidermal cell in roots, which increases the absorptive capacity of the roots.

_____10. A covering of cells over the root tip that protects delicate meristematic tissue beneath it.

Terms:

a. Adventitious root	f. Graft	k. Prop root
b. Apoplast	g. Humus	l. Root cap
c. Casparian strip	h. Leaching	m. Root hair
d. Endodermis	i. Mycorrhizae	n. Stele
e. Fibrous root	j. Pneumatophore	o. Taproot

TABLE

Fill in the blanks.

Root Structure	Function of the Root Structure
Root tip	Protects the apical meristem, may orient the root downward
Root apical meristem	#1
#2	Absorption of water and minerals
Epidermis	#3
Cortex	#4
#5	Control mineral uptake into root xylem
#6	Gives rise to lateral roots and lateral meristems
#7	Conducts water and dissolved minerals
Phloem	#8

THOUGHT QUESTIONS
Write your responses to these questions.

1. What are the major structural and functional elements in roots?
2. How do monocot and dicot roots differ?
3. Describe how water moves through roots and upward into stems.
4. Describe various root modifications and give examples of plants that possess them.
5. Describe macronutrients, micronutrients, and essential elements.

MULTIPLE CHOICE
Place your answer(s) in the space provided. Some questions may have more that one correct answer.

_____ 1. The origin of multicellular branch roots is the
    a. cambium.
    b. Casparian strip.
    c. pericycle.
    d. parenchyma.
    e. cortex.

_____ 2. The principal ingredients in inorganic fertilizers are
    a. phosphorus.
    b. potassium.
    c. nitrogen.
    d. iron.
    e. hydrogen.

_____ 3. When water first enters a root, it usually
    a. enters parenchymal cells.
    b. is absorbed by cellulose.
    c. moves from a water negative potential to a positive water potential.
    d. moves along cell walls.
    e. enters cells in the Casparian strip.

_____ 4. Essential macronutrients include
    a. phosphorus.
    b. potassium.
    c. iron.
    d. hydrogen.
    e. magnesium.

_____ 5. The function(s) performed by all roots is/are
    a. absorption of water.
    b. absorption of minerals.
    c. food storage.
    d. anchorage.
    e. aeration.

_____ 6. Structures found in primary roots that are also found in stems include the
    a. cortex.
    b. cuticle.
    c. conducting vessels.
    d. apical meristem cap.
    e. epidermis.

_____ 7. Which of the following statements is/are accurate with respect to root hairs?
    a. They are short-lived.
    b. Occur in monocots, but not dicots.
    c. Occur in dicots, but not monocots.
    d. Occur in both monocots and dicots.
    e. Some develop into root branches.

_____ 8. Roots produced in unusual places on the plants, often as aerial roots, are called _____ roots.
    a. secondary
    b. enhancement
    c. adventitious
    d. contractile
    e. aerial water-absorbing

_____ 9. Vascular tissues in monocot roots in general
    a. form a solid cylinder.
    b. are absent.
    c. are in patches arranged in a circle.
    d. contain a vascular cambium.
    e. continue into root hairs.

____10. The principal function(s) of the root cortex is/are
a. conduction.   d. water absorption.
b. storage.   e. mineral absorption.
c. production of root hairs.

____11. Essential micronutrients include
a. phosphorus.   d. hydrogen.
b. potassium.   e. magnesium.
c. iron.

____12. The inorganic materials in soil come from
a. fertilizers.   d. the atmosphere.
b. water runoff.   e. percolating water.
c. weathered rock.

____13. The two groups of organisms in soil that are the most important in decomposition and nutrient cycles are
a. insects.   d. algae.
b. fungi.   e. bacteria.
c. worms.

VISUAL FOUNDATIONS
Color the parts of the illustration below as indicated.

RED	❑	vascular cambium
GREEN	❑	secondary phloem
YELLOW	❑	periderm
BLUE	❑	primary phloem
ORANGE	❑	secondary xylem
BROWN	❑	primary xylem
TAN	❑	pericycle
PINK	❑	epidermis

CHAPTER 35

Reproduction in Flowering Plants

All flowering plants reproduce sexually; some also reproduce asexually. Sexual reproduction involves flower formation, pollination, fertilization within the flower ovary, and seed and fruit formation. The offspring resulting from sexual reproduction exhibit a great deal of individual variation, due to gene recombination and the union of dissimilar gametes. Sexual reproduction offers the advantage of new combinations of genes that might make an individual plant better suited to its environment. Flowering plants may be pollinated by wind or animals. Obligate relationships exist between some animal pollinators and the plants they pollinate. Asexual reproduction involves only one parent; the offspring are genetically identical to the parent and each other. Asexual reproduction, therefore, is advantageous when the parent is well adapted to its environment. The stems, leaves, and roots of many flowering plants are modified for asexual reproduction. Also, in some plants, seeds and fruits are produced asexually without meiosis or the fusion of gametes.

CHAPTER OUTLINE AND CONCEPT REVIEW
Fill in the blanks.

FLOWERS ARE INVOLVED IN SEXUAL REPRODUCTION

POLLINATION IS THE FIRST STEP TOWARD FERTILIZATION
Flowering plants and their animal pollinators have affected one another's evolution
Some flowering plants depend on wind to disperse pollen

FERTILIZATION OCCURS, FOLLOWED BY SEED AND FRUIT DEVELOPMENT
A unique double fertilization process occurs in flowering plants
Embryonic development in seeds is orderly and predictable
The mature seed contains an embryonic plant and storage materials
Fruits are mature, ripened ovaries

1 The four basic types of fruits are _____
_____.

2 One type of fruit, the (a)_____ fruit, develops from a single ovary. It may be fleshy or dry. Fruits that are fleshy throughout, like those in tomatoes and grapes, are called (b)_____, while fleshy fruits that have a hard pit are known as (c)_____. Of fruits that split open at maturity, some, like the milkweed (d)_____, split along *one* seam, while (e)_____ split along *two* seams, and (f)_____ split along *multiple* seams.

3 (a)_____ fruits result from the fusion of several developing ovaries in a *single* flower. The raspberry is an example. Similarly, (b)_____ fruits form from the fusion of several ovaries of many flowers that are in close proximity. The pineapple is an example.

4 (a)_____ fruits contain other plant tissues in addition to ovaries. When one eats a strawberry, for example, he or she is eating the fleshy (b)_____ of the flower, and when

eating apples and pears she or he is consuming the (c)_____ that surrounds the ovary.

Seed dispersal is highly varied in flowering plants

5 Flowering plant seeds and fruits are adapted for various means of dispersal, among which four common means are _____

_____.

ASEXUAL REPRODUCTION IN FLOWERING PLANTS MAY INVOLVE MODIFIED STEMS, LEAVES, OR ROOTS

6 Various vegetative structures may be involved in asexual reproduction. Some of them, for example, are (a)_____, like those in bamboo and grasses that have horizontal underground stems; (b)_____, underground stems greatly enlarged for food storage, like those in potatoes; (c)_____, short underground stems with fleshy storage leaves, as in onions and tulips; (d)_____, thick stems covered with papery scales; and (e)_____ or runners that are above ground stems with long internodes, like those in strawberries.

7 Roots may develop adventitious buds that form _____. These stems develop roots and may give rise to new plants.

Apomixis is the production of seeds without the sexual process

8 _____ is the production of seeds and fruits without sexual reproduction.

BUILDING WORDS

Use combinations of prefixes and suffixes to build words for the definitions that follow.

Prefixes	The Meaning
endo-	within
hypo-	under, below

Prefix	Suffix	Definition
_____	-sperm	1. Nutritive tissue within seeds.
_____	-cotyl	2. The part of a plant embryo or seedling below the point of attachment of the cotyledons.

MATCHING

For each of these definitions, select the correct matching term from the list that follows.

____ 1. A collective term for the petals of a flower

____ 2. A ripened ovary.

____ 3. The outermost parts of a flower, usually leaf-like in appearance, that protect the flower as a bud.

____ 4. A type of reproduction in which fruits and seeds are formed asexually.

____ 5. A fruit that develops from a single flower with many separate carpels, such as a raspberry.

____ 6. A plant reproductive body composed of a young, multicellular plant and nutritive tissue.

____ 7. The male gametophyte in plants.

____ 8. An above-ground, horizontal stem with long internodes.

____ 9. The female reproductive unit of a flower.

____10. A simple fleshy fruit in which the inner wall of the fruit is hard and stony; such as a plum or peach.

Terms:

a. Aggregate fruit   f. Drupe   k. Seed
b. Apomixis   g. Fruit   l. Sepal
c. Carpel   h. Ovary   m. Stolon
d. Corolla   i. Ovule
e. Cotyledon   j. Pollen

TABLE
Fill in the blanks.

Examples of Fruit	Main Type of Fruit	Subtype	Description
Tomato	Simple	Fleshy	Berry: fleshy and soft throughout, usually with many seeds
Strawberry	#1	None	Other plant tissues and ovary tissue comprise the fruit
Acorn	Simple	#2	#3
Raspberry	#4	None	#5
Peach	#6	#7	#8
Bean	Simple	Dry: opens to release seeds	#9
Corn	Simple	Dry: does not open	#10
Pineapple	#11	None	#12

THOUGHT QUESTIONS
Write your responses to these questions.

1. Diagram a perfect flower and describe the function(s) of each of its major parts.
2. What is pollination? What is fertilization? What processes occur from the moment of pollination to the moment of fertilization?
3. Describe the development processes that occur between the fertilization and the production of a seed.
3. What logical conclusions can one draw, if any, about the relationship between the structure of a flower and the structure and habits of a hummingbird?
4. Describe the different types of fruits and give an example of a plant for each.
5. Describe at least eight ways that plants reproduce asexually. Give an example of each.

MULTIPLE CHOICE

Place your answer(s) in the space provided. Some questions may have more that one correct answer.

_____ 1. A fruit that forms from many separate carpels in a single flower is a/an
    a. follicle.   d. aggregate fruit.
    b. achene.   e. accessory fruit.
    c. drupe.

_____ 2. The process by which an embryo develops from a diploid cell in the ovary without fusion of haploid gametes is called
    a. lateral bud generation.   d. apomixis.
    b. dehiscence.   e. spontaneous generation.
    c. asexual reproduction by means of suckers.

_____ 3. The tomato is actually a
    a. drupe.   d. fruit.
    b. berry.   e. mature ovary.
    c. fleshy lateral bud.

_____ 4. Which of the following are somehow related to dry dehiscent fruits?
    a. drupe   d. capsule
    b. legume   e. green bean
    c. grain

_____ 5. The "eyes" of a potato are
    a. diploid gametes.   d. lateral buds.
    b. parts of a stem.   e. capable of producing complete plants.
    c. rhizomes.

_____ 6. A fruit composed of ovary tissue and other plant parts is a
    a. follicle.   d. aggregate fruit.
    b. achene.   e. accessory fruit.
    c. drupe.

_____ 7. Horizontal, asexually reproducing stems that run above ground are
    a. diploid gametes.   d. corms.
    b. rhizomes.   e. stolons.
    c. tubers.

_____ 8. Raspberries and blackberries are examples of
    a. follicles.   d. aggregate fruits.
    b. achenes.   e. multiple fruits.
    c. drupes.

VISUAL FOUNDATIONS
Color the parts of the illustration below as indicated.

RED	☐	ovule
GREEN	☐	sepal
YELLOW	☐	ovary (and derived from ovary)
BLUE	☐	stamen
ORANGE	☐	style
BROWN	☐	stigma
TAN	☐	seed
PINK	☐	floral tube (and derived from floral tube)
VIOLET	☐	petal

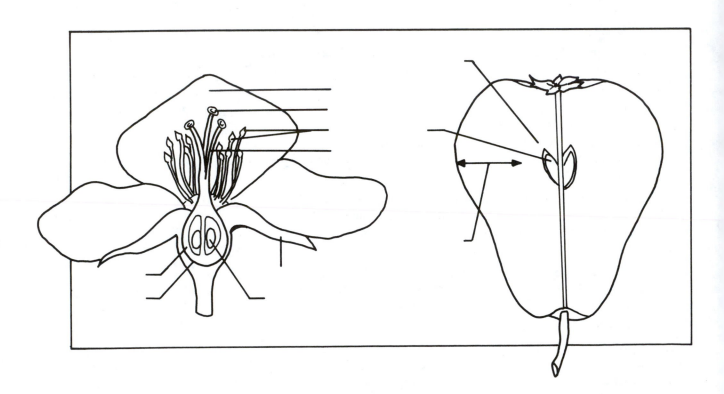

CHAPTER 36

❏

Growth Responses and Regulation of Growth

Environmental cues exert an important influence on all aspects of plant growth and development. Initiation of sexual reproduction, for example, is often under environmental control. In some plants, it is induced by changes in duration of daylight, and in others, by changes in temperature. Plants display a variety of growth movements and responses, varying from gradual to rapid. The more gradual movements and responses include the tropisms and circadian rhythms. The more rapid include the turgor movements. All aspects of plant growth and development are affected by hormones, chemical messengers produced in one part of the plant and transported to another part where they have their effect. Five different hormones interact in complex ways with one another to produce a variety of responses in plants. These are the auxins which are involved in cell elongation, phototropism, gravitropism, apical dominance, and fruit development; the gibberellins which are involved in stem elongation, flowering, and seed germination; the cytokinins which promote cell division and differentiation, delay senescence, and interact with auxins in apical dominance; ethylene which has a role in the ripening of fruits and leaf abscission; and abscisic acid which is involved in stomatal closure due to water stress, and bud and seed dormancy.

CHAPTER OUTLINE AND CONCEPT REVIEW
Fill in the blanks.

BOTH EXTERNAL AND INTERNAL FACTORS AFFECT SEED GERMINATION AND EARLY GROWTH
Dicots and monocots exhibit characteristic patterns of early growth

ENVIRONMENTAL CUES AFFECT FLOWERING AND OTHER PLANT RESPONSES
Phytochrome detects varying periods of day length

Phytochrome is involved in many other responses to light

Temperature may also affect reproduction

A BIOLOGICAL CLOCK INFLUENCES MANY PLANT RESPONSES

1 _____ are regular daily rhythms in growth or activities of the plant.

CHANGES IN TURGOR CAN INDUCE TEMPORARY PLANT MOVEMENTS

2 Some plants respond to external stimuli by changes in turgor in special cells in an organ at the base of a petiole called the (a)_____. A mechanical stimulus may initiate an electrical "impulse" that causes changes in cell membrane permeability. When this occurs, (b)_____ ions flow out of the affected cells, resulting in intracellular hypotonicity and the net movement of water out of these cells. This sudden change in water volume causes the leaf to move. Such movements, known as (c)_____, are temporary.

3 The ability of plant parts to align with the sun as it moves across the sky is known as
_____.

A TROPISM IS GROWTH IN RESPONSE TO AN EXTERNAL STIMULUS FROM A SPECIFIC DIRECTION

4 Tropisms are categorized according to the stimulus that causes them to occur. For example, growth or movement initiated by light is (a)_____, response to gravity is (b)_____, and movement caused by a mechanical stimulus is (c)_____.

THIGMOMORPHOGENESIS IS A GROWTH RESPONSE TO MECHANICAL STRESS

HORMONES ARE CHEMICAL MESSENGERS THAT REGULATE GROWTH AND DEVELOPMENT

5 The five major hormones that regulate responses in plants are _____
_____.

Auxin promotes cell elongation

6 By definition, auxins are compounds that stimulate phototropic (a)_____ in oat coleoptiles, the principal one in plants generally being (b)_____
_____. These hormones always move in one direction, specifically from the (c)_____ toward the roots. Such unidirectional movement is called (d)_____. Auxins are also involved in phototropism, gravitropism, apical dominance, and fruit development.

Gibberellins promote stem elongation

7 In addition to influencing stem elongation, gibberellins are also involved in _____
_____.

Cytokinins promote cell division

8 Cytokinins mainly promote (a)_____. They also delay (b)_____(aging), and they interact with auxins in apical dominance.

Ethylene stimulates abscission and fruit ripening

9 Ethylene is involved in many aspects of aging. Among them two principal influences are on
_____.

Abscisic acid promotes bud and seed dormancy

10 Abscisic acid is known as the stress hormone. Among other things, it is involved in stomatal closure due to _____ stress.

Florigen is a hypothetical hormone that promotes flowering

OTHER CHEMICALS HAVE BEEN IMPLICATED IN PLANT GROWTH AND DEVELOPMENT

BUILDING WORDS

Use combinations of prefixes and suffixes to build words for the definitions that follow.

Prefixes	The Meaning		Suffixes	The Meaning
photo-	light		-chrom(e)	color
phyto-	plant		-tropism	turn, turning

Prefix	Suffix	Definition
_____	_____	1. The growth response of an organism to light; usually the turning toward or away from the light source.
gravi-	_____	2. The growth response of an organism to gravity; usually the turning toward or away from the direction of gravity.
_____	_____	3. A blue-green, proteinaceous pigment involved in photoperiodism and a number of other light-initiated physiological responses of plants.

_____ periodism   4. The physiological response of organisms to variations of light and darkness.

MATCHING
For each of these definitions, select the correct matching term from the list that follows.

____ 1. The inhibition of lateral buds by the stem apical meristem.

____ 2. A special structure at the base of the petiole that functions in leaf movement by changes in turgor.

____ 3. A plant hormone involved in dormancy and responses to stress.

____ 4. Promotion of flowering in certain plants by exposing them to a cold period.

____ 5. The aging process in plants.

____ 6. A protective sheath that encloses the stem in certain monocots.

____ 7. A plant hormone involved in apical dominance and cell elongation.

____ 8. A plant hormone that promotes fruit ripening.

____ 9. Plant growth in response to contact with a solid object.

____10. A plant hormone that promotes rapid cell division and is involved in other aspects of plant growth and development.

Terms:

a. Abscisic acid
b. Apical dominance
c. Auxin
d. Coleoptile
e. Cytokinin

f. Ethylene
g. Imbibition
h. Nastic movements
i. Pulvinus
j. Senescence

k. Statolith
l. Thigmotropism
m. Vernalization

TABLE
Fill in the blanks.

Physiological Effect	Hormone	Site of Production
Promotes bud and seed dormancy	Abscisic acid	Older leaves, root cap, stem
Promotes apical dominance	#1	#2
Delays senescence	#3	Roots
Promotes germination	#4	#5
Promotes changes in plant tissues that are stressed	#6	#7
Promotes leaf abscission	#8	Stem nodes, ripening fruit, senescing tissue
Promotes cell division	Cytokinin	#9
Regulates growth by promoting cell elongation	#10	Shoot apical meristem, young leaves, seeds

THOUGHT QUESTIONS
Write your responses to these questions.

1. Describe the genetic and environmental factors that influence plant growth and development.
2. What environmental factors can cause a plant to turn or move? How do they move? What are tropisms?
3. What is periodicity? What are circadian rhythms? How does photoperiodicity affect flowering?
4. Discuss the role(s) of the hormones auxins, gibberellins, cytokinins, ethylene, and abscisic acid in the processes of germination, dormancy, elongation, ripening, and abscission. Are there other functions for these hormones?

MULTIPLE CHOICE
Place your answer(s) in the space provided. Some questions may have more that one correct answer.

_____ 1. Which of the following is/are correct about hormones?
   a. They are effective in very small amounts.   d. The effects of different hormones overlap.
   b. They are organic compounds.   e. Each plant hormone has multiple effects.
   c. They are produced in one part of the plant and transported to other parts.

_____ 2. The hormone(s) that trigger(s) changes in plants that are exposed to unfavorable environmental conditions is/are
   a. auxin.   d. ethylene.
   b. gibberellin.   e. abscisic acid.
   c. cytokinin.

_____ 3. The hormone(s) principally responsible for cell division and differentiation is/are
   a. auxin.   d. ethylene.
   b. gibberellin.   e. abscisic acid.
   c. cytokinin.

_____ 4. If a plant that touches your house continues to grow toward and attach itself to the house, the plant is exhibiting a
   a. tropism.   d. thigmotropism.
   b. phototropism.   e. turgor movement.
   c. gravitropism.

_____ 5. If potassium ions and tannins leave special cells at the base of a petiole when the plant's leaf is touched, the leaf may move. This is an example of a
   a. tropism.   d. thigmotropism.
   b. phototropism.   e. turgor movement.
   c. gravitropism.

_____ 6. The plant hormone(s) about which Charles Darwin gathered information is/are
   a. auxin.   d. ethylene.
   b. gibberellin.   e. abscisic acid.
   c. cytokinin.

_____ 7. The hormone(s) with a five ring structure that is/are involved in rapid stem elongation just prior to flowering is/are
   a. auxin.   d. ethylene.
   b. gibberellin.   e. abscisic acid.
   c. cytokinin.

_____ 8. The hormone(s) principally responsible for the growth of a coleoptile toward light is/are

a. auxin.

b. gibberellin.

c. cytokinin.

d. ethylene.

e. abscisic acid.

_____ 9. If a plant that normally opens its stomates in daylight and closes its stomates during the night is placed in total darkness for a week, one would expect that it would

a. open its stomates all week.

b. close its stomates all week.

c. assume a different periodicity for opening and closing stomates.

d. continue opening and closing stomates as usual.

e. soon exhibit a hormone imbalance.

VISUAL FOUNDATIONS

Color the parts of the illustration below as indicated.

RED ☐ coleoptile

GREEN ☐ shoot and leaves

YELLOW ☐ cotyledons

BLUE ☐ hook

ORANGE ☐ primary root

BROWN ☐ seed coat

TAN ☐ shriveled cotyledon

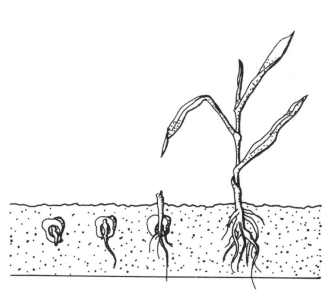

PART VII

Structures and Life Process in Animals

❏

The Animal Body: Introduction to Structure and Function

In most animals, cells are organized into tissues, tissues into organs, and organs into organ systems. The principal animal tissues are epithelial, connective, muscular, and nervous. Epithelial tissues are characterized by tight-fitting cells and the presence of a basement membrane. Covering body surfaces and lining cavities, they function in protection, absorption, secretion, and sensation. Epithelial tissues are classified on the basis of cell shape and arrangement of cell layers. Connective tissue joins together other tissues, supports the body and its organs, and protects underlying organs. There are many different types of connective tissues, consisting of a variety of cell types. For example, there is loose and dense connective tissue, elastic connective tissue, reticular connective tissue, adipose tissue, cartilage, bone, and blood. Muscle tissue is composed of cells that are specialized to contract. There are three major types of muscle tissue — cardiac, smooth, and skeletal. Nervous tissue is composed of cells that are specialized for conducting impulses and those that support and nourish the conducting cells. Organs are comprised of two or more kinds of tissues. Complex animals have many organs and ten organ systems. The organ systems work together to maintain the body's homeostasis.

CHAPTER OUTLINE AND CONCEPT REVIEW
Fill in the blanks.

INTRODUCTION

1 New cells formed by cell division remain associated in multicellular animals. The size of an animal is determined by the _____ of cells that make up its body, not the size of the individual cells.

2 A group of cells that carry out a specific function is called a (a)_____, and these groups associate to form (b)_____, which in turn are grouped into the (c)_____ of the body.

EPITHELIAL TISSUES COVER THE BODY AND LINE ITS CAVITIES

3 Epithelial tissues cover body surfaces and cavities, forming (a)_____ of tightly joined cells that are attached to underlying tissues by a fibrous noncellular (b)_____.

4 Four of the major functions of epithelial cells include _____
_____.

5 Epithelial tissues vary in the number of cell layers and the shapes of their cells. For example, (a)_____ epithelium is made up of only one layer of cells and (b)_____ epithelium has two or more cell layers. (c)_____ cells are flat, (d)_____ cells resemble dice, and (e)_____ cells are tall, slender cells shaped like cylinders.

CONNECTIVE TISSUES JOIN AND SUPPORT OTHER BODY STRUCTURES

6 Connective tissues join, support, and protect other tissues. There are many kinds of connective tissues, the main types of which are _____

_____
_____.

7 Characteristically, connective tissues contain very few cells, an (a)_____
   _____ in which cells and (b)_____ are embedded, and a (c)_____ that is
   secreted by the cells.

8 The _____ associated with connective tissue is largely
   responsible for the nature and function of the tissue.

Connective tissue contains collagen, reticular, and elastic fibers

9 There are three types of fibers in connective tissues. Collagen fibers are numerous, strong fibers
   composed of the protein (a)_____. Elastic fibers are composed of (b)_____
   and can stretch. Reticular fibers, composed of (c)_____,
   form delicate networks of connective tissue.

Connective tissue contains specialized cells

10 Fibroblast cells produce (a)_____ in connective tissues.
   (b)_____ are phagocytic scavengers.

11 The (a)_____ in connective tissue release histamines, and (b)_____
   secrete antibodies.

Loose connective tissue is widely distributed

12 The most abundant tissue in the body is called loose connective tissue or (a)_____ tissue.
   It is found in spaces between body parts, around muscles, nerves, and blood vessels, and under the
   skin where it is called the (b)_____.

Dense connective tissue consists mainly of fibers

13 Dense connective tissues are predominantly composed of (a)_____ fibers. If these fibers
   are arranged in a specific pattern, the tissue is referred to as (b)_____
   connective tissue. These strong tissues form "cablelike" cords called (c)_____.
   (d)_____ connective tissue consists of fibers arranged in
   various directions.

Elastic tissue is found in structures that must expand

14 The walls of large _____ contain elastic connective tissue.

Reticular connective tissue provides support

15 Reticular connective tissue forms an internal structural support network for soft organs such as the
   _____.

Adipose tissue stores energy

16 Two of the most important functions of adipose tissue are _____
   _____.

Cartilage and bone provide support

17 All vertebrates have an internal supporting structure called the _____ that is composed of
   cartilage and/or bone.

18 The (a)_____ that provides support in the vertebrate embryo is largely replaced
   by (b)_____ in all vertebrate adults except (c)_____.

19 Cartilage cells called (a)_____ are surrounded by a rubbery (b)_____ in
   which collagen fibers are embedded. "Imprisoned" by their own secretions, the cells are isolated in
   holes called (c)_____. Nutrients diffuse through the matrix to the cells.

20 Bone is similar to cartilage in that it too has a (a)_____ dotted with lacunae in which cells called (b)_____ are "imprisoned." Unlike cartilage however, bone is highly (c)_____, a term that refers to their abundant supply of blood vessels.

21 The matrix of bone contains (a)_____, a material that inhibits the diffusion of nutrients through the matrix. Consequently, osteocytes communicate with one another through small channels called (b)_____.

22 The structural unit of bone is called an (a)_____. At its center is a (b)_____ containing blood vessels and nerves. Concentric layers of osteocytes, known as (c)_____, surround the canal.

Blood and lymph are circulating tissues

23 The noncellular component of blood is the _____.

24 Red blood cells, or (a)_____ as they are called, contain the respiratory pigment (b)_____. Their shape is generally described as a (c)_____.

25 The white blood cells, collectively known as the _____, have important roles in defending the body against disease-causing organisms.

26 Platelets are small cell fragments that originate in _____. In some vertebrates they assist in blood clotting.

MUSCLE TISSUE IS SPECIALIZED TO CONTRACT

27 Skeletal muscle cells are called _____.

28 Contractile proteins called (a)_____, or "thin filaments," and (b)_____, or "thick filaments," are contained within the elongated (c)_____ of muscle cells.

29 Vertebrates have three kinds of muscles: (a)_____ muscles that attach to bones and cause body movements, (b)_____ muscle found in internal organs and the walls of blood vessels, and heart or (c)_____ muscle. Both (d)_____ muscles are under involuntary control, while (e)_____ muscle control is voluntary.

NERVOUS TISSUE CONTROLS MUSCLES, GLANDS, AND OTHER ORGANS

30 Cells in nervous tissue that conduct impulses are (a)_____, and support cells in this tissue are called (b)_____ cells.

31 Neurons communicate with one another at cellular junctions called _____.

32 A _____ is a collection of neurons bound together by connective tissue.

33 Neurons generally contain three functionally and anatomically distinct regions: the (a)_____ which contains the nucleus, the (b)_____ that receive incoming impulses, and the (c)_____ that carry away outgoing impulses.

COMPLEX ANIMALS HAVE ORGANS AND ORGAN SYSTEMS

34 The ten major organ systems of complex animals are the _____ _____ _____ systems.

ORGAN SYSTEMS WORK TOGETHER TO MAINTAIN HOMEOSTASIS

Homeostatic mechanisms are feedback systems
Homeostatic mechanisms regulate blood-sugar level

BUILDING WORDS

Use combinations of prefixes and suffixes to build words for the definitions that follow.

Prefixes	The Meaning		Suffixes	The Meaning
chondro-	cartilage		-blast	embryo
fibro-	fiber		-cyte	cell
homeo-	similar, "constant"		-phage	eat, devour
inter-	between, among		-stasis	equilibrium
macro-	large, long, great, excessive			
multi-	many			
myo-	muscle			
osteo-	bone			
pseudo-	false			

Prefix	Suffix	Definition
_____	-cellular	1. Composed of many cells.
_____	-stratified	2. An arrangement of epithelial cells in which the cells falsely appear to be stratified.
_____	_____	3. A cell, especially active in developing ("embryonic") tissue and healing wounds, that produces connective tissue fibers.
_____	-cellular	4. Situated between or among cells.
_____	_____	5. A large cell, common in connective tissues, that phagocytizes ("eats") foreign matter including bacteria.
_____	_____	6. A cartilage cell.
_____	_____	7. A bone cell.
_____	-fibril	8. A small longitudinal contractile fiber inside a muscle cell.
_____	_____	9. Maintaining a constant internal environment.

MATCHING

For each of these definitions, select the correct matching term from the list that follows.

____ 1. Voluntary or striated muscle of vertebrates.

____ 2. Tissue in which fat is stored, or the fat itself.

____ 3. Spindle-shaped unit of bone composed of concentric layers of osteocytes.

____ 4. Flexible skeletal tissue of vertebrates.

____ 5. A differentiated part of the body made up of tissues and adapted to perform a specific function or group of functions.

____ 6. Glands that secrete products directly into the blood or tissue fluid instead of into ducts.

____ 7. The muscle of the heart.

____ 8. Supporting cell of central nervous tissue.

____ 9. General term for a body cell or organ specialized for secretion.

____10. A protein in connective tissue fibers.

Terms:

a.   Adipose tissue
b.   Cardiac muscle
c.   Cartilage
d.   Collagen
e.   Endocrine gland

f.   Gland
g.   Glial cell
h.   Matrix
i.   Organ
j.   Organ system

k.   Osteon
l.   Skeletal muscle
m.   Stressor

TABLE
Fill in the blanks.

Principle Tissue	Tissue	Location	Function
Epithelial tissue	Stratified squamous epithelium	Skin, mouth lining, vaginal lining	Protection; outer layer continuously sloughed off and replaced from below
Epithelial tissue	#1	#2	Allows for transport of materials, especially by diffusion
#3	Pseudostratified epithelium	#4	#5
#6	Adipose tissue	Subcutaneous layer, pads certain internal organs	#7
Connective tissue	#8	Forms skeletal structure in most vertebrates	#9
#10	Blood	#11	#12
Muscle tissue	Cardiac muscle	#13	Contraction of the heart
Muscle tissue	#14	Attached to bones	Movement of the body
#15	Nervous tissue	Brain, spinal cord, nerves	Respond to stimuli, conduct impulses

THOUGHT QUESTIONS
Write your responses to these questions.

1. What are the adaptive advantages of multicellularity, if any?
2. What is a tissue? An organ? A tissue system?
3. Describe the structure and function of each of the three types of muscle tissue.
4. Describe the structure and function of cells in the nervous tissue.
5. What are the names and principal functions of the major organ systems in the typical mammal?
6. What is homeostasis? Describe in detail how it works in two different body systems. Does it seem to be a paradox that a homeostatic system is dynamic (changing) and still in equilibrium? Explain your answer.
7. How are the cells of a neoplasm different from ordinary cells? What is/are the difference(s) between a malignant neoplasm and a benign neoplasm?

MULTIPLE CHOICE

Place your answer(s) in the space provided. Some questions may have more that one correct answer.

_____ 1. Chondrocytes are
   a. part of cartilage.
   b. found in bone.
   c. cells that produce the matrix in cartilage.
   d. eventually found in the lacunae of a matrix.
   e. embryonic osteocytes.

_____ 2. Cartilage
   a. supports vertebrate embryos.
   b. comprises the shark's skeleton.
   c. is largely replaced by bone in most vertebrates.
   d. is derived from embryonic adipose tissue.
   e. is produced by chondrocytes.

_____ 3. Blood is one type of
   a. endothelium.
   b. collagen.
   c. mesenchyme.
   d. connective tissue.
   e. cardiac tissue.

_____ 4. Large muscles attached to bones are
   a. skeletal muscles.
   b. smooth muscles.
   c. composed largely of actin and myosin.
   d. nonstriated.
   e. multinucleated.

_____ 5. Cells that nourish and support nerve cells are known as
   a. neurons.
   b. glial cells.
   c. neuroblasts.
   d. synaptic cells.
   e. neurocytes.

_____ 6. A group of closely associated cells that carries out a specific function is a/an
   a. organ.
   b. system.
   c. organ system.
   d. tissue.
   e. clone.

_____ 7. Cells lining internal cavities that secrete a lubricating mucus are
   a. goblet-cells.
   b. pseudostratified.
   c. part of glandular tissue.
   d. epithelial.
   e. part of connective tissue.

_____ 8. Platelets are
   a. bone marrow cells.
   b. fragments of cells.
   c. a type of white blood cells.
   d. a type of RBCs.
   e. derived from bone marrow cells.

_____ 9. Epithelium located in the ducts of glands would most likely be
   a. simple columnar.
   b. simple cuboidal.
   c. simple squamous.
   d. stratified squamous.
   e. stratified cuboidal.

_____10. The portion of the neuron that is specialized to receive a nerve impulse is the
   a. cell body.
   b. synapse.
   c. dendrite.
   d. axon.
   e. glial body.

_____11. The connective tissue matrix is
    a. noncellular.
    b. mostly lipids.
    c. a gel.
    d. a polysaccharide.
    e. fibrous.

_____12. Collagen is
    a. part of blood.
    b. part of bone.
    c. part of connective tissue.
    d. composed of fibroblasts.
    e. fibrous.

_____13. Haversian canals
    a. contain nerves.
    b. run through cartilage.
    c. run through bone.
    d. are matrix lacunae.
    e. are synonymous with canaliculi.

_____14. Adipose tissue is a type of
    a. cartilage.
    b. epithelium.
    c. connective tissue.
    d. modified blood tissue.
    e. marrow.

_____15. The nucleus of a neuron is typically found in the
    a. cell body.
    b. synapse.
    c. dendrite.
    d. axon.
    e. glial body.

_____16. Endothelium
    a. lines ducts.
    b. lines blood vessels.
    c. lines lymph vessels.
    d. is derived from mesenchyme.
    e. is a form of sensory epithelium.

_____17. Myosin and actin are the main components of
    a. cartilage.
    b. blood.
    c. collagen.
    d. muscle.
    e. adipose tissue.

_____18. The basement membrane
    a. lies beneath epithelium.
    b. contains polysaccharides.
    c. has the typical fluid mosaic membrane structure.
    d. is synonymous with plasma membrane.
    e. is noncellular.

_____19. The major classes of animal tissues include
    a. epithelial.
    b. cardiovascular.
    c. muscular.
    d. skeletal.
    e. connective.

_____20. Epithelium cells that are flattened and thin are called
    a. columnar.
    b. cuboidal.
    c. squamous.
    d. stratified.
    e. endothelium.

VISUAL FOUNDATIONS
Color the parts of the illustration below as indicated.

RED   ☐ blood vessel
GREEN   ☐ lacuna
YELLOW   ☐ compact bone
BLUE   ☐ osteocyte
ORANGE   ☐ haversian canal
BROWN   ☐ osteon
TAN   ☐ spongy bone

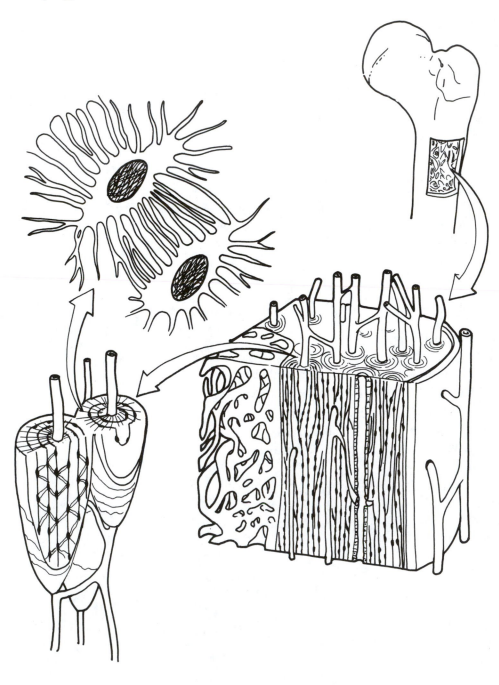

CHAPTER 38

Protection, Support, and Movement: Skin, Skeleton, and Muscle

Three organ systems — the integumentary, the skeletal, and the muscular — are closely interrelated in function and significance. Epithelium covers all external and internal body surfaces. In invertebrates, the external epithelium may contain secretory cells that produce a protective cuticle, secrete lubricants or adhesives, produce odorous or poisonous substances, or produce threads for nests or webs. In vertebrates, specifically humans, the external epithelium (skin) includes nails, hair, sweat glands, oil glands, and sensory receptors. In other vertebrates, it may include feathers, scales, mucous, and pigmentation. The skeleton supports and protects the body and transmits mechanical forces generated by contractile cells. Among the invertebrates are found hydrostatic skeletons (cnidarians and annelids) and exoskeletons (mollusks and arthropods). Exoskeletons, composed of nonliving material above the epidermis, prevent growth, necessitating periodic molting. Endoskeletons on the other hand, extensive in echinoderms and chordates, are composed of living tissue that can grow. Whereas some bones in the body develop from a noncartilage connective tissue model, most bones of the body develop from cartilage replicas. All animals have the ability to move. Muscle tissue is found in most invertebrates and all vertebrates. As muscle contracts, it moves body parts by pulling on them. Muscle contraction, fueled by ATP, involves a complex sequence of events. Muscle action is antagonistic — the movement produced by one muscle can be reversed by another.

CHAPTER OUTLINE AND CONCEPT REVIEW
Fill in the blanks.

INTRODUCTION

1 Muscle can contract. When it is anchored to the _____ system, muscle contraction causes movement (locomotion).

OUTER COVERINGS PROTECT THE BODY

2 The external and internal surfaces of the animal body are covered by _____ tissue.

The protective epithelial covering of invertebrates may function in secretion or gas exchange

3 Epithelial tissue in invertebrates may contain _____ cells that produce a protective cuticle, secrete lubricants or adhesives, produce odorous or poisonous substances, or produce threads for nests or webs.

The vertebrate skin functions in protection and temperature regulation

4 Human _____ includes nails, hair, sweat glands, oil glands, and sensory receptors.

5 Feathers, scales, mucus coverings, and poison glands are all structures associated with the _____ of vertebrates.

6   The outer layer of skin is called the (a)_____. It consists of several sublayers, or (b)_____.

7   Cells in the _____ continuously divide. As they are pushed upward, these cells mature, produce keratin, and eventually die.

8   Epithelial cells produce an insoluble, elaborately coiled protein called (a)_____, which functions in the skin for (b)_____.

9   The dermis consists of dense, fibrous (a)_____ resting on a layer of (b)_____ tissue composed largely of fat.

SKELETONS ARE IMPORTANT IN LOCOMOTION, PROTECTION, AND SUPPORT

10 The skeleton transmits mechanical forces generated by _____ and also supports and protects the body.

11 The support system in some organisms is an _____ consisting of a non-living substance overlying the epidermis.

In hydrostatic skeletons, body fluids are used to transmit force

12 Many invertebrates (e.g., hydra) have a (a)_____ in which fluid is used to transmit forces generated by contractile cells or muscle. In these animals, contractile cells are arranged in two layers, an outer (b)_____ oriented layer and an inner (c)_____ arranged layer. When the (d)_____ (outer or inner?) layer contracts, the animal becomes shorter and thicker, and when the (e)_____ (outer or inner?) contracts, the animal becomes longer and thinner.

13 Annelid worms have sophisticated hydrostatic skeletons. The body cavity is divided by transverse partitions called _____, creating isolated, fluid-filled segments that can operate independently.

Mollusks and arthropods have nonliving exoskeletons

14 The main function of the mollusk exoskeleton is for _____.

15 Arthropod exoskeletons, composed mainly of chiton, are jointed for flexibility. This nonliving skeleton prevents growth, necessitating periodic _____.

Internal skeletons are capable of growth

16 Internal skeletons, called (a)_____, are extensively developed only in the echinoderms and the (b)_____.

17 The endoskeletons of _____ are internal "shells" formed from spicules and plates composed of calcium salts.

18 The two main divisions of the vertebrate endoskeleton are the (a)_____ skeleton along the long axis of the body and the (b)_____ skeleton comprised of the bones of the limbs and girdles.

19 Components of the axial skeleton include the (a)_____, which consists of cranial and facial bones; the (b)_____, made up of a series of vertebrae; and the (c)_____, consisting of the sternum and ribs. Vertebrae have a bony, weight-bearing (d)_____ and a dorsal (e)_____ around the spinal cord.

20 Components of the appendicular skeleton include the (a)_____, consisting of clavicles and scapulas; the (b)_____ which consists of large, fused hipbones; and the (c)_____, each of which terminates in digits. Digits can be specialized, as, for example, with the (d)_____ of humans and other primates that enables them to grasp and manipulate objects.

21 The long bones of the limbs, such as the radius, have characteristic structures. They are covered by a connective tissue layer called the (a)_____. The ends of these bones are called the (b)_____, and the shaft is the (c)_____. A cartilaginous "growth center" in children called the (d)_____ becomes an (e)_____ in adults. (f)_____, made up of osteons, is a thin covering over bone. The epiphyseal ends of long bones are filled with spongy, or (g)_____, bone tissue.

22 Long bones develop from cartilage replicas; this is (a)_____ bone formation. Other bones develop from noncartilage connective tissue replica; this is (b)_____ bone development.

23 Junctions between bones are called (a)_____. They are classified according to the degree of their movement: (b)_____, such as sutures, are tightly bound by fibers; (c)_____, like those between vertebrae; and the most common type of joint, the (d)_____.

MUSCLE IS THE CONTRACTILE TISSUE THAT PERMITS MOVEMENT IN MOST ANIMALS

24 (a)_____ is a ubiquitous contractile protein in eukaryotic cells. It makes up the thin microfilaments that function in such cellular processes as skeletal muscle contraction, ameboid movement, and cell division. In many cells, as, for example, in skeletal muscle myofibrils, it functions in association with the contractile protein (b)_____.

25 Most animals move. Most invertebrates have specialized muscles for movements. For example, bivalve mollusks have both smooth and striated muscle. The (a)_____ muscle act to hold their shells closed for long periods of time, whereas the (b)_____ muscle enables them to shut their shells rapidly.

A vertebrate muscle may consist of thousands of muscle fibers

26 Each skeletal muscle of a vertebrate consists of bundles of muscle fibers called _____.

27 Striations of skeletal muscle fibers reflect the interdigitations of (a)_____ filaments. A unit of actin and myosin filaments makes up a (b)_____.

Contraction occurs when actin and myosin filaments slide past each other

28 During muscle contractions, actin filaments move _____(inward or outward?) between the myosin filaments.

29 A motor neuron releases (a)_____ into the myoneural cleft, where it combines with receptors on the surface of the muscle fiber, depolarizing the sarcolemma and initiating an (b)_____.

30 An impulse spreads through T tubules and stimulates calcium release. (a)_____ initiate a process that uncovers the binding sites of the (b)_____ filaments. Cross bridges of the (c)_____ filaments attach to the binding sites.

31 Cross bridges flex and reattach to new binding sites so that the filaments are pulled past one another, _____ the muscle.

ATP powers muscle contraction

32 (a)_____ is the immediate energy source for muscle contraction. (b)_____ is an energy storage compound. (c)_____ is the fuel stored in muscle fibers.

Skeletal muscle action depends on muscle pairs that work antagonistically

33 Skeletal muscles pull on cords of connective tissue called (a)_____, which then pull on bones. Usually, muscles cross a joint and are attached to opposite sides of the joint. At most joints,

muscles act (b)_____, which means that the action of one muscle
(the agonist) is opposed by the action of the other muscle, called the (c)_____.

Smooth, cardiac, and skeletal muscle respond in specific ways

34 (a)_____ muscle is capable of slow, sustained contractions and (b)_____ muscle
contracts rhythmically. (c)_____ muscle contains two sets of specialized fibers: the
(d)_____ fibers that are specialized for slow responses and the
(e)_____ fibers specialized for fast responses.

BUILDING WORDS

Use combinations of prefixes and suffixes to build words for the definitions that follow.

Prefixes	The Meaning		Suffixes	The Meaning
endo-	upon, over, on		-blast	embryo, "formative cell"
epi-	within		-chondr(al)	cartilage
myo-	muscle		-dermis	skin
osteo-	bone		-oste(um)	bone
peri-	about, around, beyond			

Prefix	Suffix	Definition
_____	_____	1. The outermost layer of skin, resting on the dermis.
_____	_____	2. Connective tissue membrane on the surface of bone that is capable of forming bone.
_____	_____	3. Pertains to the occurrence or formation of "something" within cartilage.
_____	_____	4. A bone-forming cell.
_____	-clast	5. Cells that break down bone.
_____	-filament	6. Subunit of myofibril consisting of either actin or myosin.

MATCHING

For each of these definitions, select the correct matching term from the list that follows.

_____ 1. All the skeletal muscle fibers that are stimulated by a single motor neuron.

_____ 2. A horny, water-insoluble protein that is a constituent of intermediate filaments in animal cells and is found in the epidermis of vertebrates; in nails, feathers, hair, horns, scales, and claws.

_____ 3. A segment of a striated muscle cell located between adjacent Z-lines that serves as a unit of contraction.

_____ 4. The deepest layer of the epidermis, consisting of cells that continuously divide.

_____ 5. A connective tissue cord or band that connects bones to each other or holds other organs in place.

_____ 6. The skull, vertebral column, sternum and rib cage.

_____ 7. The hard exterior covering of certain invertebrates.

_____ 8. A connective tissue structure that joins a muscle to another muscle, or a muscle to a bone.

_____ 9. The oxygen necessary to metabolize the lactic acid produced during strenuous exercise.

_____10. The protein of which myofilaments are composed.

Terms:

a. Actin
b. Axial skeleton
c. Endoskeleton
d. Exoskeleton
e. Joint

f. Keratin
g. Ligament
h. Motor unit
i. Oxygen debt
j. Sarcomere

k. Skeletal muscle
l. Stratum basale
m. Tendon
n. Vertebral column

TABLE
Fill in the blanks.

Organism	Skeletal Material	Kind of Skeleton	Characteristics
Many invertebrates	Fluid	Hydrostatic	Contractile tissue generates force that moves the body
Arthropods	#1	#2	#3
#4	#5	Endoskeleton	Internal shell, some have spines that project to the outer surface
Chordates	#6	#7	#8

THOUGHT QUESTIONS
Write your responses to these questions.

1. Describe the major support systems in animals, including the basic structures and the similarities and differences of a hydrostatic skeleton, an exoskeleton, and an endoskeleton.
2. Sketch a typical long bone and label the parts.
3. Sketch a sarcomere and label the parts, then relate the "sliding filament theory" of muscle contraction to each part of the sarcomere.
4. How are glycogen, creatine phosphate, and ATP involved in muscle contraction?

MULTIPLE CHOICE
Place your answer(s) in the space provided. Some questions may have more that one correct answer.

_____ 1. The human skull is part of the
    a. cervical complex.
    b. girdle.
    c. axial skeleton.
    d. appendicular skeleton.
    e. atlas.

_____ 2. Most of the mechanical strength of bone is due to
    a. spongy bone.
    b. cancellous bone.
    c. intramembranous bone.
    d. endochondral bone.
    e. marrow.

_____ 3. You perceive touch through sense organs in your
    a. epidermis.
    b. stratum basale.
    c. stratum corneum.
    d. epithelium.
    e. dermis.

_____ 4. Myofilaments are composed of
   a. myofibrils.   d. myosin.
   b. fibers.   e. sarcoplasmic reticulum.
   c. actin.

_____ 5. The outer layer of a vertebrate's skin is the
   a. epidermis.   d. epithelium.
   b. stratum basale.   e. dermis.
   c. stratum corneum.

_____ 6. An exoskeleton would not work for an organism the size of a walrus because
   a. muscle attachments can't carry the load.   d. exoskeletons weigh too much.
   b. their eggs are too small.   e. invertebrate muscles are not strong enough.
   c. the number of ecdyses is limited.

_____ 7. The tissue around bones that lays down new layers of bone is the
   a. metaphysis.   d. periosteum.
   b. epiphysis.   e. endosteum.
   c. marrow.

_____ 8. The human axial skeleton includes the
   a. ulna.   d. femur.
   b. shoulder blades.   e. breastbone.
   c. centrum.

_____ 9. Cartilaginous growth centers in children are called
   a. metaphysis.   d. periosteum.
   b. epiphysis.   e. endosteum.
   c. marrow.

_____10. The human appendicular skeleton includes the
   a. ulna.   d. femur.
   b. shoulder blades.   e. breastbone.
   c. centrum.

_____11. Vertebrate appendages are connected to
   a. the cervical complex.   d. the appendicular skeleton.
   b. girdles.   e. the atlas.
   c. the axial skeleton.

_____12. Skeletons that operate entirely by hydrostatic pressure are found in
   a. annelids.   d. lobsters and crayfish.
   b. echinoderms.   e. vertebrates.
   c. hydra.

_____13. An opposable digit is found in
   a. lobsters.   d. some insects.
   b. great apes.   e. humans.
   c. crayfish.

_____14. A snail's slime track, the sticky substance on a fly's feet, and a spider's web are all produced by
   a. the stratum basale.   d. specialized dermal cells.
   b. epithelial cells.   e. stratum corneum.
   c. epicuticle.

____15. Internal skeletons are found in
a. annelids.
b. echinoderms.
c. hydra.
d. lobsters and crayfish.
e. vertebrates.

VISUAL FOUNDATIONS

Color the parts of the illustration below as indicated

RED ☐ nucleus
GREEN ☐ T tubule
YELLOW ☐ sarcoplasmic reticulum
BLUE ☐ mitochondria
ORANGE ☐ myofibrils
BROWN ☐ plasma membrane
TAN ☐ sarcomere
PINK ☐ Z line

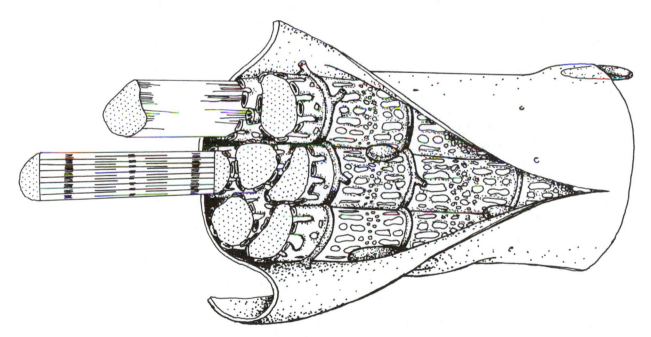

Neural Control: Neurons

Two organ systems regulate behavior and physiological processes in animals — the endocrine system and the nervous system. Endocrine regulation is slow and long lasting, whereas nervous regulation is rapid. Changes within an animal's body or in the outside world are detected by receptors and transmitted in the form of rapidly moving electrical signals or impulses to the central nervous system (CNS). The CNS sorts and interprets the information and determines an appropriate response. Impulses are then transmitted to muscle or glands where the response occurs. The structural and functional unit of the nervous system that carries the impulses is the neuron. The typical neuron consists of a cell body, dendrites, and an axon. The cell body houses the nucleus and has branched cytoplasmic projections called dendrites that receive stimuli. The axon is a single, long structure that transmits impulses from the cell body to an adjacent neuron, muscle, or gland. Transmission along a neuron is an electrochemical process involving changes in ion distribution between the inside and outside of the neuron. Transmission from one neuron to the next generally involves the release of chemicals — neurotransmitters — into the space between neurons. The neurotransmitters diffuse across the space and bring about a response in the adjacent neuron (or muscle or gland). Many types of neurotransmitters have been identified. Because every neuron networks with hundreds of other neurons, hundreds of messages may arrive at a single neuron at the same time. These messages must be integrated before the neuron can respond. After the neuron has completed its integration, it may or may not initiate an impulse along its axon. Most neural integration occurs in the CNS. Neurons are organized into complex neural pathways or circuits.

CHAPTER OUTLINE AND CONCEPT REVIEW
Fill in the blanks.

INTRODUCTION

1 In complex animals, two systems are responsible for the regulation of physiological processes and behavior. Regulation by the (a)_____ system is slow and long-lasting, whereas regulation by the (b)_____ system is rapid.

2 The ability of an organism to survive is largely dependent on its ability to detect and respond appropriately to _____, which are defined as changes in either the internal or external environments.

3 Information flow through the nervous system begins with the detection of a stimulus by a "sensory device," a process called (a)_____. The "sensory device" is associated with sensory or (b)_____ neurons that (c)_____ or relay the information to the central nervous system (CNS). Association neurons (also called (d)_____) in the CNS "manage" the information, sorting and interpreting the information in a process called (e)_____. Once the appropriate response is selected, the information is transmitted from the CNS by motor or (f)_____ neurons to muscles and/or glands (the (g)_____) that carry out the response.

THE CELL TYPES OF THE NERVOUS SYSTEM ARE NEURONS AND GLIAL CELLS

4 The _____ is the structural and functional unit of the nervous system.

Glial cells support and protect neurons

5 _____, meaning "nerve glue," are unique cells that protect, insulate, and support neurons.

A typical neuron consists of a cell body, dendrites, and an axon

6 Neurons are highly specialized cells with a distinctive structure consisting of a cell body and two types of cytoplasmic extensions. Numerous, short (a)_____ receive impulses and conduct them to the (b)_____ which integrates impulses. Once integrated, the impulse is conducted away by the (c)_____, a long process that terminates at another neuron or effector.

7 Branches at the ends of axons, called (a)_____, form tiny (b)_____ that release a transmitter chemical called a (c)_____.

8 Many axons outside of the CNS are covered by sheaths produced by cells called (a)_____ that wrap around these axons. The outer (b)_____, or neurilemma, is important in the regeneration of neurons. The inner (c)_____ acts as an insulator, due largely to the lipid-rich material called (d)_____ in the plasma membranes of sheath-producing cells.

9 A (a)_____ is a bundle of axons outside the CNS, whereas a (b)_____ is a bundle of axons within the CNS.

10 Masses of cell bodies outside the CNS are called (a)_____, whereas masses of cell bodies within the CNS are called (b)_____.

NEURONS CONVEY INFORMATION BY TRANSMITTING RAPIDLY MOVING ELECTRICAL IMPULSES

The resting potential is the difference in electrical charge across the plasma membrane

11 The steady potential difference that exists across the plasma membrane of a nonconducting neuron is the (a)_____. Its value is expressed as (b)_____.

12 When a neuron is not conducting an electrical impulse, the _____(inner or outer?) surface of the plasma membrane is negatively charged compared to the interstitial fluid.

13 In most cells, potassium ion concentration is highest (a)_____ (inside or outside?) the cell and sodium ion concentration is highest) (b)_____ (inside or outside?) the cell.

14 The ionic balance across the plasma membrane of neurons is a result of several factors. The (a)_____ is an active transport system that moves sodium out of the cell and potassium into the cell. (b)_____ are part of a facilitated diffusion system that allows sodium and potassium ions to follow their gradients. Finally, the cytoplasm of the neuron is negatively charged due to the presence of large (c)_____.

The nerve impulse is an action potential

15 If a stimulus is strong enough, it will provoke a response from the neuron called an (a)_____, which is an alteration of the resting potential caused by an increase in membrane permeability to (b)_____. The potential value for this response is expressed as (c)_____.

16 (a)_____ in the plasma membrane of the neuron open when they detect a critical level of voltage change in the membrane potential. This value, known as the (b)_____, is generally given as about (c)_____.

17 A (a)_____ is a sharp rise and fall of an action potential. The sudden rise is largely caused by the opening of (b)_____.

18 When depolarization reaches threshold level, an action potential may be generated. An action potential is a _____ that spreads along the axon.

19 As the action potential moves down the axon, _____ occurs behind it.

Saltatory conduction is rapid

20 In saltatory conduction in myelinated neurons, depolarization skips along the axon from one _____ to the next.

The neuron obeys an all-or-none law

21 Once a stimulus depolarizes a neuron to the threshold, the neuron will respond fully with an action potential. All action potentials have the same strength. A response that always occurs at a given value, or it won't occur at all, is referred to as an _____ response.

Certain substances affect excitability

SYNAPTIC TRANSMISSION OCCURS BETWEEN NEURONS

22 A junction between two neurons, or between a neuron and an effector, is called a (a)_____ _____. There are two kinds of neuroeffector junctions: (b)_____ junctions between neurons and gland cells and (c)_____ between neurons and muscle cells.

23 The neuron that terminates at a synapse is known as the (a)_____ neuron and the neuron that begins at that synapse is called the (b)_____ neuron.

24 (a)_____ are gap junctions that allow rapid communication between neurons. Most synapses are (b)_____ that involve (c)_____ _____ that cross the physical space or (d)_____ between neurons.

Neurotransmitters affect postsynaptic neurons

25 The neurotransmitter diffuses across the synaptic cleft and combines with specific (a)_____ on the postsynaptic neuron. This opens (b)_____ that permit either depolarization or hyperpolarization.

Signals may be excitatory or inhibitory

26 A depolarization of the postsynaptic membrane that brings the neuron closer to firing is called an (a)_____. A hyperpolarization of the postsynaptic membrane that reduces the probability that the neuron will fire is called an (b)_____.

Graded potentials vary in magnitude

27 Local responses in the postsynaptic membrane that vary in magnitude, fade over distance, and can be summated are called _____. EPSPs and IPSPs are examples.

28 Graded potentials can be added together in a process called (a)_____. When a second EPSP occurs before the depolarization caused by the first EPSP has decayed, (b)_____ _____ occurs. In (c)_____, several synapses in the same area generate EPSPs simultaneously. Adding EPSPs together can bring the neuron to threshold.

Many types of neurotransmitters are known

29 Cells that release (a)_____ are called cholinergic neurons. This neurotransmitter functions at (b)_____ junctions and some synapses in the autonomic nervous system.

30 (a)_____ release norepinephrine, a neurotransmitter that, along with epinephrine and dopamine, belongs to the (b)_____ class.

NERVE FIBERS MAY BE CLASSIFIED IN TERMS OF SPEED OF CONDUCTION

31 Large, heavily myelinated neurons conduct impulses fastest. In general, the _____ (further apart or closer?) the nodes of Ranvier, the faster the axon conducts.

NEURAL IMPULSES MUST BE INTEGRATED

32 Integration of EPSPs and IPSPs occurs in the _____ of the postsynaptic neuron.

NEURONS ARE ORGANIZED INTO CIRCUITS

33 Complex neural pathways are possible because of such neuronal associations as
_____.

BUILDING WORDS

Use combinations of prefixes and suffixes to build words for the definitions that follow.

Prefixes	The Meaning		Suffixes	The Meaning
inter-	between, among		-neuro(n)	nerve
multi-	many, much, multiple			
neur(i)(o)-	nerve			
post-	behind, after			
pre-	before, prior to, in advance of			

Prefix	Suffix	Definition
_____	_____	1. A nerve cell that carries impulses to other nerves, and is between an effector and sense receptor but not directly associated with either.
_____	-glia	2. Connecting and supporting cells in the central nervous system surrounding the neurons.
_____	-lemma	3. The cellular sheath formed by Schwann cells surrounding the axons of certain nerve cells.
_____	-transmitter	4. Substance used by neurons to transmit impulses across a synapse.
_____	-polar	5. Pertains to a neuron with more than two (often many) processes or projections.
_____	-synaptic	6. Pertains to a neuron that begins after a specific synapse.
_____	-synaptic	7. Pertains to a neuron that ends before a specific synapse.

MATCHING

For each of these definitions, select the correct matching term from the list that follows.

_____ 1. One of many projections from a nerve cell that conducts a nerve impulse toward the cell body.

_____ 2. A nerve cell; a conducting cell of the nervous system which typically consists of a cell body, dendrites, and an axon.

____ 3. A knot-like mass of neuron cell bodies located outside the central nervous system.

____ 4. The long, tubular extension of the neuron that transmits nerve impulses away from the cell body.

____ 5. A large bundle of axons (or dendrites) wrapped in connective tissue that convey impulses between the central nervous system and some other part of the body.

____ 6. The brief period of time that must elapse after the response of a neuron or muscle fiber, during which it cannot respond to another stimulus.

____ 7. The electrical activity developed in a muscle or nerve cell during activity.

____ 8. A fatty insulating covering around axons of certain neurons.

____ 9. The effect of a nerve impulse on a postsynaptic neuron, which brings the postsynaptic neuron closer to firing.

____ 10. The junction between two neurons or between a neuron and an effector.

Terms:

a. Action potential
b. Axon
c. Convergence
d. Dendrite
e. Facilitation

f. Ganglion
g. Integration
h. Myelin sheath
i. Nerve
j. Neuron

k. Refractory period
l. Schwann cell
m. Summation
n. Synapse
o. Threshold level

TABLE
Fill in the blanks.

Kind of Potential	mV	Membrane Response	Ion Channel Activity
Membrane potential	-70mV	Stable	Sodium-potassium pump active, sodium and potassium channels open
Resting potential	#1	#2	#3
#4	Over -55mV	Depolarization	#5
Action potential	#6	#7	Rise: voltage-activated sodium ion channels and potassium ion channels open Fall: voltage-activated sodium channels start to close
EPSP	Increasingly positive	#8	#9
IPSP	Increasingly negative	#10	Chemically activated potassium or chloride channels may open

THOUGHT QUESTIONS
Write your responses to these questions.

1. Sketch the path followed through parts of the nervous system during a typical response to an outside stimulus. Include descriptions of reception, transmission, integration, and response.
2. Illustrate and describe a resting potential, then show how a stimulus transforms that resting potential into an action potential. Show how an impulse is transmitted along a neuron and how it is carried from one neuron to the next.
3. Contrast continuous conduction and saltatory conduction.
4. Illustrate and describe convergence, divergence, facilitation, and reverberation.

MULTIPLE CHOICE
Place your answer(s) in the space provided. Some questions may have more that one correct answer.

_____ 1. The inner surface of a resting neuron is _____ compared with the outside.
  a. positively charged   d. 70 mV
  b. negatively charged   e. −70 mV
  c. polarized

_____ 2. The part of the neuron that transmits an impulse from the cell body to an effector cell is the
  a. dendrite.   d. collateral.
  b. axon.   e. hillock.
  c. Schwann body.

_____ 3. The structural and functional unit of the nervous system is the
  a. cell body.   d. nerve.
  b. nerve cell.   e. synapse.
  c. neuron.

_____ 4. Saltatory conduction
  a. occurs between nodes of Ranvier.   d. requires less energy than continuous conduction.
  b. occurs only in the CNS.   e. involves depolarization at nodes of Ranvier.
  c. is more rapid than the continuous type.

_____ 5. The nodes of Ranvier are
  a. on the cell body.   d. gaps between adjacent Schwann cells.
  b. on the dendrites.   e. insulated with myelin.
  c. on the axon.

_____ 6. A neuron that begins at a synaptic cleft is called a
  a. presynaptic neuron.   d. neurotransmitter.
  b. postsynaptic neuron.   e. acetylcholine releaser.
  c. synapsing neuron.

_____ 7. Which of the following correctly expresses the movement of ions by the sodium-potassium pump?
  a. sodium out, potassium in   d. more sodium out than potassium in
  b. sodium in, potassium out   e. less sodium out than potassium in
  c. sodium and potassium both in and out, but in different amounts

_____ 8. The neurilemma is
  a. a cellular sheath.   d. found primarily in neurons of the CNS.
  b. the neuron membrane.   e. associated with nodes of Ranvier.
  c. composed of Schwann cells.

____ 9. A nerve pathway in the CNS is
   a. one neuron.
   b. a group of nerves.
   c. a bundle of axons.
   d. a ganglion.
   e. a bundle of cell bodies.

____10. An axon cannot transmit an action potential no matter how great a stimulus is applied when it is
   a. hyperpolarized.
   b. depolarized.
   c. in the absolute refractory period.
   d. in the relative refractory period.
   e. in the resting state.

____11. If a neurotransmitter hyperpolarizes a postsynaptic membrane, the change in potential is referred to as
   a. spatial summation.
   b. temporal summation.
   c. a threshold level impulse.
   d. IPSP.
   e. EPSP.

____12. Aside from the sodium-potassium pump, the resting potential is due mainly to
   a. inward diffusion of chloride ions.
   b. inward diffusion of sodium ions.
   c. outward diffusion of potassium ions.
   d. open sodium channels in the membrane.
   e. large protein anions inside the cell.

____13. Branchlike extensions of the cell body involved in receiving stimuli are
   a. dendrites.
   b. axons.
   c. Schwann bodies.
   d. collaterals.
   e. hillocks.

VISUAL FOUNDATIONS
Color the parts of the illustration below as indicated.

RED ☐ synaptic knob
GREEN ☐ axon terminal
YELLOW ☐ myelin sheath
BLUE ☐ axon
ORANGE ☐ cellular sheath
PINK ☐ cell body and dendrite
VIOLET ☐ nucleus

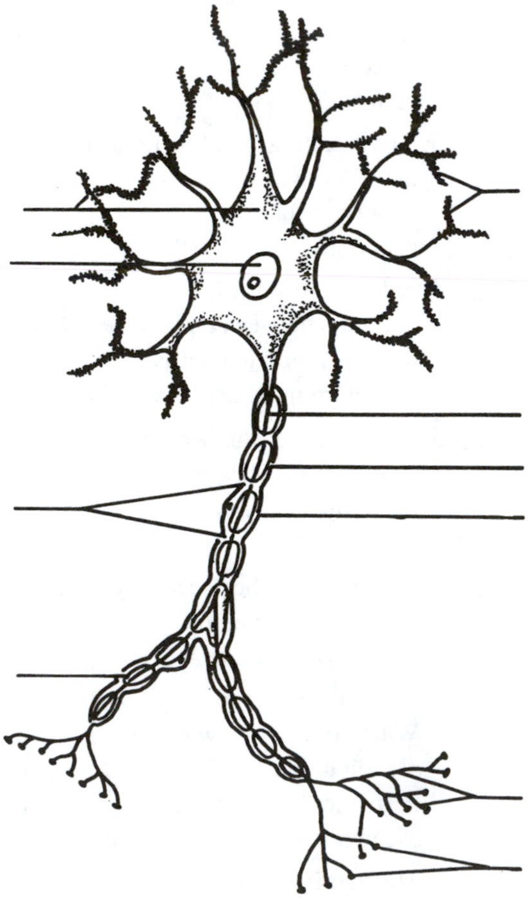

Color the parts of the illustration below as indicated.

RED	❑	sodium ion
GREEN	❑	potassium ion
YELLOW	❑	plasma membrane
BLUE	❑	arrows indicating diffusion into the cell
ORANGE	❑	sodium channel
BROWN	❑	potassium channel
TAN	❑	cytoplasm
PINK	❑	extracellular fluid
VIOLET	❑	arrows indicating diffusion out of the cell

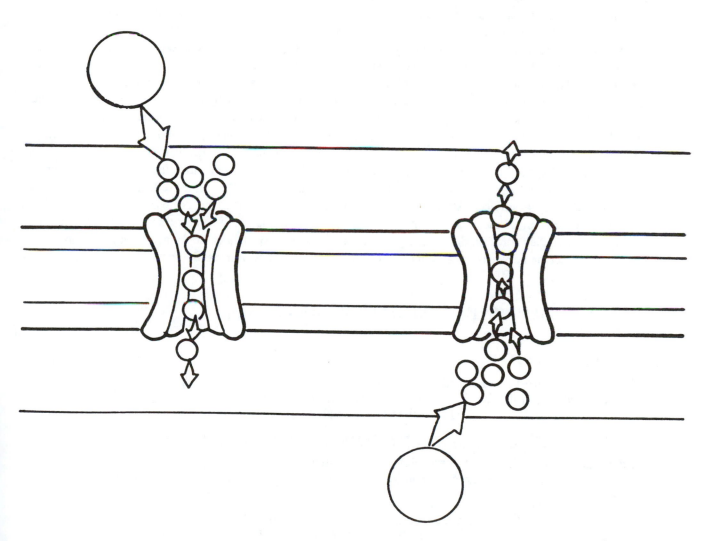

CHAPTER 40

Neural Regulation: Nervous Systems

The simplest organized nervous system is the nerve net found in cnidarians. A nerve net consists of nerve cells scattered throughout the body; there is no central control organ and no definite nervous pathways. The nervous system of echinoderms is more complex, with a nerve ring and nerves that extend into various parts of the body. Bilaterally symmetric animals have more complex nervous systems. The vertebrate nervous system has two main divisions: a central nervous system (CNS) consisting of a complex tubular brain that is continuous with a tubular nerve cord, and a peripheral nervous system (PNS). The CNS provides centralized control, integrating incoming information and determining appropriate responses. The PNS consists of sensory receptors and cranial and spinal nerves. It functions to regulate the organism's internal environment and to help it adjust to its external environment. The spinal cord transmits impulses to and from the brain and controls many reflex activities. The outer layer of the cerebrum, the largest subdivision of the brain, is functionally divided into sensory areas that receive incoming signals from the sense organs, motor areas which control voluntary movement, and association areas that link the sensory and motor areas and are responsible for thought, learning, language, memory, judgment, and personality. The reticular activating system is responsible for maintaining wakefulness. The limbic system influences the emotional aspects of behavior, sexual behavior, biological rhythms, autonomic responses, and motivation. Brain waves, electrical potentials given off by the brain, reflect a person's state of relaxation and mental activity. Sleep is a state of unconsciousness marked by decreased electrical activity. Learning is a function of the brain, involving the storage of information and its retrieval. Memory storage appears to involve the formation of new synaptic connections. Memory circuits are formed throughout the brain. Many drugs affect the nervous system, some by changing the levels of neurotransmitters within the brain.

CHAPTER OUTLINE AND CONCEPT REVIEW
Fill in the blanks.

INTRODUCTION

1 In general, the lifestyle of an animal is closely related to the design and complexity of its nervous system. For example, the simplest nervous system, called a (a)_____, is found in (b)_____, which are sessile animals like *Hydra*.

MANY INVERTEBRATES HAVE COMPLEX NERVOUS SYSTEMS

2 Echinoderms have a modified nerve net. In this system, a _____ surrounds the mouth, and large, radial nerves extend into each arm. Branches from these nerves coordinate the animal's movements.

3 Planarian worms have a (a)_____ nervous system with a (b)_____ in the head region that serves as a simple brain.

4 Each body segment in annelids contains a pair of (a)_____ which are connected by ventrally located (b)_____.

KEY FEATURES OF THE VERTEBRATE NERVOUS SYSTEM ARE THE HOLLOW DORSAL NERVE CORD AND THE WELL-DEVELOPED BRAIN

5 The vertebrate nervous system is divided into a (a)_____ system consisting of the brain and spinal cord, and a (b)_____ system consisting of sense receptors, nerves, and ganglia. This latter system is further divided into the somatic or voluntary portion and the autonomic portion that regulates the internal environment. The autonomic system has two efferent pathways made up of (c)_____ nerves.

THE EVOLUTION OF THE VERTEBRATE BRAIN IS MARKED BY INCREASING COMPLEXITY

6 The (a)_____ of the vertebrate embryo differentiates anteriorly into the brain and posteriorly into the spinal cord. Three bulges in the anterior end develop into the (b)_____.

The hindbrain develops into the medulla, pons, and cerebellum

7 The hindbrain, or _____, divides to form the metencephalon and myelencephalon.

8 The (a)_____, which coordinates muscle activity, and the (b)_____, which connects the spinal cord and medulla with higher brain centers, are formed from the (c)_____. The (d)_____, made up of nerve tracks and nuclei, is formed from the (e)_____.

9 The _____ is made up of the medulla, pons, and midbrain.

The midbrain is most prominent in fishes and amphibians

10 The midbrain is the largest part of the brain in fish and amphibians. It is their main _____ area, linking sensory input and motor output.

11 In mammals, the midbrain consists of the (a)_____ which regulate visual reflexes and the (b)_____ which regulate some auditory reflexes. It also contains the (c)_____ that integrates information about muscle tone and posture.

The forebrain gives rise to the thalamus, hypothalamus, and cerebrum

12 The forebrain, or (a)_____, differentiates into the diencephalon and the telencephalon, which in turn develop into the (b)_____ and (c)_____ respectively.

13 Cavities called (a)_____ are located within the cerebrum. They communicate with the (b)_____, a cavity in the diencephalon.

14 The (a)_____ is a relay center for motor and sensory signals in all vertebrate classes. Located below this center, the (b)_____ regulates autonomic responses and links the nervous and endocrine systems. The (c)_____ gland is connected to the hypothalamus.

15 Olfactory bulbs develop from the telencephalon, as does the _____, a structure in birds associated with innate behavioral patterns.

16 The right and left cerebral (a)_____ are predominantly comprised of axons, which are collectively referred to as (b)_____. Mammals and most reptiles have a layer of grey matter called the (c)_____ that covers the cerebrum. In complex mammals the surface area of the cortex is increased by ridges called (d)_____, shallow folds called (e)_____, and deep fissures.

THE HUMAN CENTRAL NERVOUS SYSTEM IS THE MOST COMPLEX BIOLOGICAL MECHANISM KNOWN

17 The human central nervous system consists of the brain and spinal cord. Both are protected by bone and by three meninges, the (a)_____, and both are bathed by cerebrospinal fluid produced in a specialized capillary bed called the (b)_____.

The spinal cord transmits impulses to and from the brain and controls many reflex activities

18 The spinal cord extends from the base of the brain to the _____ vertebra.

19 Grey matter in the spinal cord is shaped like the letter "H." It is surrounded by (a)_____ that contains nerve pathways or (b)_____.

20 The spinal cord consists of (a)_____ that transmit information to the brain, and (b)_____ that transmit information from the brain.

21 A relatively fixed reaction pattern in response to a simple stimulus is called a (a)_____. A simple reflex, such as the withdrawal reflex, involves three neurons: a (b)_____ afferent neuron that receives information and transmits it to the CNS, an (c)_____ neuron in the CNS, and a (d)_____ efferent neuron that transmits the information to an effector.

The largest, most prominent part of the human brain is the cerebrum

22 The human cerebral cortex consists of three functional areas. (a)_____ areas receive incoming sensory information, (b)_____ areas control voluntary movement, and (c)_____ areas link the other two areas.

23 Each hemisphere of the human cerebrum is divided into lobes. (a)_____ contain primary motor areas and are separated from the primary sensory areas in the (b)_____ lobes by a groove called the (c)_____. (d)_____ lobes contain centers for hearing and the (e)_____ lobes are associated with vision.

The brain integrates information

24 The reticular activating system (RAS) is responsible for maintaining consciousness. It is located within the _____.

25 The limbic system affects the emotional aspects of behavior, motivation, sexual behavior, autonomic responses, and biological rhythms. It consists of parts of the _____.

26 Electrical activity in the form of brain waves can be detected by a device that produces brain wave tracings called an (a)_____. The slow (b)_____ waves are associated with sleep, while (c)_____ waves are associated with relaxation, and (d)_____ waves are generated by an active, thinking brain.

27 There are two main stages of sleep: (a)_____ sleep which is associated with dreaming, and (b)_____ sleep which is characterized by delta waves.

28 According to current theory, there are three levels of memory. (a)_____ memory is very short, intense, and requires attention; (b)_____ memory stores information for a few seconds, and this information may get processed into more-or-less permanent (c)_____ memory.

29 Research shows that early stimulation can enhance the development of motor areas, result in a thicker cerebral _____ and stimulate memory.

THE PERIPHERAL NERVOUS SYSTEM INCLUDES SOMATIC AND AUTONOMIC SUBDIVISIONS

The somatic system helps the body adjust to the external environment

30 The somatic nervous system consists of (a)_____ that detect stimuli,
(b)_____ that transmit information to the CNS, and
(c)_____ that carry information from the CNS to effectors.

31 Neurons are organized into nerves. The 12 pairs of (a)_____ nerves connect to the brain
and are largely involved with sense receptors. Thirty-one pairs of (b)_____ nerves connect
to the spinal cord.

32 Each spinal nerve splits into two roots: a (a)_____ root that carries only sensory afferent
neurons and a (b)_____ root that carries only motor efferent neurons.

The autonomic system regulates the internal environment

33 The efferent component of the autonomic nervous system is divided into two systems. The
(a)_____ system enables the body to respond to stressful situations. The
(b)_____ system influences organs to conserve and restore energy.

MANY DRUGS AFFECT THE NERVOUS SYSTEM

BUILDING WORDS

Use combinations of prefixes and suffixes to build words for the definitions that follow.

Prefixes	The Meaning
hypo-	under, below
para-	beside, near
post-	behind, after
pre-	before, prior to, in advance of

Prefix	Suffix	Definition
_____	-thalamus	1. Part of the brain located below the thalamus; principal integration center for the regulation of the viscera.
_____	-ganglionic	2. Pertains to a neuron located distal to (after) a ganglion.
_____	-ganglionic	3. Pertains to a neuron located proximal to (before) a ganglion.
_____	-vertebral	4. Pertains to structures located beside the vertebral column.

MATCHING

For each of these definitions, select the correct matching term from the list that follows.

_____ 1. An action system of the brain that plays a role in emotional responses.

_____ 2. The outer layer of the cerebrum, composed of gray matter and consisting of densely-packed nerve cells.

_____ 3. The receptors and nerves that lie outside of the central nervous system.

_____ 4. The deeply-convoluted subdivision of the brain lying beneath the cerebrum; concerned with coordination of muscular movements.

_____ 5. The membranes that envelop the brain and spinal cord.

_____ 6. The portion of the peripheral nervous system that controls the visceral functions of the body.

_____ 7. A large bundle of nerve fibers interconnecting the two cerebral hemispheres.

_____ 8. The white bulge that is the part of the brainstem between the medulla and midbrain; connects various parts of the brain.

_____ 9. The nervous system consisting of the brain and spinal column.

_____10. A division of the autonomic nervous system concerned primarily with the control of the internal organs; functions to conserve or restore energy.

Terms:

a. Autonomic nervous system
b. Central nervous system
c. Cerebellum
d. Cerebral cortex
e. Corpus callosum
f. Gyrus

g. Limbic system
h. Meninges
i. Parasympathetic nervous system
j. Peripheral nervous system
k. Pons

l. Sensory areas
m. Somatic nervous system
n. Thalamus
o. Brain ventricle

TABLE
Fill in the blanks.

Human Brain Structure	Location of Structure	Function of Structure
Medulla	Myelencephalon	Contains vital centers and other reflex centers
#1	Metencephalon	Connects various parts of the brain
#2	Mesencephalon	#3
Thalamus	#4	#5
#6	#7	Controls autonomic functions; links nervous and endocrine systems, controls temperature, appetite and fluid balance; involved in some emotional and sexual responses
Cerebellum	#8	Responsible for muscle tone, posture and equilibrium
#9	Telencephalon	Complex association functions
#10	Brain stem and thalamus	Arousal system
#11	Certain structures of the cerebrum, diencephalon	Affect emotional aspects of behavior, motivation, sexual behavior, autonomic responses and biological rhythms

THOUGHT QUESTIONS
Write your responses to these questions.

1. Describe the similarities and differences in nerve nets, radial nervous systems, and bilateral nervous systems of invertebrates and vertebrates. Name animal groups that possess each type.
2. Make labelled sketches of the brains of insects, fish, amphibians, reptiles, birds, and mammals.
3. In humans, what are the principal functions of the medulla, pons, midbrain, thalamus, hypothalamus, cerebellum, and the parts of the cerebrum?
4. What is the reticular activating system? Compare it to the limbic system.
5. How do you think? How does one learn? How are motor abilities affected by repetitive experiences?
6. Briefly outline the major components of the vertebrate nervous system, including the CNS and PNS, the somatic and autonomic systems, and the sympathetic and parasympathetic divisions.
7. Name the basic functional types of drugs, give examples of each type, and describe how they affect the nervous system and behavior.

MULTIPLE CHOICE
Place your answer(s) in the space provided. Some questions may have more that one correct answer.

_____ 1. Regulation of body temperature under ordinary circumstances is under the control of the
    a. autonomic n.s.   d. cranial nerve VI.
    b. sympathetic n.s.   e. dorsal root ganglion.
    c. parasympathetic n.s.

_____ 2. The principal integration center for the regulation of viscera is the
    a. cerebellum.   d. hypothalamus.
    b. cerebrum.   e. red nucleus.
    c. thalamus.

_____ 3. In non-REM sleep, as compared to REM sleep, there is/are
    a. more delta waves.   d. more dream consciousness.
    b. faster breathing.   e. more norepinephrine released.
    c. lower blood pressure.

_____ 4. When you are eating one of your favorite foods, the cranial nerve(s) directly involved in your perception of this pleasurable experience is/are
    a. facial.   d. glossopharyngeal.
    b. VII.   e. XI.
    c. X.

_____ 5. In higher mammals that engage in complex association functions, there would likely be
    a. a convoluted cerebral cortex.   d. a well-developed neopallium.
    b. a small amount of gray matter.   e. fissures.
    c. an expanded cortical surface area.

_____ 6. Cerebrospinal fluid
    a. is produced by choroid plexuses.   d. is within layers of the meninges.
    b. circulates through ventricles.   e. is in the subarachnoid space.
    c. is between the arachnoid and pia mater.

_____ 7. The brain and spinal cord are wrapped in connective tissue called
    a. gray matter.   d. sulcus.
    b. meninges.   e. neopallium.
    c. gyri.

_____ 8. The part of the brain that coordinates muscular activity is the
    a. cerebrum.   d. medulla oblongata.
    b. cerebellum.   e. midbrain.
    c. myelencephalon.

_____ 9. If you are having strong sexual feelings one minute and enraged over little or nothing the next minute, chances are very good that you have just had a very active
    a. frontal lobe.   d. reticular activating system (RAS).
    b. limbic system.   e. cerebellum.
    c. thalamus.

_____10. The part of the mammalian brain that integrates information about posture and muscle tone is the
    a. cerebrum.   d. medulla oblongata.
    b. cerebellum.   e. midbrain.
    c. myelencephalon.

_____11. When you are very sleepy, there's a good chance that
    a. the meninges are damaged.   d. your reticular activating system is not very active.
    b. beta waves are diminishing.   e. your brain lacks gray matter.
    c. you have convolutions, but not gyri.

_____12. If only one lateral temporal lobe is damaged, one might expect
    a. blindness in one eye.   d. partial loss of hearing in both ears.
    b. total blindness.   e. inability to detect odors.
    c. loss of hearing in one ear.

_____13. Skeletal muscles are controlled by the
    a. parietal lobes.   d. temporal lobes.
    b. frontal lobes.   e. cerebral ganglion.
    c. central sulcus.

_____14. In vertebrates, the cerebrum is typically
    a. divided into two hemispheres.   d. mostly white matter.
    b. mostly gray matter.   e. mainly axons connecting parts of the brain.
    c. mainly cell bodies and some sensory neurons.

VISUAL FOUNDATIONS
Color the parts of the illustration below as indicated.

RED	☐	cerebral hemisphere
GREEN	☐	cerebellum
YELLOW	☐	medulla
BLUE	☐	olfactory bulb, olfactory tract, olfactory lobe
ORANGE	☐	optic lobe
BROWN	☐	epiphysis
TAN	☐	corpus striatum

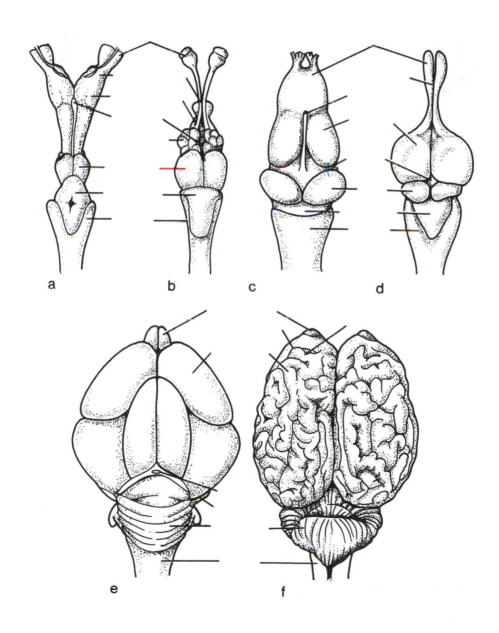

a   b   c   d

e   f

Color the parts of the illustration below as indicated.

RED	☐	sensory neuron
GREEN	☐	association neuron
YELLOW	☐	receptor
BLUE	☐	motor neuron
ORANGE	☐	nerve cell body of sensory neuron
BROWN	☐	muscle
TAN	☐	central nervous system

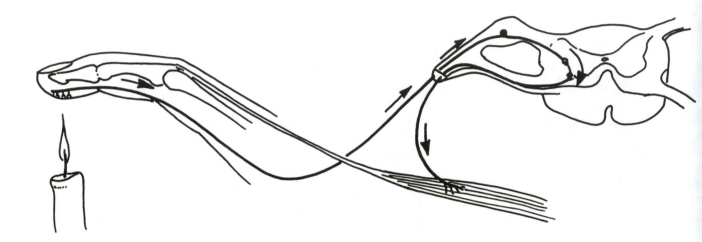

Color the parts of the illustration below as indicated.

RED ☐ diencephalon

GREEN ☐ pituitary

YELLOW ☐ spinal cord

BLUE ☐ ventricle

ORANGE ☐ medulla and pons

BROWN ☐ midbrain

TAN ☐ cerebellum

PINK ☐ corpus callosum and fornix

VIOLET ☐ cerebrum

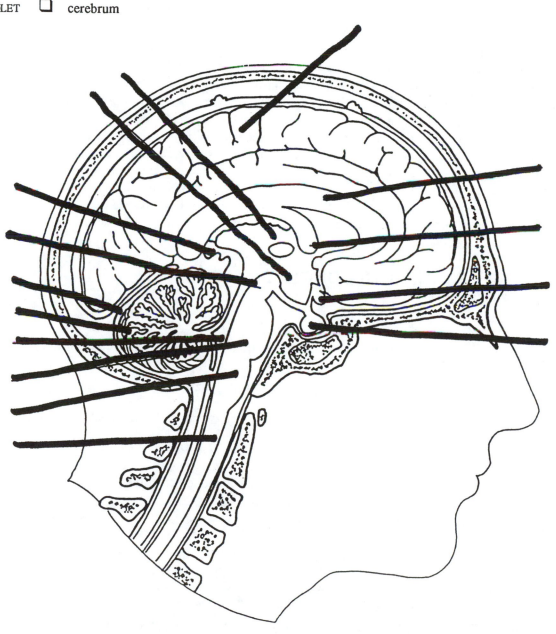

CHAPTER 41

Sensory Reception

Sense organs are specialized structures with receptor cells that detect changes in the internal or external environment and transmit the information to the nervous system. Receptor cells may be neuron endings or specialized cells in contact with neurons. Sense organs are classified according to the location of the stimuli and according to the types of energy to which they respond. They work by absorbing energy and converting it into electrical energy that depolarizes the receptor cell. When the state of depolarization reaches a threshold level, an impulse is generated in the axon, thereby transmitting the information to the CNS. Many receptors do not continue to respond at the initial rate even if the stimulus maintains its intensity. Such adaptation permits an animal to ignore persistent unpleasant or unimportant stimuli. Sense organs that respond to mechanical energy include the various tactile receptors in the skin, the lateral line organs in fish, the receptors that respond continuously to tension and movement in muscles and joints, the receptors that enable organisms to maintain their orientation, and the receptors responsible for hearing. Sense organs that respond to chemical energy are responsible for the sense of taste and smell. Those that respond to heat and cold provide important cues about body temperature, and help some animals locate a warm-blooded host. Those that detect electrical energy are found in some fish; they can detect electric currents in water, and in some species can generate a shock for defense or to stun prey. Sense organs that respond to light energy are responsible for vision.

CHAPTER OUTLINE AND CONCEPT REVIEW
Fill in the blanks.

INTRODUCTION

1 The receptor cells in a sense organ detect changes in the environment. These cells may be specialized for the task or merely _____ endings.

2 The five human senses that have been recognized traditionally are (a)_____
_____. (b)_____
_____ are now also recognized as human senses.

SENSORY RECEPTORS CAN BE CLASSIFIED ACCORDING TO THE STIMULI TO WHICH THEY RESPOND

3 Sense receptors can be classified by their location. Those that detect stimuli in the external environment are called (a)_____; (b)_____ are located within joints, muscles, and tendons; and (c)_____ detect change in the internal environment.

4 Classification of sense organs can also be based on the type of stimuli detected. For example,
(a)_____ respond to gravity, pressure, and stretching;
(b)_____ respond to electrical energy; heat and cold are detected by
(c)_____; and chemicals provoke responses in (d)_____.

RECEPTOR CELLS WORK BY PRODUCING RECEPTOR POTENTIALS

5 Various forms of energy are transduced into (a)_____ energy by receptor cells that then produce a (b)_____ potential.

6 A receptor potential is a kind of (a)_____ potential. If it can (b)_____ the axon hillock to threshold, it can initiate an (c)_____ that will travel along the (d)_____ afferent neuron to the central nervous system.

7 A strong stimulus will generate _____(more or less?) action potentials than a weak stimulus.

Sensation depends on transmission of a "coded" message

8 Sensation is determined by the part of the _____ that receives the message from the sense organ.

9 Impulses from sense organs are encoded into messages based upon the (a)_____ _____ of neurons (or fibers) carrying the message and the (b)_____ of action potentials carried by the neuron (fiber).

Receptors adapt to stimuli

10 When a receptor continues to receive a constant stimulus, yet the sensory neuron becomes less responsive to that stimulus, the receptor is demonstrating _____.

MECHANORECEPTORS RESPOND TO TOUCH, PRESSURE, GRAVITY, STRETCH, OR MOVEMENT

11 Mechanoreceptors are activated when they are mechanically _____.

Touch receptors are located in the skin

12 The tactile receptors in the skin are _____ that respond to mechanical displacement of hairs or of the receptor cells themselves.

13 In human skin, (a)_____ are responsible for the perception of deep pressure, (b)_____ are responsible for touch, and free nerve endings perceive (c)_____.

Many invertebrates have gravity receptors called statocysts

14 (a)_____ are tiny granules that stimulate (b)_____ when pulled down by gravity. They occur in sense organs called statocysts.

Lateral line organs supplement vision in fish

15 Lateral line organs consist of a (a)_____ lined with (b)_____ that runs the length of the animal's lateral surface. Water disturbances move the (c)_____ on the end of hairs in the receptor cells, generating an electrical response.

Proprioceptors help coordinate muscle movement

16 _____ are proprioceptors that respond continuously to tension and movement in the muscles and joints.

The vestibular apparatus of the vertebrate ear maintains equilibrium

17 Gravity detectors in the form of ear stones called (a)_____ are housed in the (b)_____ of the vestibular apparatus.

18 Turning movements, referred to as (a)_____, cause movement of a fluid called (b)_____ in the semicircular canals, which in turn stimulate the receptor cells of (c)_____.

Auditory receptors are located in the cochlea

19 Auditory receptors in an inner ear structure of birds and mammals, called the (a)_____, contain (b)_____ that can detect pressure waves.

20 In terrestrial vertebrates, sound waves first initiate vibrations in the eardrum, or
    (a)_____. Three tiny earbones, the (b)_____
    _____, transmit the vibration to fluids in the inner ear through an opening called the
    (c)_____.

21 Fluids in the cochlea, in response to vibrations from the oval window, initiate vibrations in the
    (a)_____ which in turn cause stimulation of hair cells in the
    (b)_____. The hair cells initiate impulses in the (c)_____.

CHEMORECEPTORS ARE ASSOCIATED WITH THE SENSES OF TASTE AND SMELL

22 (a)_____ refers to the sense of taste, and (b)_____ applies to the
    sense of smell, both of which are highly sensitive (c)_____ systems.

Taste buds detect dissolved food molecules

23 The four basic tastes are _____.

The olfactory epithelium is responsible for the sense of smell

24 The (a)_____ of terrestrial vertebrates is located on the roof of the
    nasal cavity. It reacts to as many as (b)_____ different odors.

THERMORECEPTORS ARE SENSITIVE TO HEAT

25 In two types of snakes, the _____, thermoreceptors are used to locate prey.

26 Thermoreceptors that detect internal temperature changes in mammals are found in the
    _____ of the brain.

ELECTRORECEPTORS DETECT ELECTRICAL CURRENTS IN WATER

PHOTORECEPTORS USE PIGMENTS TO ABSORB LIGHT

27 Cephalopods, arthropods, and vertebrates all have photosensitive pigments called
    _____ in their eyes.

Eyespots, simple eyes, and compound eyes are found among invertebrates

28 Planaria have _____, which are photoreceptive organs capable of differentiating
    the intensity of light.

29 Photoreceptors in simple eyes detect (a)_____, but these eyes do not form images effectively.
    Effective image formation and interpretation is called (b)_____, and it requires a
    (c)_____ capable of focusing.

30 The compound eye in insects and crustaceans consists of _____, which collectively
    form a mosaic image.

Vertebrate eyes form sharp images

31 The tough outer coat of the mammalian eye, called the (a)_____, helps maintain the
    (b)_____ of the eyeball. The anterior, transparent part of this coat, which allows the entry of
    light, is called the (c)_____.

32 Light is focused by the (a)_____ on the (b)_____, which contains the photoreceptive
    cells. (c)_____, containing the pigment (d)_____, are concentrated in the
    periphery of the retina. They function best in dim light and perceive black and white.

33 (a)_____ are responsible for color vision and are densest in the (b)_____ in the
    center of the retina.

BUILDING WORDS

Use combinations of prefixes and suffixes to build words for the definitions that follow.

Prefixes	The Meaning		Suffixes	The Meaning
endo-	within		-lith	stone
oto-	ear			
proprio-	one's own			

Prefix	Suffix	Definition
_____	-ceptor	1. Sense organs within muscles, tendons, and joints that enable the animal to perceive the position of its own body parts.
_____ _____		2. Calcium carbonate "stone" in the inner ear of vertebrates.
_____	-lymph	3. Fluid within the semicircular canals of the vertebrate ear.

MATCHING

For each of these definitions, select the correct matching term from the list that follows.

_____ 1. A light-sensitive pigment found in rod cells of the vertebrate eye.

_____ 2. The structure of the inner ear of mammals that contains the auditory receptors.

_____ 3. A sensory receptor that responds to mechanical energy.

_____ 4. The pigmented portion of the vertebrate eye.

_____ 5. Sensory receptor that provides information about body temperature.

_____ 6. A mechanoreceptor that is sensitive to deep pressure touches.

_____ 7. The "ear drum."

_____ 8. The transparent anterior covering of the eye.

_____ 9. A sense organ or sensory cell that responds to chemical stimuli.

_____ 10. A conical photoreceptive cell of the retina that is particularly sensitive to bright light, and, by light of various wave lengths, mediates color vision.

Terms:

a.	Chemoreceptor	f.	Exteroceptor	k.	Pacinian corpuscle
b.	Cochlea	g.	Fovea	l.	Rhodopsin
c.	Cone	h.	Interoceptor	m.	Statocyst
d.	Cornea	i.	Iris	n.	Thermoreceptor
e.	Electroreceptor	j.	Mechanoreceptor	o.	Tympanic membrane

TABLE
Fill in the blanks.

Stimuli	Receptor Classification: Type of Stimuli	Receptor Classification: Location	Example
Light touch	Mechanoreceptor	Exteroceptor	Meissner's corpuscle in skin
Electrical currents in water	#1	#2	Electric organ in rays
#3	#4	Exteroceptor	Organ of Corti in birds
Change in internal body temperature	Thermoreceptor	#5	#6
#7	Mechanoreceptor	#8	Muscle spindle in human Skeletal muscle
Heat from prey	#9	Exteroceptor	#10
#11	#12	Exteroceptor	Ocelli of flatworms
Ischemia	Pain receptor	#13	Pain receptor in heart
Food dissolved in saliva	#14	Exteroceptor	#15
Pheromones	#16	#17	Olfactory epithelia of terrestrial vertebrates

THOUGHT QUESTIONS
Write your responses to these questions.

1. What are exteroceptors, proprioceptors, and interoceptors? Give specific examples of each.
2. Describe the interactions that occur between five different types of stimuli and the receptors in sense organs that can respond to those stimuli.
3. Describe the events that occur between receptor response and cognition in the human eye.
4. Describe the events that occur between auditory receptor response and cognition in the human.
5. Describe the events that occur between olfactory receptor response and cognition in the human.
6. Describe the events that occur between gustatory receptor response and cognition in the human.
7. Distinguish between tactile receptors, statocysts, lateral line organs, and proprioceptors.

MULTIPLE CHOICE
Place your answer(s) in the space provided. Some questions may have more that one correct answer.

_____ 1. The increasing inability with age to focus light from near objects is due to
    a. deformity of the retina.　　　　d. loss of lens elasticity.
    b. deterioration of rhodopsin.　　　e. decreasing sensitivity of rods and/or cones.
    c. inability of the lens to become roundish.

_____ 2. Movement in ligaments is detected by
    a. muscle spindles.   d. proprioceptors.
    b. joint receptors.   e. tonic sense organs.
    c. Golgi tendon organs.

_____ 3. Variations in the quality of sound are recognized by the
    a. number of hair cells stimulated.   d. intensity of stimulation.
    b. pattern of hair cells stimulated.   e. amplitude of response.
    c. frequency of nerve impulses.

_____ 4. Receptors within muscles, tendons, and joints that perceive position and body orientation are
    a. exteroceptors.   d. mechanoreceptors.
    b. proprioceptors.   e. electroreceptors.
    c. interoceptors.

_____ 5. One means of coding sensory perceptions may be related to the frequency of the action potentials in a neuron. This mechanism of coding is known as
    a. temporal patterning.   d. fiber specificity.
    b. spatial localization.   e. adaptation.
    c. cross-fiber patterning.

_____ 6. Sense organs that detect changes in pH, osmotic pressure, and temperature within body organs are
    a. exteroceptors.   d. mechanoreceptors.
    b. proprioceptors.   e. electroreceptors.
    c. interoceptors.

_____ 7. The state of depolarization in a receptor neuron that is caused by a stimulus is the
    a. depolarization potential.   d. action potential.
    b. resting potential.   e. threshold potential.
    c. receptor potential.

_____ 8. Receptors that respond only when stimulated by motion are known as
    a. mechanoreceptors.   d. phasic receptors.
    b. tonic receptors.   e. proprioceptors.
    c. labyrinths.

_____ 9. The membrane at the opening of the inner ear that is in contact with the stirrup is in the
    a. tectorial membrane.   d. eardrum.
    b. basilar membrane.   e. round window.
    c. oval window.

Visual Foundations on next page ➤

VISUAL FOUNDATIONS
Color the parts of the illustrations below as indicated.

RED ☐ hair cells   ORANGE ☐ cilia

GREEN ☐ cochlear duct   BROWN ☐ tympanic canal and vestibular canal

YELLOW ☐ cochlear nerve   TAN ☐ bone

BLUE ☐ tectorial membrane   PINK ☐ basilar membrane

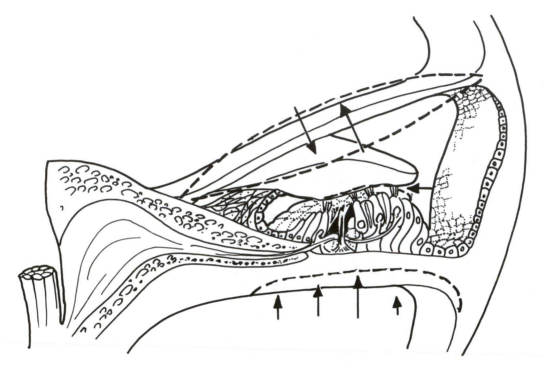

Color the parts of the illustrations below as indicated.

RED ☐ rod cell

GREEN ☐ cone cell

YELLOW ☐ light rays

BLUE ☐ optic nerve fibers

ORANGE ☐ vitreous body

BROWN ☐ pigmented epithelium

TAN ☐ choroid layer and sclera

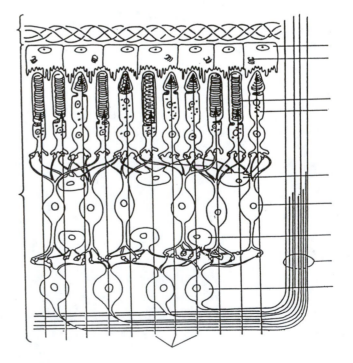

Color the parts of the illustrations below as indicated.

RED ☐ ommatidia

GREEN ☐ facets

YELLOW ☐ optic nerve

BLUE ☐ receptor cell

ORANGE ☐ optic ganglion

BROWN ☐ cornea

TAN ☐ lens

PINK ☐ crystalline cone

VIOLET ☐ rhabdome

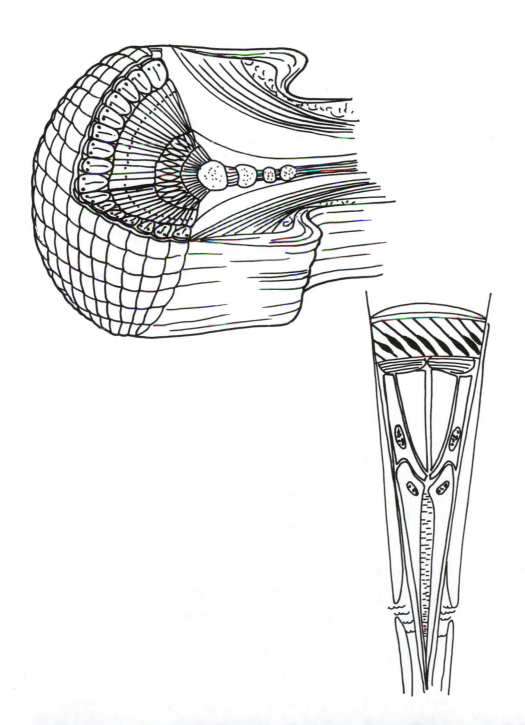

CHAPTER 42

Internal Transport

Some animals are so small that diffusion alone is effective at transporting materials to and from their cells. Larger animals, however, require a circulatory system, not only to transport materials, but also to help maintain fluid balance, to defend the body against invading microorganisms, and, in some animals, to regulate body temperature. Some invertebrates have an open circulatory system in which blood is pumped from a heart into vessels that have open ends. Blood spills out of the vessels into the body cavity and baths the tissues directly. The blood then passes back into the heart, either directly through openings in the heart (arthropods) or indirectly, passing first through open vessels that lead to the heart (some mollusks). Other animals have a closed circulatory system in which blood flows through a continuous circuit of blood vessels. The walls of the smallest vessels are thin enough to permit exchange of materials between the vessels and the extracellular fluid that baths the tissue cells. The vertebrate circulatory system consists of a muscular heart that pumps blood into a system of blood vessels: the arteries that carry blood away from the heart to the capillaries, the capillaries where the exchange of materials between the blood and tissues occurs, and the veins that carry the blood back to the heart. Vertebrate blood consists of liquid plasma in which red blood cells, white blood cells, and platelets are suspended. The red blood cells transport oxygen, the white blood cells defend the body against disease organisms, and the platelets function in blood clotting. The vertebrate heart consists of one or two chambers that receive blood, and one or two that pump blood into the arteries. In birds and mammals, there are two circuits of blood vessels. One pumps blood from the heart to the lungs, and the other, from the heart to every tissue in the body. The lymphatic system, a subsystem of the circulatory system, collects extracellular fluid and returns it to the blood. It also functions to defend the body against disease organisms and to absorb lipids from the digestive tract.

CHAPTER OUTLINE AND CONCEPT REVIEW
Fill in the blanks.

INTRODUCTION

1 _____ are animal groups that have no specialized circulatory structures.

2 The bodies of complex animals need a (a)_____ system to transport materials because simple (b)_____, which distributes materials adequately in thin-bodied, small animals, cannot supply all cells in larger animals.

3 Components of a circulatory system include a fluid, the (a)_____, that is usually pumped by a (b)_____ through a system of spaces or blood (c)_____.

4 An (a)_____ system consists of a blood cavity, or (b)_____, a heart, and open-ended vessels. In a (c)_____ system, blood is contained within the heart and a continuous system of blood vessels.

SOME INVERTEBRATES HAVE NO CIRCULATORY SYSTEM

5 The _____ of cnidarians serves as both a circulatory organ and a digestive organ.

MANY INVERTEBRATES HAVE AN OPEN CIRCULATORY SYSTEM

6 Blood is called (a)_____ in animals with open circulatory systems because blood is indistinguishable from (b)_____ fluid.

7 (a)_____ have an open circulatory system in which hemolymph flows into large spaces, called (b)_____, of the hemocoel.

8 (a)_____, a blood pigment that imparts a bluish color to hemolymph in some invertebrates, contains the metal (b)_____.

SOME INVERTEBRATES HAVE A CLOSED CIRCULATORY SYSTEM

9 Annelids have a (a)_____ blood vessel that conducts blood anteriorly, a (b)_____ blood vessel that conducts blood posteriorly, and (c)_____(#?) "hearts" that connect these two vessels.

10 Hemoglobin is found in the _____ of earthworm blood.

THE CLOSED CIRCULATORY SYSTEM OF VERTEBRATES IS ADAPTED TO CARRY OUT A VARIETY OF FUNCTIONS

11 The circulatory system of all vertebrates is a/an (a)_____(open or closed?) system with a muscular (b)_____ that pumps (c)_____ through blood vessels.

12 The vertebrate circulatory system transports (a)_____. Other functions include (b)_____, as well as thermoregulation in (c)_____.

VERTEBRATE BLOOD CONSISTS OF PLASMA, BLOOD CELLS, AND PLATELETS

13 Human blood consists of (a)_____ suspended in (b)_____.

Plasma is the fluid component of blood

14 Plasma is in dynamic equilibrium with (a)_____ fluid, which bathes cells, and with (b)_____ fluid within cells.

15 (a)_____, a protein involved in blood clotting, (b)_____, which are involved in immunity, and albumin are all (c)_____. (d)_____ in plasma transport triglycerides and cholesterol.

Red blood cells transport oxygen

16 Red blood cells (erythrocytes) are specialized to transport (a)_____. They are produced in (b)_____, contain the respiratory pigment (c)_____, and live about (d)_____ days.

White blood cells defend the body against disease organisms

17 White blood cells (leukocytes) called (a)_____ are the principle phagocytic cells in blood. These cells, together with (b)_____ are called (c)_____ because of the distinctive granules in their cytoplasm.

18 Agranular leukocytes are (a)_____ which secrete antibodies, and (b)_____ which develop into phagocytic (c)_____.

Platelets function in blood clotting

19 Platelets, also known as (a)_____, play an important role in the control of bleeding, or (b)_____.

VERTEBRATES HAVE THREE MAIN TYPES OF BLOOD VESSELS

20 (a)_____ carry blood away from the heart; (b)_____ return blood to the heart.

21 The wall of blood vessels is made up of the (a)_____, which contains endothelium; the (b)_____ with smooth muscle; and the (c)_____, which contains many elastic and collagen fibers.

22 The exchange of nutrients and waste products takes place across the thin wall of _____.

ADAPTATIONS OF THE HEART AND CIRCULATION HAVE EVOLVED IN EACH VERTEBRATE CLASS

23 Chambers in the heart that pump blood into arteries are called (a)_____, and chambers that receive blood from veins are called (b)_____.

24 Fish hearts have (a)_____(#?) atrium(ia) and (b)_____ ventricle(s). In this structure, veins first empty into the (c)_____, an accessory chamber that then pumps blood into the atrium. Blood leaving the ventricle empties into the (d)_____, from which it passes into the ventral aorta.

25 In the (a)_____(#?)-chambered amphibian heart, the (b)_____ pumps venous blood into the right atrium, and the (c)_____ helps to separate oxygen-rich blood from oxygen-poor blood.

26 Because oxygen-rich blood can be isolated from oxygen-poor blood in a four chambered heart, tissues receive (a)_____(more or less?) oxygen, and can maintain a (b)_____(higher or lower?) metabolic rate.

THE HUMAN HEART IS WELL ADAPTED FOR PUMPING BLOOD

27 The human heart is enclosed by a (a)_____, creating a (b)_____ filled with fluid that serves to reduce friction.

28 The (a)_____ septum separates the two ventricles, whereas the (b)_____ separates the two atria in a four-chambered heart.

29 (a)_____(#?) valves control the flow of blood in the human heart. The (b)_____ prevents backflow into the atria during ventricular contractions. The valve on the right is also known as the (c)_____, and the valve on the left is known as the (d)_____. The (e)_____ "guard" the exits from the heart.

Each heartbeat is initiated by a pacemaker

30 The conduction system of the heart contains a pacemaker called the (a)_____, an (b)_____ which links the atria to the (c)_____, which divides, sending a branch into each ventricle.

31 The portion of the (a)_____ cycle in which contraction occurs is called (b)_____, and the portion in which relaxation occurs is called (c)_____.

Two main heart sounds can be distinguished

32 Of the two main heart sounds, the (a)_____ occurs first and is associated with closure of the (b)_____, and the (c)_____ is produced by the closure of the (d)_____, which marks the beginning of ventricular (e)_____.

The electrical activity of the heart can be recorded

33 The _____ is a written record of the electrical activity of the heart.

Cardiac output varies with the body's need

34 (a)_____ is the volume of blood pumped by one ventricle in one minute. It is determined by multiplying the number of ventricular beats per minute, called the

(b)_____, times the amount of blood pumped by one ventricle per beat, a value referred to as the (c)_____. The normal volume per minute value is about (d)_____, although it can vary dramatically.

Stroke volume depends on venous return

35 Stroke volume is determined primarily by (a)_____ return, although (b)_____ mechanisms also have an effect.

Heart rate is regulated by the nervous system

36 Although the heart can beat independently, it is regulated by _____ in the medulla.

BLOOD PRESSURE DEPENDS ON BLOOD FLOW AND RESISTANCE TO BLOOD FLOW

37 High blood pressure is called (a)_____. It can be caused by an (b)_____(increase or decrease?) in blood volume such as frequently occurs with a high dietary intake of (c)_____.

38 The most important factor in determining peripheral resistance to blood flow is the

_____.

Blood pressure is highest in arteries

39 Veins are low-pressure vessels that contain _____ to prevent blood backflow.

Blood pressure is carefully regulated

40 (a)_____ are sense organs in the walls of some arteries that detect (b)_____ and send the perceived information to (c)_____ in the medulla.

IN BIRDS AND MAMMALS BLOOD IS PUMPED THROUGH A PULMONARY AND A SYSTEMIC CIRCUIT

41 A double circulatory system consists of a (a)_____ between the heart and lungs, and (b)_____ between the heart and body.

The pulmonary circulation oxygenates the blood

42 Pulmonary veins carry oxygen-(a)_____(rich or poor?) blood to the (b)_____ atrium of the heart.

The systemic circulation delivers blood to the tissues

43 Arteries in the systemic circuit branch from the (a)_____, the largest artery in the body, and serve major body areas. For example, the carotid arteries feed the (b)_____, subclavian arteries supply the (c)_____, and iliac arteries feed the (d)_____.

44 Veins returning blood from the head and neck empty into the large (a)_____ _____ vein, while venous return from the lower body empties into the (b)_____ vein.

The coronary circulation delivers blood to the heart
Four arteries deliver blood to the brain
The hepatic portal system delivers nutrients to the liver

THE LYMPHATIC SYSTEM IS AN ACCESSORY CIRCULATORY SYSTEM

45 Considered an accessory circulatory system in vertebrates, the lymphatic system returns (a)_____ to the blood. It is also involved with immune mechanisms and absorbs (b)_____ from the digestive tract.

The lymphatic system consists of lymphatic vessels and lymph tissue

46 (a)_____, the fluid contained within lymphatic vessels, is formed from (b)_____ _____, filtered by (c)_____, and emptied into (d)_____ veins by the (e)_____ duct on the left side and the (f)_____ duct on the right.

The lymphatic system plays an important role in fluid homeostasis

47 _____, the excessive accumulation of interstitial fluid, can result if lymph vessels are obstructed.

BUILDING WORDS

Use combinations of prefixes and suffixes to build words for the definitions that follow.

Prefixes	The Meaning	Suffixes	The Meaning
baro-	pressure	-cardium	heart
erythro-	red	-coel	cavity
hemo-	blood	-cyte	cell
leuk(o)-	white (without color)	-lunar	moon
neutro-	neutral	-phil	loving, friendly, lover
peri-	about, around, beyond		
semi-	half		
vaso-	vessel		

Prefix	Suffix	Definition
_____	_____	1. The blood cavity that comprises the open circulatory system of arthropods and some mollusks.
_____	-cyanin	2. The copper-containing blood pigment in some mollusks and arthropods.
_____	_____	3. Red blood cell.
_____	_____	4. General term for all of the body's white blood cells.
_____	_____	5. The principal phagocytic cell in the blood that has an affinity for neutral dyes.
eosino-	_____	6. WBC with granules that have an affinity for eosin.
baso-	_____	7. WBC with granules that have an affinity for basic dyes.
_____	-emia	8. A form of cancer in which WBCs multiply rapidly within the bone marrow.
_____	-constriction	9. Constriction of a blood vessel.
_____	-dilation	10. Relaxation of a blood vessel.
_____	_____	11. The tough connective tissue sac around the heart.
_____	_____	12. Shaped like a half-moon.
_____	-receptor	13. Pressure receptor.

MATCHING

For each of these definitions, select the correct matching term from the list that follows.

_____ 1. Network of fluid-carrying vessels and associated organs that participate in immunity and in the return of tissue fluid to the main circulation.

_____ 2. The bicuspid heart valve located between the left atrium and the left ventricle.

_____ 3. The fluid in lymphatic vessels.

_____ 4. The fluid portion of the blood consisting of a pale, yellowish fluid.

_____ 5. A contracting chamber of the heart which forces blood into the ventricle.

_____ 6. The rhythmic expansion of an artery that may be felt with the finger.

_____ 7. The smallest arteries, which carry blood to the capillary beds.

_____ 8. Contraction of the heart muscle, especially that of the ventricle, during which the heart pumps blood into arteries.

_____ 9. Largest blood vessel in the body through which blood leaves the heart and enters the systemic circulation.

_____10. A blood vessel that carries blood from the tissues toward the heart.

Terms:

a.	Aorta	f.	Interstitial fluid	k.	Pulse
b.	Arterioles	g.	Lymph	l.	Systole
c.	Artery	h.	Lymphatic system	m.	Vein
d.	Atrium	i.	Mitral valve		
e.	Blood pressure	j.	Plasma		

TABLE

Fill in the blanks.

Specific Protein or Cell Type	Blood Component	Function
Fibrinogen	Plasma	Involved in blood clotting
#1	#2	Transport triglycerides, cholesterol
Erythrocytes	#3	#4
Thrombocytes	Cellular	#5
#6	#7	Defense, are phagocytic in tissue
#8	#9	Contribute to osmotic pressure of blood, therefore are important in blood volume regulation
Neutrophils	Cellular	#12

THOUGHT QUESTIONS
Write your responses to these questions.

1. How is internal transport accomplished in a single cell? In a nonvascular plant? In a vascular plant? In a hydra? In an insect? In a human?
2. Make sketches that show the basic structures of RBCs, WBCs, and platelets and describe the principal functions of each.
3. Describe the cellular and molecular events involved in blood clotting.
4. Trace the path of a red blood cell from one of the vena cavas to the aorta. Include descriptions of the structural and physiological events that occur in each step, including valve operations, coordinated chamber contractions, gas exchange at the lungs, initiation of contractions, etc.
5. What is atherosclerosis? Can it be prevented? Explain the relationship between atherosclerosis and angina pectoris and myocardial infarction.
6. Describe the functions of the lymphatic system and interactions of the lymphatic system with the "blood system."

MULTIPLE CHOICE
Place your answer(s) in the space provided. Some questions may have more that one correct answer.

_____ 1. Prothrombin
    a. is produced in liver.
    b. is a precursor to thrombin.
    c. catalyzes conversion of fibrinogen to fibrin.
    d. requires vitamin K for its production.
    e. is a toxic by-product of metabolism.

_____ 2. The thrombocytes of mammals are
    a. WBCs.
    b. RBCs.
    c. called platelets.
    d. anucleate.
    e. derived from megakaryocytes.

_____ 3. Erythrocytes are
    a. also called RBCs.
    b. spherical.
    c. produced in bone marrow.
    d. one kind of leucocyte.
    e. carriers of oxygen.

_____ 4. The white blood cells that contain an anticlotting chemical
    a. contain histamine.
    b. carry oxygen.
    c. stain blue with basic dyes.
    d. transport carbon dioxide.
    e. contain hemoglobin.

_____ 5. Monocytes
    a. are RBCs.
    b. are WBCs.
    c. are produced in the spleen.
    d. are large cells.
    e. can become macrophages.

_____ 6. An antibody is
    a. a plasma lipid.
    b. in serum.
    c. a protein.
    d. a gamma globulin.
    e. involved in clotting.

_____ 7. Fibrinogen is
    a. a plasma lipid.
    b. in serum.
    c. a protein.
    d. a gamma globulin.
    e. involved in clotting.

_____ 8. Very small vessels that deliver oxygenated blood to capillaries are called
    a. arterioles.   d. veins.
    b. venules.   e. setum vessels.
    c. arteries.

_____ 9. The blood-filled cavity of an open circulatory system is called the
    a. lymphocoel.   d. sinus.
    b. hemocoel.   e. ostium.
    c. atracoel.

_____10. Blood enters the right atrium from the
    a. right ventricle.   d. pulmonary artery.
    b. inferior vena cava.   e. jugular vein.
    c. superior vena cava.

_____11. Pulmonary arteries carry blood that is
    a. low in oxygen.   d. high in CO_2.
    b. low in CO_2.   e. on its way to the lungs.
    c. high in oxygen.

_____12. Phenomena that tend to be associated with hypertension include
    a. obesity.   d. decrease in ventricular size.
    b. increased vascular resistance.   e. deteriorating heart function.
    c. increased workload on the heart.

_____13. Ventricles receive stimuli for contraction directly from the
    a. SA node.   d. atria.
    b. AV node.   e. intercalated disks.
    c. Perkinje fibers.

_____14. A heart murmur may result when
    a. the heart is punctured.   d. diastole is too low.
    b. systole is too high.   e. blood backflows into ventricles.
    c. semilunar valves don't close tightly.

_____15. If a person's heart rate is 65 and one ventricle pumps 75 ml with each contraction, what is that person's cardiac output in liters?
    a. 3.575   d. 4.875
    b. 4.000   e. 5.350
    c. 4.525

_____16. The semilunar valve(s) is/are found between a ventricle and
    a. an atrium.   d. the pulmonary vein.
    b. another ventricle.   e. the pulmonary artery.
    c. the aorta.

_____17. The pericardium surrounds
    a. atria only.   d. the heart.
    b. ventricles only.   e. clusters of some blood cells.
    c. blood vessels.

_____18. One would expect an ostrich to have a heart most like a
    a. lizard.   d. human.
    b. crocodile.   e. guppy.
    c. earthworm.

____19. Three chambered hearts generally consist of the following number of atria/ventricles:
      a. 2/1.   d. 0/3.
      b. 1/2.   e. 3/0.
      c. 1/1 and an accessory chamber.

____20. In general, blood circulates through vessels in the following order:
      a. veins, venules, capillaries, arteries, arterioles.   d. arteries, arterioles, capillaries, venules, veins.
      b. venules, veins, capillaries, arterioles, arteries.   e. arteries, arterioles, capillaries, veins, venules.
      c. arteries, arterioles, venules, capillaries, veins.

VISUAL FOUNDATIONS
Color the parts of the illustration below as indicated.

RED ❑ artery and arrows indicating flow of blood away from the heart

GREEN ❑ capillary bed

YELLOW ❑ lymph capillaries

BLUE ❑ vein and arrows indicating flow of blood to the heart

ORANGE ❑ lymphatic

BROWN ❑ lymph node

PINK ❑ arteriole

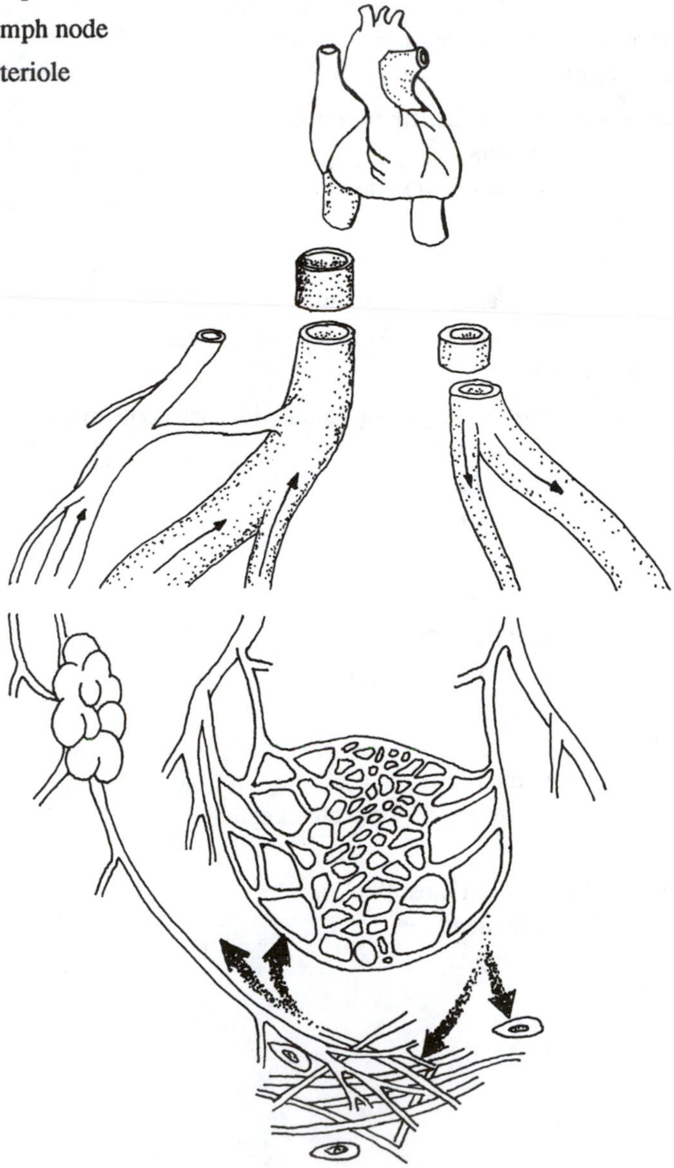

Color the parts of the illustration below as indicated.

RED ☐ smooth muscle

GREEN ☐ outer coat

YELLOW ☐ endothelium

Label artery, vein, and capillary.

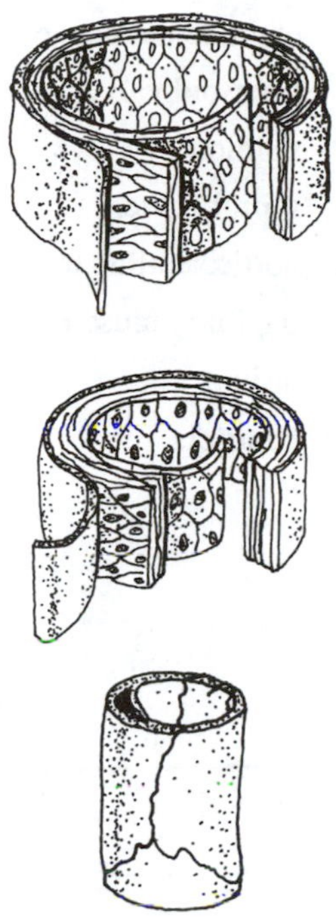

Color the parts of the illustration below as indicated.

RED ☐ aorta and pulmonary artery   ORANGE ☐ valve

GREEN ☐ ventricle   BROWN ☐ partition

YELLOW ☐ atrium   TAN ☐ conus

BLUE ☐ vein from body   PINK ☐ sinus venosus

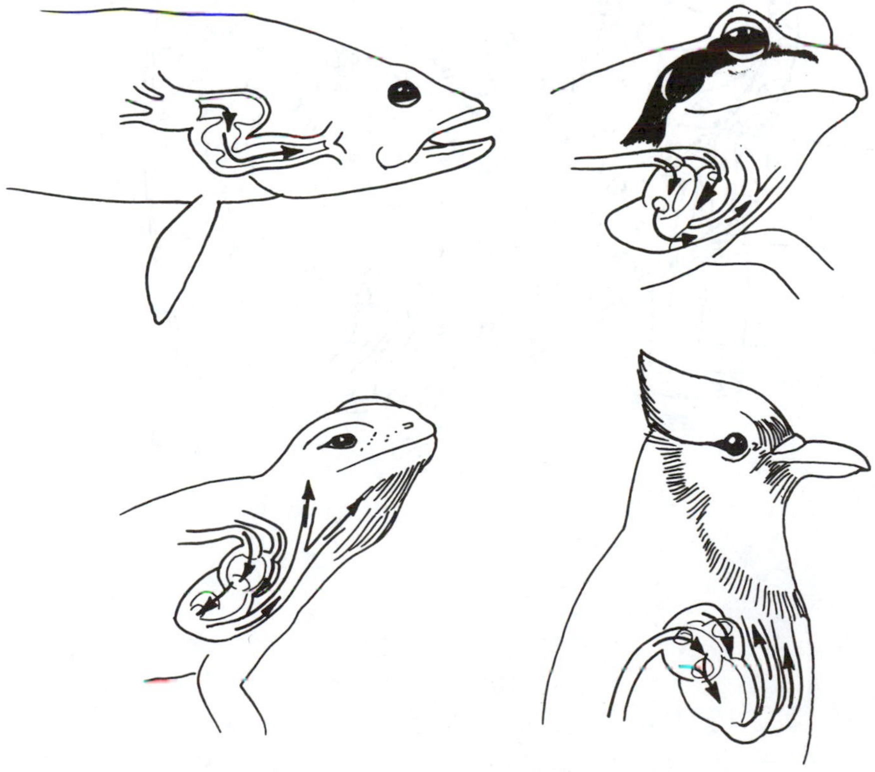

Color the parts of the illustration below as indicated.

RED ☐ chordae tendineae
GREEN ☐ tricuspid valve
YELLOW ☐ pulmonary valve
BLUE ☐ mitral valve
ORANGE ☐ aortic semilunar valve
BROWN ☐ papillary muscle
TAN ☐ atrium
PINK ☐ ventricle

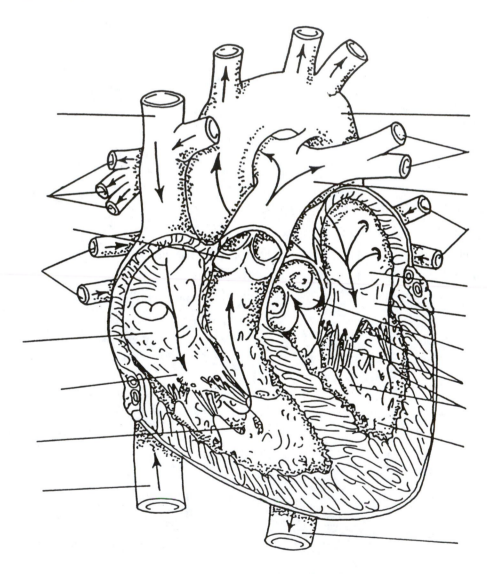

□

CHAPTER 43

Internal Defense

All animals have the ability to prevent disease-causing microorganisms (pathogens) from entering their body. Because these external defense mechanisms sometimes fail, most animals have developed internal defense mechanisms as well. Before an animal can attack an invader, it must be able to recognize its presence, to distinguish between self (its own cells) and nonself (foreign matter). Whereas most animals are capable of nonspecific responses, such as phagocytosis and the inflammatory response, only vertebrates are capable of specific responses, such as the production of antibodies that target specific pathogens. This type of internal defense is characterized by a more rapid and intense response the second time the organism is exposed to the pathogen. In this manner, the organism develops resistance to the pathogen. Resistance to many pathogens can be induced artificially by either the injection of the pathogen in a weakened, killed, or otherwise altered state, or the injection of antibodies produced by another person or animal. The immune system is constantly surveying the body for abnormal cells and destroys them whenever they arise; failure results in cancer. This same surveillance system is responsible for graft rejection; the system simply reads foreign cells as nonself cells and destroys them. Sometimes the surveillance system fails and the immune system falsely reads self cells as nonself and destroys them; the resulting conditions are known as autoimmune diseases. Allergies are another example of abnormal immune responses.

CHAPTER OUTLINE AND CONCEPT REVIEW
Fill in the blanks.

INTRODUCTION

1 A substance that stimulates an immune response is an _____.

2 The two broad categories of defense mechanisms are, first, (a)_____ _____, which are those that prevent pathogens from entering the body and attack those that do enter and, second, the "tailor-made," precisely directed (b)_____.

3 _____ is the study of specific defense mechanisms.

4 _____ are highly specific proteins that help destroy antigens.

INVERTEBRATES HAVE MAINLY NONSPECIFIC INTERNAL DEFENSE MECHANISMS

5 Most invertebrates are capable only of nonspecific responses as in "eating" invading cells, or (a)_____, and carrying out aspects of the (b)_____, which is characterized by localized heat, swelling, and discoloration.

VERTEBRATES LAUNCH BOTH NONSPECIFIC AND SPECIFIC IMMUNE RESPONSES

VERTEBRATE NONSPECIFIC DEFENSE MECHANISMS INCLUDE MECHANICAL AND CHEMICAL BARRIERS

6 The first line of defense in animals is the (a)_____. Other nonspecific defense mechanisms that prevent entrance of pathogens include the skin, (b)_____ in the stomach, and the (c)_____ of the respiratory passageways.

7 Interferons are proteins that inhibit _____.

Cytokines are important in both nonspecific and specific immune responses
Complement leads to destruction of pathogens
Inflammation is a protective mechanism

8 When pathogens invade tissues, they trigger an inflammatory response, which brings needed phagocytic cells and antibodies to the infected area. Injured cells release _____ that dilates blood vessels in the area.

9 The four clinical manifestations of inflammation are _____ _____.

10 A fever may result when leukocytes release pyrogens that act to reset the body's thermostat in the (a)_____ gland. The most potent of these regulatory proteins is (b)_____ released by macrophages.

Phagocytes destroy pathogens

11 _____ are two leukocytes that phagocytize and destroy bacteria.

SPECIFIC DEFENSE MECHANISMS INCLUDE ANTIBODY-MEDIATED IMMUNITY AND CELL-MEDIATED IMMUNITY

Cells of the immune system include lymphocytes and phagocytes

12 T cells are a type of agranular lymphocyte that originate from the (a)_____ cells that are found in (b)_____.

13 The three main types of T cells are: the (a)_____ T cells, or killer T cells that kill cells with antigens on their surfaces, the (b)_____ T cells that secrete substances that initiate or enhance the immune response, and the (c)_____ T cells that inhibit the immune response.

14 Macrophages are sometimes referred to as APCs, which stands for _____ _____.

The thymus "instructs" T cells and produces hormones

15 The thymus gland somehow "instructs" or otherwise effects immunological competence upon T cells. It also secretes _____, a hormone that is thought to stimulate mature T cells.

The major histocompatibility complex permits recognition of self
Antibody-mediated immunity is a chemical warfare mechanism

16 _____ is the process whereby one or a few selected B cells respond to a specific antigen by repeatedly dividing, thereby forming a population of virtually identical cells, all from the one or few cells.

17 Antibodies are highly specific proteins called (a)_____, abbreviated as (b)_____. These molecules "recognize" specific amino acid sequences on antigens called (c)_____. The antibody molecule is Y-shaped, the two arms functioning as (d)_____ _____.

18 Antibodies are grouped into five principle isotypes, which are abbreviated as (a)_____. Of these, about 75% are (b)_____ in humans. (c)_____ predominate in secretions, (d)_____ is an antigen receptor on the membranes of B lymphocytes, and (e)_____ is the mediator of allergic responses.

Cell-mediated immunity provides cellular warriors

19 After a cytotoxic T cell combines with an antigen on the surface of a target cell, it secretes _____ that contain a variety of lytic proteins.

20 _____ released by cytotoxic T cells are soluble proteins that are toxic to cancer cells.

21 Cell-mediated immune response can be summarized as

pathogen enters body ——> (a)_____ phagocytizes
pathogen ——> (b)_____ complex is displayed ——>
(c)_____ cell activated by complex ——> clone of
(d)_____ secretes interleukens ——>
(e)_____ develop and migrate to the infected area ——>
proteins are released ——> target cells destroyed

A secondary immune response is more rapid than a primary response

22 The principal antibody synthesized in the primary response is (a)_____. Second exposure to an antigen evokes a secondary immune response, which is more rapid and more intense than the primary response. The principal antibody synthesized in the secondary response is (b)_____.

Active immunity follows exposure to antigens

23 Active immunity develops as a result of exposure to antigens; it may occur naturally or may be artificially induced by _____.

Passive immunity is borrowed immunity

24 Immunity conferred on an individual by antibodies produced in another individual is called _____.

Normally the body effectively defends itself against cancer

25 Antibodies that combine with cancer cell antigens, thereby preventing T cells from adhering to and destroying them, are called _____.

Graft rejection is an immune response against transplanted tissue

26 Transplanted human tissues possess protein markers that stimulate graft rejection, an immune response launched mainly by _____ that destroy the transplant.

Certain sites in the body are immunologically privileged
In an autoimmune disease the body attacks its own tissues
Allergic reactions are abnormal immune responses

27 In an allergic response, an (a)_____ stimulates production of (b)_____ type antibody, which combines with the receptors on mast cells; the mast cells then release (c)_____ and other substances, causing inflammation and other symptoms of allergy.

AIDS is an immune disease caused by a retrovirus

28 AIDS is an acronym for (a)_____ which is caused by infection with the retrovirus (b)_____ _____. The virus compromises the immune system of its victims by destroying (c)_____ cells.

BUILDING WORDS

Use combinations of prefixes and suffixes to build words for the definitions that follow.

Prefixes	The Meaning	Suffixes	The Meaning
anti-	against, opposite of	-cyte	cell
auto-	self, same		
lyso-	loosening, decomposition		
mono-	alone, single, one		

Prefix	Suffix	Definition
_____	-body	1. A specific protein that acts against pathogens and helps destroy them.
_____	-histamine	2. A drug that acts against (blocks) the effects of histamine.
lympho-	_____	3. A white blood cell strategically positioned in the lymphoid tissue; the main "warrior" in specific immune responses.
_____	-clonal	4. An adjective pertaining to a single clone of cells.
_____	-immune	5. An adjective pertaining to the situation wherein the body reacts immunologically against its own tissues (against "self").
_____	-zyme	6. An enzyme that lyses bacteria by degrading the cell wall.

MATCHING

For each of these definitions, select the correct matching term from the list that follows.

____ 1. Any substance capable of stimulating an immune response; usually a protein or large carbohydrate that is foreign to the body.

____ 2. An endocrine gland that functions as part of the lymphatic system.

____ 3. Substance released from mast cells that is involved in allergic and inflammatory reactions.

____ 4. A type of white blood cell responsible for cell-mediated immunity.

____ 5. The production of antibodies or T cells in response to foreign antigens.

____ 6. A protein produced by animal cells when challenged by a virus.

____ 7. Type of lymphocyte that secretes antibodies.

____ 8. An organism capable of producing disease.

____ 9. A type of cell found in connective tissue; contains histamine and is important in allergic reactions.

____10. Temporary immunity derived from the immunoglobulins of another organism.

Terms:

a. Allergen
b. Antigen
c. Complement
d. Histamine
e. Immune response

f. Immunoglobulin
g. Interferon
h. Interleukin
i. Mast cell
j. Memory cell

k. Passive immunity
l. Pathogen
m. Plasma cell
n. T cell
o. Thymus gland

TABLE
Fill in the blanks.

Cells	Nonspecific Defense Mechanism	Specific Defense Mechanism: Antibody-Mediated Immunity	Specific Defense Mechanism: Cell-Mediated Immunity
#1	None	Activated by helper T cells and macrophages; multiply into a clone	None
#2	None	Secrete specific antibodies	None
#3	None	Long-term immunity	None
#4	None	None	Activated by helper T cells and macrophages; multiply into a clone
#5	None	Involved in B cell activation	Involved in T cell activation
#6	None	None	Chemically destroy cells infected with viruses
#7	None	None	Long-term immunity
#8	Kill virus-infected cells & tumor cells	None	Kill virus-infected cells and tumor cells
#9	Phagocytic, destroy bacteria	Antigen-presenting cells, activate helper T cells	Antigen-presenting cells; stimulates cloning
#10	Phagocytic, destroy bacteria	None	None

THOUGHT QUESTIONS
Write your responses to these questions.

1. How do the internal defense mechanisms of plants compare to those found in animals? How do the internal defense mechanisms of invertebrates compare to those found in vertebrates? Do protists have internal defense mechanisms? Prokaryotes?
2. Describe the inflammatory response and explain why it is a "nonspecific" defense mechanism.
3. Explain why the antigen-antibody response is a "specific" defense mechanism.
4. Summarize the involvement of different cell types in a typical primary immune response. Contrast those events with a secondary immune response.
5. What is an autoimmune disease?
6. What is it about HIV that ultimately reduces the effectiveness of the immune system? What does this have to do with AIDS? What do the acronyms HIV and AIDS stand for?

MULTIPLE CHOICE

Place your answer(s) in the space provided. Some questions may have more that one correct answer.

_____ 1. T cells
- a. are lymphocytes.
- b. are called LGLs.
- c. are also called T lymphocytes.
- d. are agranular and mononucleated.
- e. are involved in specific cellular immunity.

_____ 2. An allergic reaction involves
- a. killer T cells.
- b. mast cells.
- c. interaction between an allergen and mast cells.
- d. production of IgE.
- e. histocompatibility antigens.

_____ 3. The histocompatibility complex is
- a. found in cell nuclei.
- b. called HLA in humans.
- c. different in each individual.
- d. the same on individuals comprising a species.
- e. a group of proteins.

_____ 4. Which of the following is/are true of Ig?
- a. They are antibodies.
- b. They contain a C region.
- c. They are produced in response to specific antigens.
- d. They contain an antigenic determinant.
- e. They are also called immunoglobulins.

_____ 5. The millions of different types of antibodies produced by the immune system are most likely due to
- a. a one gene-one antibody ratio.
- b. gene mutations.
- c. a combination of many germ line V genes and mutations.
- d. spacers between V and C regions of DNA.
- e. many C regions.

_____ 6. The functions of antibodies are determined by their
- a. C region.
- b. V region.
- c. J region.
- d. constant segment.
- e. variable segment.

_____ 7. Which of the following immunoglobulins would you expect to be most involved in protecting you from air borne pathogens?
- a. IgG
- b. IgM
- c. IgA
- d. IgD
- e. IgE

_____ 8. Treatment with thymosin might be appropriate when the patient has
- a. an underdeveloped thymus gland.
- b. too few B cells.
- c. too many immunologically active T cells.
- d. dysfunction of the MHC system.
- e. certain types of cancer.

_____ 9. Complement
- a. is a system of several proteins.
- b. is highly antigen-specific.
- c. is stimulated into action by an antibody-antigen complex.
- d. helps destroy pathogens.
- e. is an antibody.

_____ 10. Active immunity can be artificially induced by
- a. transfusions.
- b. injecting attenuated virus.
- c. passing maternal antibodies to a fetus.
- d. injecting gamma globulin.
- e. stimulating macrophage growth.

____11. Nonspecific defense mechanisms in vertebrates include
    a. skin.   d. phagocytes.
    b. acid secretions.   e. antibody-mediated immunity.
    c. inflammation.

____12. The secondary response is due to
    a. killer T cells.   d. macrophages.
    b. memory cells.   e. helper T cells.
    c. plasma cells.

____13. B cells
    a. are granular.   d. are derived from plasma cells.
    b. are lymphocytes.   e. include many antigen-binding forms.
    c. clone after contacting its targeted antigen.

____14. T-cell receptors
    a. bind antigens.   d. are found on killer cells.
    b. have no known function. e. stimulate antibody production.
    c. are identical on all T cells.

____15. The body's thermostat is reset during fever by the action of
    a. a peptide.   d. a substance released by macrophages.
    b. phagosomes.   e. interleukin1.
    c. prostaglandins.

____16. Which of the following is true of AIDS?
    a. HIV infects helper T cells.   d. HIV can be transmitted by sharing needles.
    b. AIDS is not spread by casual contact. e. There is currently no cure for AIDS.
    c. Both heterosexuals and homosexuals are at risk.

VISUAL FOUNDATIONS

Color the parts of the illustration below as indicated. Label variable portion and constant portion.

RED ☐ antigenic determinants
GREEN ☐ antigen
YELLOW ☐ antibody-heavy chain
BLUE ☐ bonding sites
ORANGE ☐ antibody-light chain
TAN ☐ antigen-antibody complex

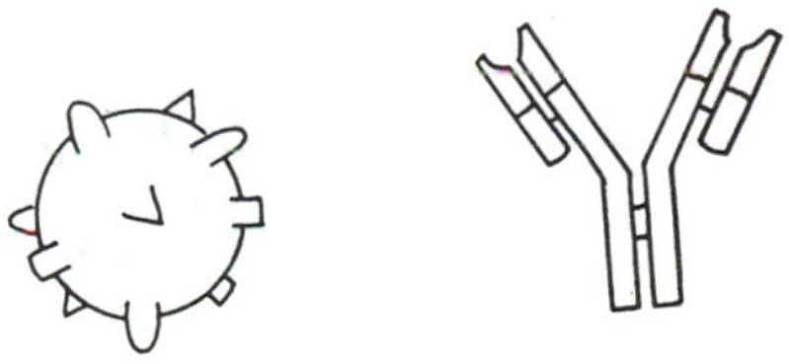

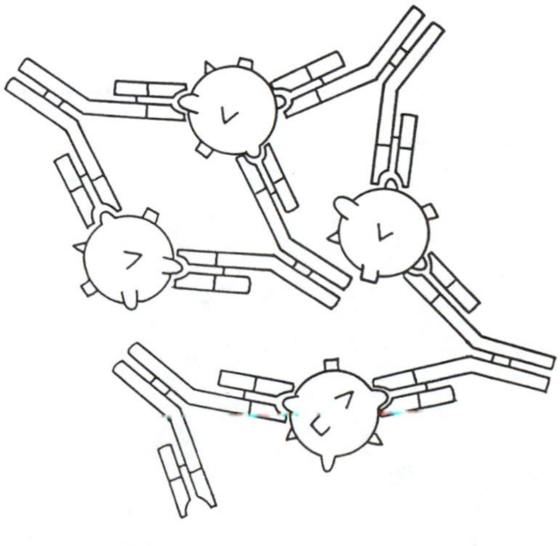

Color the parts of the illustration below as indicated. Label cell-mediated immunity, antibody-mediated immunity

RED	❑	bone marrow
GREEN	❑	antigen stimulation
YELLOW	❑	T cell
BLUE	❑	migration to lymph node
ORANGE	❑	thymus
BROWN	❑	natural killer cell
TAN	❑	B cell
PINK	❑	plasma cell

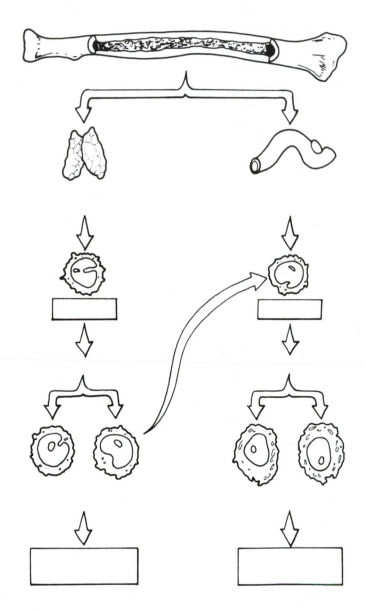

CHAPTER 44

Gas Exchange

Gas exchange with air is more efficient than gas exchange with water. This is because air contains more oxygen than water, and oxygen diffuses more rapidly in air than in water. In small, aquatic organisms, gases diffuse directly between the environment and all body cells. In larger, more complex organisms, specialized respiratory structures, such as tracheal tubes, gills, and lungs, are required. Respiratory pigments, such as hemoglobin and hemocyanin, greatly increase respiratory efficiency by increasing the capacity of blood to transport oxygen. The human respiratory system consists of two lungs and a system of tubes through which air reaches them. The largest tube, the trachea, branches into two bronchi, one extending into each lung. Each bronchus branches repeatedly, giving rise ultimately to tiny bronchioles. The bronchioles, in turn, branch repeatedly, giving rise to clusters of alveoli. Oxygen diffuses from the alveoli into pulmonary capillaries, and carbon dioxide diffuses from the blood into the alveoli, each gas moving from a region of its greater concentration to a region of its lower concentration. Although breathing is an involuntary process controlled by respiratory centers in the brain, it can be consciously influenced. Respiration can be adversely affected by hyperventilation, flying, diving, smoking, and air pollution.

CHAPTER OUTLINE AND CONCEPT REVIEW
Fill in the blanks.

INTRODUCTION
1 The exchange of gases between an organism and the medium in which it lives is called
 (a)_____, while actively moving air or water over the respiratory surface is
 (b)_____.

RESPIRATORY STRUCTURES ARE ADAPTED FOR GAS EXCHANGE IN AIR OR WATER
2 Primarily because of water's greater (a)_____, aquatic animals expend 10 to
 20 times more energy than air breathers to move their medium over their respiratory surfaces. Air
 contains (b)_____(more of less?) oxygen than water, and oxygen diffuses (c)_____(faster
 or slower?) in air than in water.

FOUR MAIN TYPES OF SURFACES FOR GAS EXCHANGE HAVE EVOLVED
3 Specialized respiratory structures must have (a)_____ to facilitate the transfer of
 adequate volumes of gases, (b)_____ surfaces that can dissolve oxygen and carbon
 dioxide, and a rich supply of (c)_____ to transport respiratory gases.

4 _____ are the four
 principal types of respiratory structures.

 The body surface may be adapted for gas exchange
5 Animals that rely entirely on gas exchange across the body surface are small and therefore have a
 large (a)_____ ratio and a low (b)_____.

Tracheal tube systems of arthropods deliver air directly to the cells

6 Arthropods have respiratory system consisting of a network of (a)_____ that air enters through small openings called (b)_____. The respiratory network terminates in small fluid-filled (c)_____ where gas exchange takes place.

Gills of aquatic animals are specialized respiratory surfaces

7 The gills of bony fish have many thin (a)_____ that extend out into the water, within which are numerous capillaries. Blood flows through these structures in a direction opposite to water movement, an arrangement called the (b)_____, a system that maximizes the difference in oxygen concentrations between the animal's blood and the water.

Terrestrial vertebrates exchange gases through lungs

8 Lungs are respiratory structures that develop as ingrowths of the (a)_____ or from the wall of a (b)_____ such as the pharynx. For example, the (c)_____ of spiders are within an inpocketing of the abdominal wall.

RESPIRATORY PIGMENTS INCREASE CAPACITY FOR OXYGEN TRANSPORT

9 The respiratory pigment of most vertebrates is hemoglobin, a compound containing a (a)_____ group bound to the protein (b)_____.

THE HUMAN RESPIRATORY SYSTEM IS TYPICAL OF AIR-BREATHING VERTEBRATES

The airway conducts air into the lungs

10 The cough reflex occurs when the (a)_____ fails to close off the (b)_____ from the esophagus.

Gas exchange occurs in the alveoli of the lungs

11 The lungs are located in the thoracic cavity. The right lung has (a)_____(#?), and the left lung has (b)_____(#?) lobes.

12 The pathway of air as it passes through the human's "breathing apparatus" can be summarized as

   nose and mouth ——> (a)_____ (throat region) ——>
   (b)_____ (voice box) ——> (c)_____ (wind pipe) ——>
   (d)_____ (tubes leading to lungs) ——> (e)_____
   (small branching tubes) ——> (f)_____ (air sacs)

Ventilation is accomplished by breathing

13 Breathing involves inhalation, or (a)_____, and exhalation, or (b)_____.

The quantity of air respired can be measured

14 (a)_____ is the volume of air that is inhaled and exhaled during "normal" breathing. It averages about (b)_____(volume?). (c)_____ is the maximum volume of air that can be expelled.

Gas exchange takes place in the air sacs

15 The factor that determines the direction and rate of diffusion of a gas across a respiratory surface is the (a)_____ of that gas. (b)_____ _____ states that the total pressure in a mixture of gases is the sum of the pressures of the individual gases

Oxygen is transported in combination with hemoglobin

16 Oxygen combines with the element (a)_____ in the (b)_____ group of hemoglobin. This "association" can be illustrated as $Hb + O_2 \longrightarrow$ (c)_____. The resultant is a compound that carries oxygen and releases it where it is in lower concentrations.

17 HbO_2 is prone to dissociation in a (a)_____(lower or higher?) pH. A change in the normal HbO_2 dissociation curve caused by a change in pH is known as the (b)_____.

Carbon dioxide is transported mainly as bicarbonate ions

18 Most of the carbon dioxide transported in blood is dissolved in plasma as (a)_____ ions, the formation of which is catalyzed by an enzyme in RBCs called (b)_____.

Breathing is regulated by respiratory centers in the brain

19 Respiratory centers are groups of neurons in the (a)_____ that regulate the rhythm of ventilation. (b)_____ are specialized nerve endings that are sensitive to changes in hydrogen ion concentration.

Hyperventilation reduces carbon dioxide concentration
High flying or deep diving can disrupt homeostasis

20 _____ is a deficiency of oxygen that may cause drowsiness, mental fatigue, and headaches.

21 A sudden decrease in environmental pressure may cause "bubbling" in the tissues and/or blood vessels; that is, the release of dissolved nitrogen in the blood in the form of bubbles that block capillaries, causing a very painful syndrome which, in the vernacular, is called "the bends," and in the scientific community is called _____.

BREATHING POLLUTED AIR DAMAGES THE RESPIRATORY SYSTEM

22 Chronic bronchitis and emphysema are examples of COPDs, which stands for _____. COPDs are linked to smoking and air pollution.

BUILDING WORDS

Use combinations of prefixes and suffixes to build words for the definitions that follow.

Prefixes	The Meaning		Suffixes	The Meaning
hyper-	over		-ox(ia)	containing oxygen
hyp(o)-	under			

Prefix	Suffix	Definition
_____	-ventilation	1. Excessive rapid and deep breathing.
_____	_____	2. Oxygen deficiency.

MATCHING

For each of these definitions, select the correct matching term from the list that follows.

____ 1. "Windpipe."

____ 2. An air sac of the lung through which gas exchange with the blood takes place.

____ 3. A type of respiratory organ of aquatic animals.

____ 4. The membrane that lines the thoracic cavity and envelopes the lungs.

____ 5. One of the branches of the trachea and its immediate branches within the lung.

_____ 6. The throat region in humans.

_____ 7. Tiny air ducts of the lung that branch to form the alveoli.

_____ 8. The dome-shaped muscle that forms the floor of the thoracic cavity.

_____ 9. The bony plate covering the gills in fish.

_____10. The organ at the upper end of the trachea that contains the vocal cords.

Terms:

a. Alveolus
b. Bronchioles
c. Bronchus
d. Diaphragm
e. Epiglottis

f. Gill
g. Larynx
h. Lung
i. Operculum
j. Pharynx

k. Pleural cavity
l. Pleural membrane
m. Respiratory center
n. Trachea

TABLE
Fill in the blanks.

Organism	Type of Gas Exchange Surface
Polychaete worms	#1
Nudibranch mollusks	#2
Insects	#3
Sea stars	#4
Clams	#5
Fish	#6
Spiders	#7
Reptiles	#8
Birds	#9
Mammals	#10

THOUGHT QUESTIONS
Write your responses to these questions.

1. What are the basic requirements for an effective gas exchange system? What can you say about the surface area? What about the consistency of the exchange surface — can it be dry? What about relationship between the exchange surface and the internal transport system? What about "carrier molecules?" Explain how each of these essential elements is satisfied when gas is exchanged by means of an organism's body surface, or tracheal tubes, or gills, or lungs.

2. Trace the path of an oxygen molecule from the atmosphere into a human red blood cell. Name all of the structures involved and their respective roles.
3. Describe how oxygen and carbon dioxide are carried by the circulatory system and how they are exchanged at the respiratory surface and in body tissues.
4. What is hyperventilation? The bends? Sudden decompression?

MULTIPLE CHOICE

Place your answer(s) in the space provided. Some questions may have more that one correct answer.

_____ 1. The diaphragm and rib muscles alternately contract and relax, these actions causing respectively
   a. inspiration and expiration.　　　d. exhalation and inhalation.
   b. inhalation and exhalation.　　　e. oxygen intake and carbon dioxide output.
   c. expiration and inspiration.

_____ 2. If the PO_2 in the tissue of our pet dog is 10, atmospheric PO_2 is 150, and arterial PO_2 is 110, one would expect the dog to (units are mm Hg)
   a. function normally.　　　d. die.
   b. accumulate carbon dioxide.　　　e. become dizzy from too much oxygen.
   c. have a serious, but not lethal, oxygen deficit.

_____ 3. If you shared the same parameters as those given for your dog (question 2) and your tissues were consuming 2.5 liters of oxygen per minute, it's likely that
   a. you are asleep.　　　d. you are exercising vigorously.
   b. you are resting.　　　e. you and your dog will both die.
   c. you and your dog are out for a leisurely walk.

_____ 4. If the parameters given in questio3 applied, the color(s) of blood in your arteries and veins respectively would be
   a. pink and light blue.　　　d. distinctly red and blue.
   b. both blue.　　　e. distinctly blue and red.
   c. both red.

_____ 5. We humans do not have to gulp air like a frog because we have
   a. alveoli.　　　d. lower oxygen requirements.
   b. lungs.　　　e. a larger surface to volume ratio.
   c. a diaphragm.

_____ 6. The Heimlich maneuver may be necessary if
   a. the epiglottis does not close.　　　d. air contacts surfactants.
   b. food enters the esophagus.　　　e. we have severe pleurisy.
   c. a foreign object sticks in the larynx.

_____ 7. Most of the mucus produced by epithelial cells in the nasal cavities is disposed of by means of
   a. a large handkerchief.　　　d. absorption in surrounding lymph ducts.
   b. nose picking.　　　e. swallowing.
   c. evaporation.

_____ 8. Characteristics that all respiratory surfaces share in common include
   a. moist surfaces.　　　d. large surface to volume ratio.
   b. thin walls.　　　e. alveoli or spiracles.
   c. specialized structures such as tracheal tubes, gills, or lungs.

_____9. We detect odors once they have entered our respiratory systems through the
   a. nostrils.
   b. nasal cavities.
   c. buccal cavity.
   d. external nares.
   e. olfactory os.

_____10. Organismic respiration in a swordfish relies on
   a. dermal gills.
   b. filaments.
   c. internal gills along edges of slits in the pharynx.
   d. spiracles and tracheal tubes.
   e. countercurrents.

_____11. A cockroach obtains oxygen for tissues located deep in its body by means of
   a. book gills.
   b. pseudolungs.
   c. circulation of hemolymph.
   d. a countercurrent exchange system.
   e. branching tubes.

_____12. A respiratory pigment that contains protein but no porphyrin group is called
   a. hemoglobin.
   b. hemolymph.
   c. erythrolymph.
   d. hemocyanin.
   e. erythrocyanin.

_____13. Air passes through structures of a mammal's respiratory system in the following order:
   a. trachea, pharynx, bronchus, bronchioles.
   b. larynx, trachea, bronchioles, bronchus.
   c. pharynx, larynx, bronchus, bronchioles.
   d. larynx, trachea, bronchus, bronchioles.
   e. trachea, larynx, bronchus, bronchioles.

_____14. With respect to organismic respiration, some of the advantages of life on land as compared to an aquatic life pertain to
   a. energy conservation.
   b. facilitated diffusion.
   c. maintaining ion homeostasis.
   d. keeping the respiratory surface moist.
   e. maintenance of a higher body temperature.

_____15. Reasonable rates of expiration and inspiration fall in the range of
   a. 25/min.
   b. 10/min.
   c. 72/hr.
   d. 0.2/sec.
   e. 650/hr.

_____16. The ability of oxygen to be released from oxyhemoglobin is affected by
   a. temperature.
   b. pH.
   c. concentration of carbon dioxide.
   d. oxygen concentration in tissues.
   e. ambient oxygen concentration.

_____17. If the action of carbonic anhydrase in RBCs increases, one would expect an increase in
   a. concentration of reduced hemoglobin.
   b. bicarbonate ions in plasma.
   c. diffusion of chloride ions out of cells.
   d. the chloride shift.
   e. bonding of carbon dioxide and hemoglobin.

_____18. If a patient is found to have inelastic air sacs, trouble expiring air, an enlarged right ventricle, and large alveoli, he/she probably
   a. has lung cancer.
   b. has emphysema.
   c. has chronic obstructive pulmonary disease.
   d. lives at a high altitude.
   e. experiences acute bronchitis periodically.

_____19. If you are SCUBA diving for a lengthy time at 200 feet, you might get the bends if
   a. you surface quickly.
   b. you are using a helium mixture.
   c. nitrogen in your blood is rapidly absorbed by your tissues.
   d. you come to the surface very slowly.
   e. you stay at 200 feet even longer.

_____20. Proper CPR procedures include
    a. placing the victim on his/her stomach.
    b. external cardiac compression.
    c. exhalation into the victim's mouth.
    d. pausing 15 seconds between breaths.
    e. extending the victim's neck.

VISUAL FOUNDATIONS
Color the parts of the illustration below as indicated.

RED   ☐   lungs
GREEN   ☐   gills
YELLOW   ☐   trachea
BLUE   ☐   body surface
ORANGE   ☐   book lung

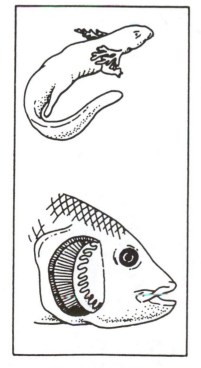
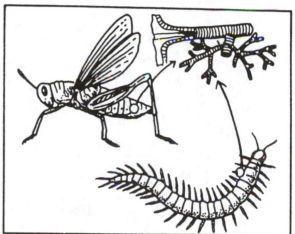
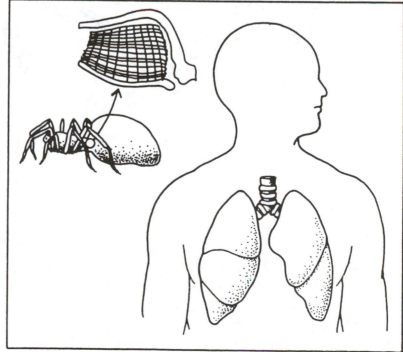
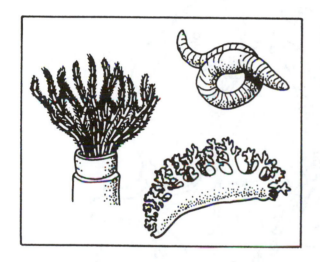

Color the parts of the illustration below as indicated.

RED   ☐ epithelial cell of alveolus

GREEN   ☐ bronchiole

YELLOW   ☐ capillary

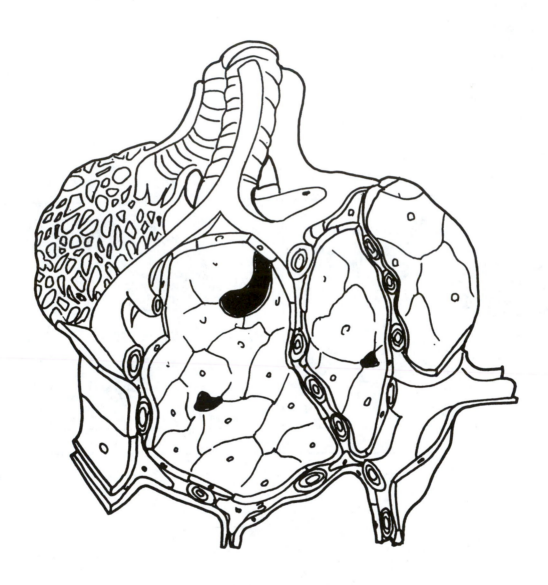

CHAPTER 45

❑

Processing Food and Nutrition

The processing of food involves several steps — taking food into the body, breaking it down into its constituent nutrients, absorbing the nutrients, and eliminating the material that is not broken down and absorbed. Animals eat either plants, animals, or both. Some animals have no digestive systems, with digestion occurring intracellularly within food vacuoles. Others have incomplete digestive systems with only a single opening for both food to enter and wastes to exit. Still other animals, most in fact, have complete digestive systems, in which the digestive tract is a complete tube with two openings, a mouth where food enters and an anus where waste is expelled. The human digestive system has highly specialized structures for processing food. All animals require the same basic nutrients — carbohydrates, lipids, proteins, vitamins, and minerals. Carbohydrates are used by the body as fuel. Lipids are also used as fuel, and additionally, as components of cell membranes, and as substrates for the synthesis of steroid hormones and other lipid substances. Proteins serve as enzymes and as structural components of cells. Vitamins and minerals are needed for many biochemical processes. Serious nutritional problems result from eating too much food, eating too little food, or not eating a balanced diet.

CHAPTER OUTLINE AND CONCEPT REVIEW
Fill in the blanks.

INTRODUCTION

1 Organisms that obtain their main source of energy from organic molecules synthesized by other organisms are known as _____.

2 Undigested material is removed from the digestive tracts of simpler animals by a process known as (a)_____, and from more complex animals by (b)_____.

ANIMALS ARE ADAPTED TO THEIR MODE OF NUTRITION

3 Primary consumers, or _____, have special digestive processes to break down the plant material they consume.

4 (a)_____ have well-developed canine teeth, digestive enzymes that break down proteins, and an overall (b)_____(shorter or longer?) gastrointestinal tract.

5 Organisms that utilize both animal and plant material are called _____.

SOME INVERTEBRATES HAVE DIGESTIVE SYSTEMS WITH A SINGLE OPENING

MOST ANIMAL DIGESTIVE SYSTEMS HAVE TWO OPENINGS

6 The vertebrate digestive system is a complete tube extending from the (a)_____ to the (b)_____. The specialized portions of this tube, in order, are the

mouth ——> (c)_____ ——> (d)_____ ——>
stomach ——> (e)_____ ——> large intestine ——> anus

THE HUMAN DIGESTIVE SYSTEM IS HIGHLY SPECIALIZED FOR PROCESSING FOOD

7   The wall of the vertebrate digestive tract consists of four layers: the (a)_____ is densely packed with blood vessels and nerves, the (b)_____ contains goblet cells that secrete mucous and a large folded surface area, the (c)_____ is the outer layer of the system, and the (d)_____ contains muscles that carry out peristalsis.

Food processing begins in the mouth

8   Ingestion and the beginning of mechanical and enzymatic breakdown of food take place in the mouth. The teeth of mammals perform varied functions: (a)_____ are designed for biting, (b)_____ for tearing and stabbing (ask Dracula!), and (c)_____ crush and grind food. Each tooth is covered by a coating of very hard (d)_____, under which is the main body of the tooth, the (e)_____, which resembles bone. The (f)_____ contains blood vessels and nerves.

9   The salivary glands of terrestrial vertebrates moisten food and release _____, an enzyme that initiates carbohydrate (starch) digestion.

The pharynx and esophagus conduct food to the stomach

10   Waves of muscular contractions called (a)_____ move a lump of food, at this point called a (b)_____, through the esophagus into the stomach.

Food is mechanically and enzymatically digested in the stomach

11   In the stomach, food is mechanically broken down, pepsin in gastric juice initiates protein digestion, and food is reduced to chyme. The stomach is lined with (a)_____ _____ cells, which contain many gastric glands and specialized cells. For example, (b)_____ cells secrete hydrochloric acid and (c)_____, a substance needed for absorption of vitamin B, and (d)_____ cells secrete pepsinogen, an enzyme precursor that converts to (e)_____.

Most enzymatic digestion takes place inside the small intestine

12   The three regions of the small intestine are (a)_____. Most enzymatic digestion of food in vertebrates takes place in the (b)_____ portion.

13   The surface area of the small intestine is increased by small fingerlike projections called (a)_____ along its surface, and by (b)_____, which are outfoldings of the plasma membranes of cells that line the intestinal lumen.

The liver secretes bile, which mechanically digests fats

14   Among the many important functions performed by the liver are the secretion of bile, maintenance of homeostasis, the conversion of glucose to the carbohydrate storage molecule (a)_____, the conversion of excess amino acids to (b)_____ and the detoxifying of drugs and other poisons.

15   Bile produced by the liver is stored in the (a)_____. Bile breaks apart fat droplets in a process called (b)_____.

The pancreas secretes digestive enzymes
Enzymatic digestion occurs as food moves through the digestive tract

16   Pancreatic juice contains several categories of enzymes, including the (a)_____ that hydrolyze proteins, the (b)_____ that break down fats, and the (c)_____ that act on polysaccharides.

17   Enzymes reduce macromolecular polymers to the small subunits that comprise them. For example, carbohydrates are digested to (a)_____, proteins are reduced to (b)_____, and fats are depolymerized to (c)_____.

Nerves and hormones regulate digestion

Absorption takes place mainly through the villi of the small intestine

18 Most digested nutrients are absorbed through the (a)_____ of the small intestine. Monosaccharides and amino acids enter the (b)_____; glycerol, fatty acids, and monoacylglycerols enter the (c)_____.

The large intestine eliminates waste

19 The large intestine absorbs sodium and water, cultures bacteria, and eliminates wastes. It is made up of seven regions, which are, in order: the (a)_____, a blind pouch near the junction of the small and large intestines; the (b)_____; the (c)_____; the (d)_____; the (e)_____; the (f)_____, the last portion of the tube; and the (g)_____, the opening at the end of the tube.

20 (a)_____ involves mainly the filtering out of wastes by the kidneys and lungs, while (b)_____ involves disposal of wastes that never participated in metabolism.

ADEQUATE AMOUNTS OF REQUIRED NUTRIENTS ARE NECESSARY TO SUPPORT METABOLIC PROCESSES

21 The amount of energy in food is given as (a)_____ per gram, which is equivalent to the unit (b)_____.

Carbohydrates are a major energy source in the human diet

22 _____ is a mixture of cellulose and other indigestible carbohydrates derived from plants.

Lipids are used as an energy source and to make biological molecules

23 Macromolecular complexes of cholesterol or triacylglycerols bound to proteins are called (a)_____. Two types of these important complexes are the (b)_____, which apparently decrease the risk of heart disease by transporting excess cholesterol to the liver; and (c)_____, which have been associated with coronary artery disease.

Proteins serve as enzymes and as structural components of cells

24 Excess amino acids are removed from the system by the (a)_____ through a process known as (b)_____.

Vitamins are organic compounds essential for normal metabolism

25 Vitamins are divided into two broad groups: the (a)_____ vitamins such as A, D, E, and K, and the (b)_____ vitamins that include (c)_____.

Minerals are inorganic nutrients required by cells

26 The essential minerals required by the body are _____ _____.

ENERGY METABOLISM IS BALANCED WHEN ENERGY INPUT EQUALS ENERGY OUTPUT

27 When energy input equals energy output, body weight remains constant. When energy input exceeds energy output, body weight (a)_____(increases or decreases?); and when energy input is less than output, the body draws on fuel reserves (fat) and body weight (b)_____(increases or decreases?).

28 (a)_____ reflects the amount of energy an organism must expend to survive. (b)_____ is the sum of basic energy required to survive *and* the energy needed to carry out daily activities.

Obesity is a serious nutritional problem
Malnutrition can cause serious health problems

BUILDING WORDS

Use combinations of prefixes and suffixes to build words for the definitions that follow.

Prefixes	The Meaning	Suffixes	The Meaning
epi-	upon, over, on	-itis	inflammation
micro-	small	-micro(n)	small, "tiny"
omni-	all	-vore	eating
sub-	under, below		

Prefix	Suffix	Definition
herbi-	_____	1. An animal that eats plants.
carni-	_____	2. An organism that eats flesh.
_____	_____	3. An organism that eats both plants and animals.
_____	-mucosa	4. A layer of connective tissue below the mucosa that binds it to the muscle layer beneath.
periton-	_____	5. Inflammation of the peritoneum.
_____	-glottis	6. A flap of tissue over the glottis (larynx opening) that prevents food and drink from entering the larynx when swallowing.
_____	-villi	7. Small projections of the cell membrane that increase the surface area of the cell.
chylo-	_____	8. Tiny droplets of lipid that are absorbed from the intestine into the lymph circulation.

MATCHING

For each of these definitions, select the correct matching term from the list that follows.

_____ 1. A cell layer that lines the digestive tract and secretes a lubricating layer of mucous.

_____ 2. The ejection of waste products, especially undigested food remnants, from the digestive tract.

_____ 3. The chief enzyme of gastric juice; hydrolyses proteins.

_____ 4. A large digestive gland located in the vertebrate abdominal cavity, having both exocrine and endocrine functions.

_____ 5. An organic compound necessary in small amounts for the normal metabolic functioning of a given organism; usually acts as a coenzyme.

_____ 6. Folds in the stomach wall, giving the inner lining a wrinkled appearance.

_____ 7. The taking up of a substance, as by the lining of the digestive tract.

_____ 8. Digestive juice produced by the liver and stored in the gallbladder.

_____ 9. Powerful, rhythmic waves of muscular contraction and relaxation in the walls of hollow tubular organs that aid in the movement of substances through the tube.

_____ 10. A minute "finger-like" projection from the surface of a membrane.

Terms:

a. Absorption
b. Adventitia
c. Bile
d. Digestion
e. Elimination

f. Gastrin
g. Mineral
h. Mucosa
i. Pancreas
j. Pepsin

k. Peristalsis
l. Rugae
m. Stomach
n. Villus
o. Vitamin

TABLE
Fill in the blanks.

Enzyme	Function	Site of Production
Salivary amylase	Enzymatic digestion of starch	Salivary glands
Maltase	#1	#2
#3	Proteins to polypeptides	#4
#5	Polypeptides to dipeptides	#6
Ribonuclease	#7	#8
#9	Degrades neutral fats	Pancreas
#10	Dipeptides to amino acids	#11
Lactase	#12	Small intestine

THOUGHT QUESTIONS
Write your responses to these questions.

1. Do plants carry out processes comparable to ingestion, digestion, absorption, and elimination in animals? What about protists? Prokaryotes? How do these functions compare in animals with incomplete digestive systems (one opening) and animals with complete digestive systems (two openings)?
2. Sketch the major components of the human digestive system and describe their principal functions.
3. Describe the structural and functional aspects of villi in the process of absorption.
4. Describe the events that transpire between ingestion of a protein and absorption of an amino acid. Do the same for a complex polysaccharide such as starch. Do the same for a lipid.
5. What roles do vitamins and minerals serve in nutrition? Give specific examples.
6. What is metabolic rate? How do basal metabolic rate and total metabolic rate differ? What is the equation that relates energy and body weight?

MULTIPLE CHOICE
Place your answer(s) in the space provided. Some questions may have more that one correct answer.

_____ 1. The principal sources of energy in the human diet are
    a. proteins.
    b. lipids.
    c. carbohydrates.
    d. starches and sugars.
    e. meat.

_____ 2. Which of the following is true concerning human nutrition?
    a. Water is essential.
    b. Excess nutrients are converted to fat.
    c. We cannot synthesize some required fatty acids.
    d. Proteins cannot be used for energy.
    e. The liver synthesizes amino acids.

_____ 3. The total metabolic rate
a. encompasses the basal metabolic rate.
b. is less than basal metabolic rate.
c. refers to metabolic rate after exercise.
d. is the rate when at rest.
e. is a nonexistent phrase.

_____ 4. A herbivore would likely have
a. well-developed claws.
b. flattened molars.
c. large, sharp canines.
d. a short gut.
e. symbiotic microorganisms.

_____ 5. Glands near the ears that initiate starch digestion are the
a. submaxillaries.
b. parietals.
c. pancreatic glands.
d. parotids.
e. sublinguals.

_____ 6. Glucose can result from
a. glycogenesis.
b. glycogenolysis.
c. gluconeogenesis.
d. glycolysis.
e. B-oxidation.

_____ 7. Nerves and blood vessels in teeth are located in the
a. pulp.
b. enamel.
c. dentin.
d. cementum.
e. canines.

_____ 8. All animals are
a. omnivores.
b. carnivores.
c. primary consumers.
d. consumers.
e. heterotrophs.

_____ 9. Bile is principally involved in the digestion of
a. carbohydrates.
b. lipids.
c. starch.
d. nucleic acids.
e. proteins.

_____10. Which of the following indicate the correct order of layers in the mammalian digestive tract?
a. muscularis, mucosa, submucosa, adventitia.
b. mucosa, submucosa, muscularis, adventitia.
c. muscularis, submucosa, mucosa, adventitia.
d. mucosa, submucosa, adventitia, muscularis.
e. adventitia, muscularis, submucosa, mucosa.

_____11. A herbivore is a/an
a. plant consumer.
b. primary consumer.
c. carnivore.
d. omnivore.
e. mutualistic symbiote.

_____12. The principal function(s) of villi is/are to
a. absorb nutrients.
b. secrete enzymes.
c. increase surface area.
d. stimulate digestion.
e. secrete hormones.

_____13. The single most crucial intermediate molecule in the metabolism of most nutrients is
a. keto acid.
b. ATP.
c. acetyl CoA.
d. NAPH.
e. glucose.

_____14. Which of the following is true regarding cellulose in the human diet?
a. It's harmful.
b. It's a source of bulk.
c. We can't digest it.
d. It's an important carbohydrate nutrient.
e. It's an important source of fiber.

_____15. A lacteal is a
- a. lymph vessel.
- b. capillary bed.
- c. digestive gland.
- d. milk protein enzyme.
- e. salivary enzyme.

_____16. Which of the following has/have a complete digestive system?
- a. birds.
- b. earthworms.
- c. sponges.
- d. jelly fish.
- e. fish.

_____17. Maltase is principally involved in the digestion of
- a. carbohydrates.
- b. lipids.
- c. starch.
- d. nucleic acids.
- e. proteins.

_____18. Goblet cells and chief cells are found in the
- a. mucosa.
- b. submucosa.
- c. epithelium.
- d. stomach.
- e. adventitia.

_____19. Peristalsis results from activity of tissues in the
- a. submucosa.
- b. mucosa.
- c. adventitia.
- d. muscularis.
- e. peritoneum.

_____20. Pepsin is principally involved in the digestion of
- a. carbohydrates.
- b. lipids.
- c. starch.
- d. nucleic acids.
- e. proteins.

_____21. Some important functions of minerals include their role in/as
- a. cofactors.
- b. nerve impulses.
- c. neurotransmitters.
- d. maintaining fluid balance.
- e. digestive hormones.

_____22. The smallest molecular units would be obtained by the action of
- a. exopeptidases.
- b. endopeptidases.
- c. dipeptidases.
- d. bile salts.
- e. amylase.

_____23. Most absorption of macromolecular subunits occurs in the
- a. stomach.
- b. rectum.
- c. large intestine.
- d. small intestine.
- e. colon.

_____24. Which of the following is true of vitamins in human nutrition?
- a. Some are needed in large quantities.
- b. Megadoses can be harmful.
- c. Antibiotics may cause a vitamin deficiency.
- d. Vitamin A can be stored in fat tissue.
- e. Vitamin D must be ingested.

_____25. Which of the following is/are essential nutrient(s) for humans?
- a. lysine.
- b. starch.
- c. cellulose.
- d. glucose.
- e. linoleic acid.

_____26. The relative amounts of nutrient types consumed by impoverished societies, as compared to affluent societies, would likely be
   a. more protein, less carbohydrate.   d. more carbohydrate, less lipid.
   b. more lipid, less carbohydrate.   e. more vegetable matter, less animal matter.
   c. more carbohydrate, less protein.

_____27. Lipids are used to
   a. maintain body temperature.   d. produce hormones.
   b. build membranes.   e. transport vitamins.
   c. provide energy.

_____28. Which of the following would likely be used for energy as a last resort?
   a. carbohydrate   d. glycogen
   b. lipid   e. fatty acid
   c. protein

_____29. Products of complete fat digestion include
   a. fatty acids.   d. glycerol.
   b. amino acids.   e. triacylglycerol.
   c. chylomicrons.

VISUAL FOUNDATIONS

Color the parts of the illustration below as indicated. Label the mucosa, submucosa, muscle layer, and peritoneum.

RED   ☐ artery
GREEN   ☐ lymph
YELLOW   ☐ nerve
BLUE   ☐ vein
ORANGE   ☐ goblet cell
BROWN   ☐ intestinal gland
TAN   ☐ villus
PINK   ☐ epithelial cell of villus

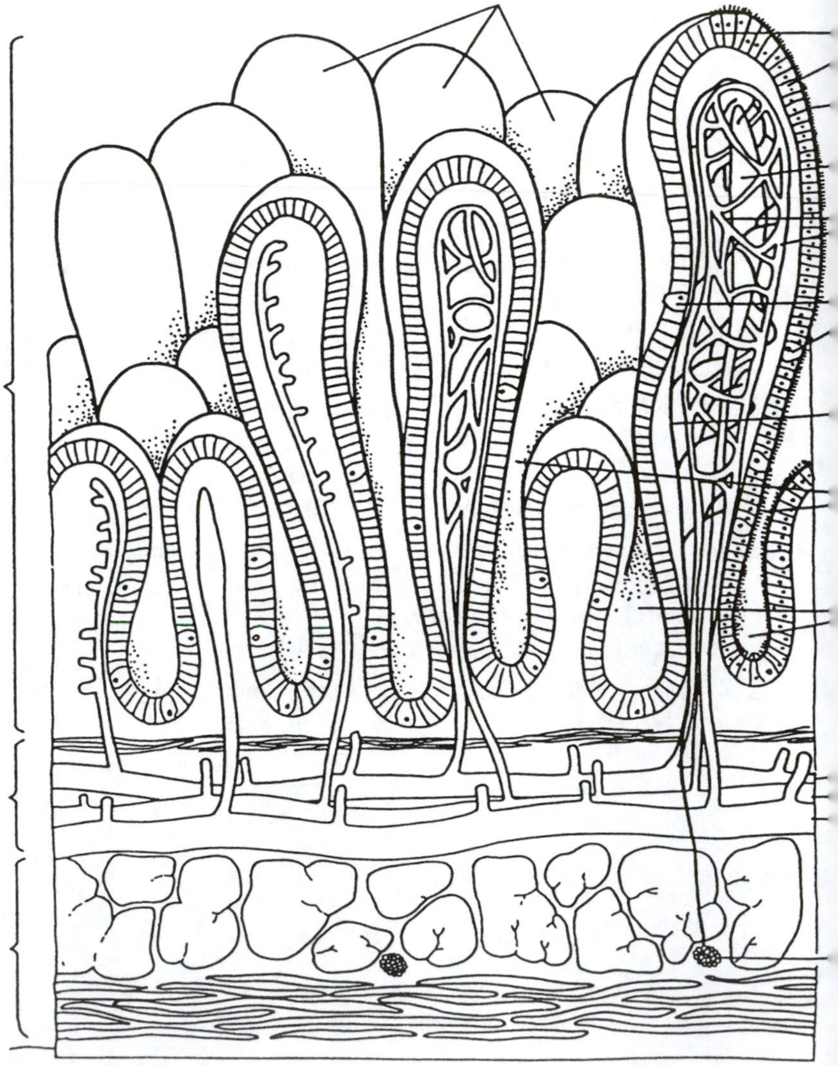

Color the parts of the illustration below as indicated.

RED	☐	liver
GREEN	☐	gall bladder
YELLOW	☐	esophagus
BLUE	☐	pancreas
ORANGE	☐	stomach
BROWN	☐	large intestine
TAN	☐	small intestine
PINK	☐	rectum and anus
VIOLET	☐	vermiform appendix

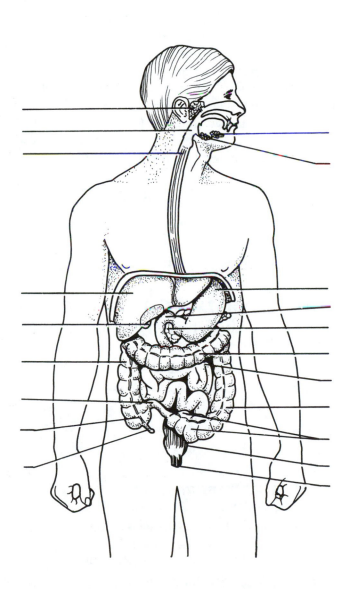

CHAPTER 46

Osmoregulation, Disposal of Metabolic Wastes, and Temperature Regulation

The water content of the animal body, as well as the concentration and distribution of ions in body fluids is carefully regulated. Most animals have excretory systems that function to rid the body of excess water, ions, and metabolic wastes. The excretory system collects fluid from the blood and interstitial fluid, adjusts its composition by reabsorbing from it the substances the body needs, and expels the adjusted excretory product, e.g., urine in humans. The principal metabolic wastes are water, carbon dioxide, and nitrogenous wastes. Excretory systems among invertebrates are diverse and adapted to the body plan and lifestyle of each species. The kidney, with the nephron as its functional unit, is the primary excretory organ in vertebrates. Animals have evolved structural, behavioral, and/or metabolic strategies for maintaining their body temperature within an optimal range. Whereas most animals are dependent on heat from the surrounding environment, birds and mammals obtain most of their body heat from their own metabolic processes.

CHAPTER OUTLINE AND CONCEPT REVIEW
Fill in the blanks.

INTRODUCTIONS

1 Two processes that maintain homeostasis of fluids in the body are _____
_____.

EXCRETORY SYSTEMS HELP MAINTAIN HOMEOSTASIS

THE PRINCIPAL METABOLIC WASTE PRODUCTS ARE WATER, CARBON DIOXIDE, AND NITROGENOUS WASTES

2 Principal metabolic wastes in most animals are water, carbon dioxide, and nitrogenous wastes in the form of _____.

3 The urea cycle is a series of steps that produces urea from the toxic nitrogenous compound
(a)_____ and (b)_____.

4 Many desert animals conserve water by excreting nitrogenous wastes as the semi-solid compound
_____.

INVERTEBRATES HAVE SOLVED PROBLEMS OF OSMOREGULATION AND METABOLIC WASTE DISPOSAL WITH A VARIETY OF MECHANISMS

5 Animals whose body fluids are at equilibrium with their environments are called
(a)_____, in contrast to (b)_____ that have adaptations enabling them to live in hypertonic or hypotonic media.

Nephridial organs are specialized for osmoregulation and/or excretion

6 The nephridial organs of many invertebrates consist of tubes that open to the outside of the body through (a)_____. In flatworms, these excretory organs are called

(b)_____, and in annelids and mollusks they are known as
(c)_____.

Malpighian tubules are an important adaptation for conserving water in insects

7 Malpighian tubules have blind ends that lie in the (a)_____. Wastes are transferred from blood to the tubule by (b)_____.

THE KIDNEY IS THE KEY VERTEBRATE ORGAN OF OSMOREGULATION AND EXCRETION

8 In most vertebrates, the _____ all aid the kidneys in ridding the body of wastes and maintaining fluid balance.

Freshwater vertebrates must rid themselves of excess water

Marine vertebrates must replace lost fluid

The mammalian kidney is important in maintaining homeostasis

THE KIDNEYS, URINARY BLADDER, AND THEIR DUCTS MAKE UP THE HUMAN URINARY SYSTEM

9 The urinary system is the principal excretory system in human beings and other vertebrates. Urine produced in the kidneys is transported to the urinary bladder through two tubes called (a)_____, then it passes from the bladder to the outside through the (b)_____.

The nephron is the functional unit of the kidney

10 The functional units of the kidneys are the nephrons. Each nephron consists of a cup-shaped (a)_____ and a (b)_____. The filtrate passes through structures in the nephron in this sequence:

   (c)_____ ——> proximal convoluted tubule ——>
   (d)_____ ——> distal convoluted tubule ——>
   (e)_____

Urine is produced by filtration, reabsorption, and secretion

11 Urine formation is accomplished by filtration of plasma, reabsorption of needed materials, and secretion of a few substances into the renal tubule. (a)_____ is not a selective process, whereas (b)_____ is highly selective.

12 Plasma is filtered out of the glomerular capillaries and into _____. Needed materials as well as wastes and excess substances become part of the filtrate.

13 Most of the filtrate is reabsorbed from the _____ back into the blood in order to return needed materials to the blood and adjust the composition of the filtrate.

14 _____ is the adjusted filtrate consisting of water, nitrogenous wastes, salts, and other substances.

Urine is concentrated as it passes through the renal tubule

15 There are two types of nephrons in kidneys. The "ordinary variety" in the cortex are called (a)_____ nephrons and those with an exceptionally long loop of Henle that extends deep into the medulla are called (b)_____ nephrons.

16 Filtrate is concentrated as it moves downward through the (a)_____ loop and diluted as it moves upward through the (b)_____ loop. This movement of filtrate in opposite directions, called the (c)_____, helps maintain a hypertonic interstitial fluid that draws water out of the collecting ducts.

17 The _____ are efferent arterioles that collect water from interstitial fluid.

Urine volume is regulated by the hormone ADH

18 Urine volume is regulated by the hormone (a)_____. It is
released by the (b)_____.

19 The "thirst center" that responds to dehydration is located in the _____.

Sodium reabsorption is regulated by the hormone aldosterone

20 Aldosterone secreted by the (a)_____ stimulates distal tubules and collecting
ducts to reabsorb more (b)_____.

Urine is composed of water, nitrogenous wastes, and salts

ANIMALS HAVE OPTIMAL TEMPERATURE RANGES

Ectotherms absorb heat from their surroundings
Endotherms derive heat from metabolic processes

BUILDING WORDS

Use combinations of prefixes and suffixes to build words for the definitions that follow.

Prefixes	The Meaning	Suffixes	The Meaning
juxta-	beside, near	-cyte	cell
podo-	foot		
proto-	first, earliest form of		

Prefix	Suffix	Definition
_____	-nephridium	1. The flame cell excretory organs of lower invertebrates and of some larval higher animals; the earliest form of specialized excretory organ.
_____	_____	2. A specialized epithelial cell possessing elongated foot processes, which cover the surfaces of most of the glomerular capillaries.
_____	-medullary	3. Pertains to nephrons situated nearest the medulla of the kidney.

MATCHING

For each of these definitions, select the correct matching term from the list that follows.

_____ 1. An animal that uses metabolic energy to maintain a constant body temperature despite variations in environmental temperature.

_____ 2. Steroid hormone produced by the vertebrate adrenal cortex which governs the excretion or retention of sodium and potassium ions.

_____ 3. The knot of capillaries at the proximal end of a nephron; enclosed by the Bowman's capsule.

_____ 4. The tube that conducts urine from the bladder to the outside of the body.

_____ 5. The principal nitrogenous excretory product of insects, birds and reptiles; a relatively insoluble end product of protein metabolism.

_____ 6. The active regulation of the osmotic pressure of body fluids.

_____ 7. One of the paired tubular structures that conducts urine from the kidney to the bladder.

_____ 8. The principal nitrogenous excretory product of mammals; one of the water-soluble end products of protein metabolism.

_____ 9. The functional, microscopic unit of the vertebrate kidney.

_____10. A hormone secreted by the posterior lobe of the pituitary which controls the rate of water reabsorption by the kidney.

Terms:

a.   Aldersterone
b.   Antidiuretic hormone
c.   Bowman's capsule
d.   Ectotherm
e.   Endotherm

f.   Glomerulus
g.   Malpighian tubule
h.   Metanephridium
i.   Nephron
j.   Osmoregulation

k.   Reabsorption
l.   Urea
m.   Ureter
n.   Urethra
o.   Uric acid

TABLE
Fill in the blanks.

Organism	Excretory Mechanism/Structure
Marine sponges	#1
Flatworms	#2
Earthworms	#3
Insects	#4
Vertebrates	#5

THOUGHT QUESTIONS
Write your responses to these questions.

1. What adaptive advantages are gained from excretion of uric acid? From urea? From ammonia?
2. Compare the evolutionary adaptations for osmoregulation in freshwater organisms, marine organisms, and terrestrial organisms.
3. Sketch the main components of the mammalian excretory system and describe the function of each part. Include an expanded view of a nephron and associated blood vessels.
4. Describe the events that occur during urine formation in a mammal beginning at the glomerulus and ending in collection in the ureters.

MULTIPLE CHOICE
Place your answer(s) in the space provided. Some questions may have more that one correct answer.

_____ 1. The amount of needed substances that can be reabsorbed from renal tubules is a function of
   a. Tm.
   b. blood pH.
   c. tubular transport maximum.
   d. concentration of urea in forming urine.
   e. renal threshold.

_____ 2. The principal functional unit(s) in the vertebrate kidneys is/are
   a. nephridia.
   b. nephrons.
   c. Bowman's capsules.
   d. antennal glands.
   e. Malpighian tubules.

_____ 3. Water is reabsorbed by interstitial fluids when osmotic concentration in interstitial fluid is increased by
a. salts from filtrate.
d. concentration of filtrate.
b. urea from filtrate.
e. deamination in kidney cells.
c. free ions.

_____ 4. Urea is a principal nitrogenous waste product produced by
a. amphibians.
d. mammals.
b. the kidneys.
e. the liver.
c. nephrons.

_____ 5. Most reabsorption of filtrate in the kidney takes place at the
a. ureter.
d. urethra.
b. loop of Henle.
e. collecting duct.
c. proximal convoluted tubule.

_____ 6. Urea is synthesized
a. from uric acid.
d. in the urea cycle.
b. in kidneys.
e. in aquatic invertebrates.
c. from ammonia and carbon dioxide.

_____ 7. Relative to sea water, fluids in the bodies of marine organisms
a. are isotonic.
d. lose water.
b. are hypotonic.
e. gain water.
c. are hypertonic.

_____ 8. The duct in humans that leads from the urinary bladder to the outside is the
a. ureter.
d. urethra.
b. loop of Henle.
e. collecting duct.
c. proximal convoluted tubule.

_____ 9. Which of the following statements most accurately describes changes in the concentration of filtrate in the two portions of the loop of Henle?
a. decreases in both
d. decreases in descending/increases in ascending
b. increases in both
e. remains essentially the same in both
c. increases in descending/decreases in ascending

_____10. A potato bug would excrete wastes by means of
a. nephridia.
d. green glands.
b. nephrons.
e. Malpighian tubules.
c. a pair of kidney-like structures.

_____11. Relative to fresh water, fluids in the bodies of aquatic organisms
a. are isotonic.
d. lose water.
b. are hypotonic.
e. gain water.
c. are hypertonic.

_____12. Excretion is specifically defined as
a. maintaining water balance.
d. the elimination of undigested wastes.
b. homeostasis.
e. the concentration of nitrogenous products.
c. removal of metabolic wastes from the body.

_____13. The main way(s) that *any* excretory system maintains homeostasis in the body is/are to
a. excrete metabolic wastes.
d. regulate salt and water.
b. eliminate undigested food.
e. concentrate urea in urine.
c. regulate body fluid constituents.

_____14. The group(s) of animals that have no specialized excretory systems include
    a. insects.   d. sponges.
    b. cnidarians.   e. sharks and rays.
    c. annelids.

_____15. The principal forms of nitrogenous waste products in various animal groups include(s)
    a. uric acid.   d. ammonia.
    b. carbon dioxide.   e. urea.
    c. amino acids.

_____16. The principal nitrogenous waste product(s) in human urine is/are
    a. uric acid.   d. ammonia.
    b. carbon dioxide.   e. urea.
    c. amino acids.

_____17. The first step in the catabolism of amino acids
    a. is conversion of ammonia to uric acid.   d. is removal of the amino group.
    b. is deamination.   e. is accomplished with peptidases.
    c. produces ammonia.

_____18. The type(s) of excretory organ(s) found in animals that collect wastes in flame cells is/are
    a. nephridia.   d. green glands.
    b. nephrons.   e. Malpighian tubules.
    c. branching tubes that open to the outside through pores.

_____19. Conservation of body fluids in a monarch butterfly is accomplished by
    a. ingestion of large volumes of fluids.   d. reabsorption of water into the hemocoel.
    b. concentrating urea in wastes.   e. reabsorption of water from excretory tubules.
    c. reabsorption of water from the digestive tract.

VISUAL FOUNDATIONS
Color the parts of the illustration below as indicated.

RED ☐ abdominal aorta
GREEN ☐ adrenal gland
YELLOW ☐ ureter and renal pelvis
BLUE ☐ inferior vena cava
ORANGE ☐ urethra
BROWN ☐ kidney
TAN ☐ urinary bladder
PINK ☐ renal artery
VIOLET ☐ renal vein

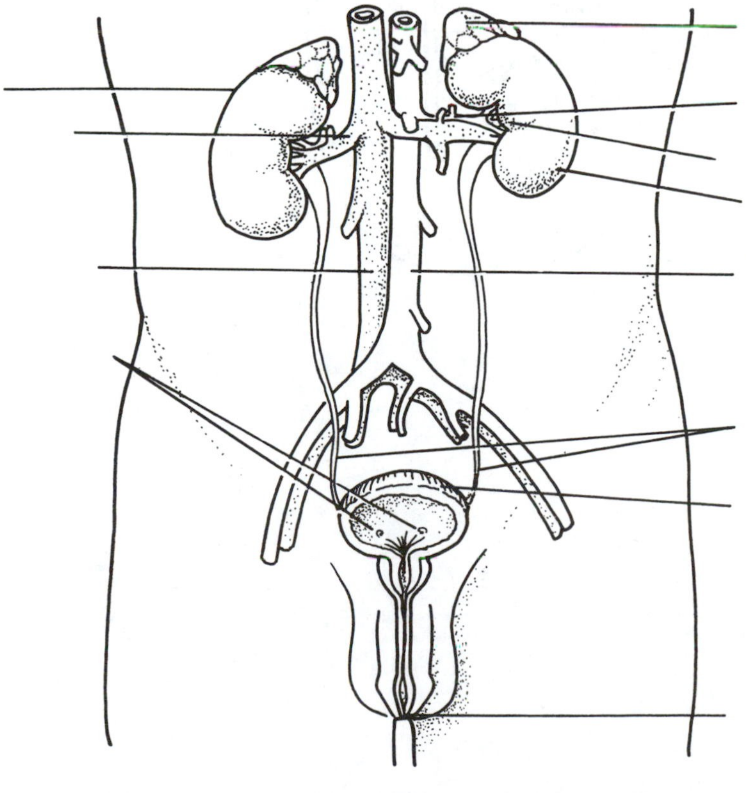

Color the parts of the illustration below as indicated.

RED	☐	artery
GREEN	☐	renal pelvis
YELLOW	☐	collecting tubule
BLUE	☐	vein
ORANGE	☐	loop of Henle
BROWN	☐	medulla
TAN	☐	cortex
PINK	☐	glomerulus
VIOLET	☐	proximal convoluted tubule and distal convoluted tubule

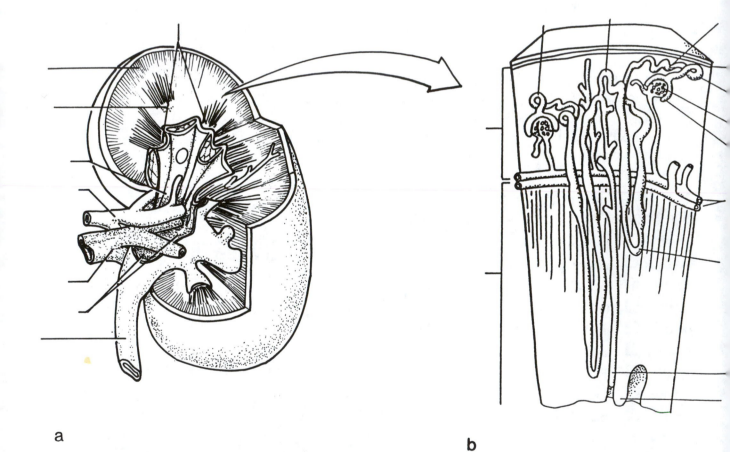

a

b

CHAPTER 47

Endocrine Regulation

The endocrine system is a diverse collection of glands and tissues that secrete hormones — chemical messengers that are transported by the blood to target tissue where they stimulate a physiological change. The endocrine system works closely with the nervous system to maintain the steady state of the body. Hormones may be steroids, peptides, proteins, or derivatives of amino acids or fatty acids. They diffuse from the blood into the interstitial fluid and then combine with receptor molecules on or in the cells of the target tissue. Some hormones activate genes that lead to the synthesis of specific proteins. Others activate a second messenger that relays the hormonal message to the appropriate site within the cell. Most invertebrate hormones are secreted by neurons rather than endocrine glands. They regulate growth, metabolism, reproduction, molting, and pigmentation. In vertebrates, hormonal activity is controlled by the hypothalamus, which links the nervous and endocrine systems. Vertebrate hormones regulate growth, reproduction, salt and fluid balance, and many aspects of metabolism. The hypothalamus secretes hormones that target the pituitary gland. The pituitary, in turn, secretes hormones that regulate a variety of body functions. Hormones produced by the thyroid gland regulate metabolic rate. The parathyroid glands regulate the calcium level in the blood. The islets of the pancreas secrete insulin, which acts to lower the glucose level in the blood, and glucagon, which acts to elevate it. The adrenal glands secrete hormones that help the body cope with stress. Malfunction of any of the endocrine glands can lead to specific disorders.

CHAPTER OUTLINE AND CONCEPT REVIEW
Fill in the blanks.

INTRODUCTION

1 The endocrine system is a collection of ductless (a)_____ that works closely with the (b)_____ system.

2 Hormones are transported throughout the body by blood, however, they affect only specific areas known as _____.

3 Hormones secreted by neurons are (a)_____ and the cells that secrete them are called (b)_____ cells.

HORMONES CAN BE ASSIGNED TO FOUR CHEMICAL GROUPS

4 The four chemical groups of hormones _____
_____.

5 In vertebrates, testes and ovaries secrete steroids synthesized from (a)_____.
Steroids are classified as (b)_____ molecules.

HORMONE SECRETION IS REGULATED BY NEGATIVE FEEDBACK MECHANISMS

HORMONES COMBINE WITH SPECIFIC RECEPTOR PROTEINS IN TARGET CELLS

Some hormones enter the cell and activate genes

6 Steroid hormones pass through the plasma membrane of a target cell and combine with specific receptors in the cell. These hormone-receptor complexes combine with (a)_____ in the nucleus, which activates certain genes to synthesize (b)_____ that in turn codes for specific proteins.

Some hormones work through second messengers

7 Many protein hormones combine with receptors in the cell membrane of the target cell and act by way of a second messenger. Two common second messengers are (a)_____, which is derived from ATP, and (b)_____, which, for example, takes a role in disassembling microtubules.

8 Second messengers act by altering the functions of _____.

Prostaglandins are local chemical mediators

9 Prostaglandins may help regulate hormone action by regulating the formation of cyclic AMP. They are sometimes called _____ because they usually act only on nearby cells.

INVERTEBRATE HORMONES REGULATE GROWTH, DEVELOPMENT, METABOLISM, REPRODUCTION, MOLTING, AND PIGMENTATION

10 Most invertebrate hormones are secreted by _____ rather than by endocrine glands. They help to regulate regeneration, molting, metamorphosis, reproduction, and metabolism.

Color change in crustaceans is regulated by hormones
Insect development is regulated by hormones

11 Hormones control development in insects. Neurosecretory cells in the brain of insects produce the "brain hormone" called (a)_____, which stimulates (b)_____ glands to produce the molting hormone (c)_____.

VERTEBRATE HORMONES REGULATE GROWTH, DEVELOPMENT, FLUID BALANCE, METABOLISM, AND REPRODUCTION

Endocrine disorders may involve too little or too much hormone
The hypothalamus integrates neural and endocrine regulation

12 The hypothalamus regulates the anterior lobe of the pituitary gland by producing several _____ hormones that regulate secretion of pituitary hormones.

The posterior lobe of the pituitary gland releases two hormones

13 The hypothalamus produces (a)_____, which causes collecting ducts in the kidney to reabsorb more water. It also produces (b)_____, which stimulates the breasts of nursing mothers. Although produced in the hypothalamus, these hormones are released by the (c)_____.

The anterior lobe of the pituitary gland regulates growth and other endocrine glands

14 _____ hormones stimulate other endocrine glands.

15 _____ stimulates mammary glands to produce milk.

16 Among hormones that influence growth are growth hormone (GH), also called (a)_____, secreted by the (b)_____, and thyroid hormones, secreted by the thyroid gland. The fundamental function of both GH and the thyroid hormones at the molecular level is to promote (c)_____.

17 Release of growth hormone is regulated by a negative feedback mechanism that signals the (a)_____. When GH blood titers are high, this structure secretes (b)_____, which slows the release of GH

from the pituitary. When GH titers are low, (c)_____
is secreted, so the pituitary releases more GH.

18 Hypersecretion of GH during childhood may cause a disorder known as (a)_____.
In adulthood, hypersecretion can cause "large extremities," or (b)_____.

Thyroid hormones increase metabolic rate

19 One thyroid hormone, (a)_____, is also known as T_4 because each molecule contains
four atoms of (b)_____.

20 Thyroid secretion is regulated by a negative feedback system between the thyroid gland and the
(a)_____ gland, which releases more (b)_____
_____ when thyroid hormone blood titers are too low.

21 Hyposecretion of thyroid hormones during infancy and childhood may result in retarded
development, a condition known as _____.

The parathyroid glands regulate calcium concentration

22 Parathyroid hormone regulates calcium levels in body fluids by stimulating release of calcium from
(a)_____ and calcium reabsorption by the (b)_____. A negative
feedback system induces the (c)_____ gland to secret (d)_____, which
inhibits parathyroid hormone activity.

The islets of the pancreas regulate glucose concentration

23 Glucose concentration in the blood is regulated mainly by "islet hormones." Numerous clusters of
cells in the pancreas, named (a)_____, secrete hormones that regulate
glucose concentration. When blood glucose levels are high, (b)_____ cells release the hormone
(c)_____; when it is low, (d)_____ cells release (e)_____.

24 Insulin lowers the concentration of glucose in the blood by stimulating uptake of glucose by
(a)_____; and by promoting
(b)_____, the formation of glycogen.

25 In the endocrine metabolism disorder known as _____, cells are unable
to utilize glucose properly and turn to fat and protein for fuel.

The adrenal glands help the body adapt to stress

26 The adrenal medulla and the adrenal cortex secrete hormones that help the body cope with stress.
The adrenal medulla secretes the two hormones _____
_____ (also known as adrenaline and noradrenaline) which increase heart rate,
metabolic rate, and strength of muscle contraction, and reroute the blood to organs that require more
blood in time of stress.

27 The adrenal cortex produces three types of hormones in appreciable amounts.
(a)_____. Of
these, the (b)_____ known as (c)_____ is converted
to the masculinizing hormone testosterone.

28 Kidneys reabsorb more sodium and excrete more potassium in response to the mineralocorticoid
hormone _____.

29 The main function of _____ is to enhance gluconeogenesis in
the liver.

30 Stress stimulates secretion of CRF, which is the abbreviation for (a)_____
_____, from the hypothalamus. CRF stimulates the
anterior pituitary to release (b)_____, which regulates the
secretion of glucocorticoids and aldosterones.

Many other hormones are known

31 The thymus gland produces the hormone _____.

32 _____ from the kidneys helps regulate blood pressure.

33 The heart secretion ANF, which stands for _____, lowers blood pressure.

34 The (a)_____ gland in the brain produces the hormone (b)_____ which influences the onset of sexual maturation.

BUILDING WORDS

Use combinations of prefixes and suffixes to build words for the definitions that follow.

Prefixes	The Meaning		Suffixes	The Meaning
hyper-	over		-megaly	enlargement
hypo-	under			
neuro-	nerve			

Prefix	Suffix	Definition
_____	-hormone	1. A hormone secreted by a nerve cell.
_____	-secretory	2. Refers to a nerve cell that secretes a neurohormone.
_____	-secretion	3. An excessive secretion; over secretion.
_____	-secretion	4. A diminished secretion; under secretion.
_____	-thyroidism	5. A condition resulting from an overactive thyroid gland.
_____	-glycemia	6. An abnormally low level of glucose in the blood.
_____	-glycemia	7. An abnormally high level of glucose in the blood.
acro-	_____	8. An abnormal condition characterized by enlargement of the head, and sometimes other structures.

MATCHING

For each of these definitions, select the correct matching term from the list that follows.

_____ 1. A hormone produced by the hypothalamus and released by the posterior lobe of the pituitary; causes the uterus to contract and stimulates the release of milk from the mammary glands.

_____ 2. A hormone secreted by the thyroid gland that rapidly lowers the calcium content in the blood.

_____ 3. An endocrine gland that lies anterior to the trachea and releases hormones that regulate the rate of metabolism.

_____ 4. Paired endocrine glands, each located just superior to each kidney.

_____ 5. General term for an organic chemical produced in one part of the body and transported to another part where it affects some aspect of metabolism.

_____ 6. Cells of the pancreas that secrete insulin.

_____ 7. A gland that secretes products directly into the blood or tissue fluid instead of into ducts.

_____ 8. General term for a hormone that helps regulate another endocrine gland.

_____ 9. The principal hormone of the thyroid gland.

_____10. An endocrine gland that produces thymosin; important in the development of the immune response mechanism.

Terms:

a. Adrenal gland
b. Aldosterone
c. Beta cells
d. Calcitonin
e. Endocrine gland

f. Glucagon
g. Hormone
h. Insulin
i. Oxytocin
j. Prostaglandin

k. Thymus
l. Thyroid gland
m. Thyroxine
n. Tropic hormone

TABLE
Fill in the blanks.

Hormone	Chemical Group	Function
Testosterone	Steroid	Develops and maintains sex characteristics of males; promotes spermatogenesis
Aldosterone	#1	#2
Thyroxine	#3	#4
#5	Peptide or protein	Stimulates reabsorption of water; conserves water
#6 s	#7	Helps body adapt to long-term stress; mobilizes fat; raises blood glucose levels
#8	#9	Affects wide range of body processes; may interact with other hormones to regulate metabolic activities
ACTH	Peptide or protein	#10
Epinephrine	#11	#12
#13	#14	Stimulates uterine contraction

THOUGHT QUESTIONS
Write your responses to these questions.

1. Outline the similarities and differences in the mode of action among steroid and protein hormones.
2. Create a table that shows each of the major vertebrate hormones, the endocrine glands that produce them and their locations in the body, and the target organ(s) and function(s) of each.
3. Explain positive and negative feedback and give examples of each.
4. Describe relationships between the nervous system and the endocrine system. What serves as the interface between these systems and how does it function to serve that purpose?
5. Describe the actions of each of the following: posterior and anterior pituitary hormones, thyroid hormones, growth hormones, insulin and glucagon, adrenal gland secretions.

MULTIPLE CHOICE

Place your answer(s) in the space provided. Some questions may have more that one correct answer.

____ 1. A person with a fasting level of 750 mg glucose per 100 ml of blood

    a. is hyperglycemic.
    b. is hypoglycemic.
    c. is about normal.
    d. probably has cells that are not using enough glucose.
    e. has too much insulin.

____ 2. Diabetes in an adult with ample numbers of adequately functioning beta cells

    a. is type I diabetes.
    b. is type II diabetes.
    c. indicates that target cells are not using insulin.
    d. indicates that not enough insulin is being produced.
    e. is unusual since this condition usually occurs in children.

____ 3. Control of hormone secretions by a negative feedback mechanism generally includes

    a. hyperactivity.
    b. hypoactivity.
    c. change in a direction that tends to maintain homeostasis.
    d. increased hormone secretion with increase in hormone effect.
    e. decreased hormone secretion with increase in hormone effect.

____ 4. A three-year old cretin and an adult suffering from myxedema most likely have

    a. a low metabolic rate.
    b. Cushing's disease.
    c. Addison's disease.
    d. a low titer of thyroid hormones.
    e. a high titer of thyroid hormones.

____ 5 Small, hydrophobic hormones that form a hormone-receptor complex with intranuclear receptors include

    a. steroids.
    b. cyclic AMP.
    c. chemicals that activate genes.
    d. thyroid hormones.
    e. prostaglandins.

____ 6 A chemical produced by one cell that has a specific regulatory effect on another cell is the definition of a

    a. neurohormone.
    b. hormone.
    c. exocrine gland secretion.
    d. pheromone.
    e. prohormone.

____ 7 The hormone and target tissue that are involved in raising glucose concentration in blood by glycogenolysis and gluconeogenesis are

    a. insulin and pancreas.
    b. glucagon and liver.
    c. thyroid-stimulating hormone and thyroid gland.
    d. adrenocorticotropic hormone and adrenal cortex.
    e. thyroxine and various metabolically active cells.

____ 8 Increased skeletal growth results directly and/or indirectly from activity of

    a. aldosterone.
    b. growth hormone.
    c. hormones from the hypothalamus.
    d. somatomedins.
    e. epinephrine and/or norepinephrine.

____ 9 Activity of the anterior lobe of the pituitary gland is controlled by

    a. the hypothalamus.
    b. ADCH.
    c. epinephrine and norepinephrine.
    d. releasing hormones.
    e. release-inhibiting hormones.

VISUAL FOUNDATIONS

Color the parts of the illustration below as indicated.

RED ❏ receptor molecule

GREEN ❏ ribosome

YELLOW ❏ hormone

BLUE ❏ DNA

ORANGE ❏ mRNA

BROWN ❏ endocrine gland cell

TAN ❏ nucleus

PINK ❏ blood vessel

VIOLET ❏ protein molecule

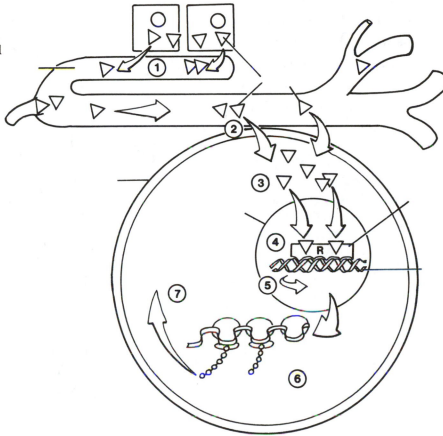

Color the parts of the illustration below as indicated.

RED ❏ receptor

GREEN ❏ cytoplasm

YELLOW ❏ hormone

BLUE ❏ extracellular fluid

ORANGE ❏ second messenger

BROWN ❏ plasma membrane

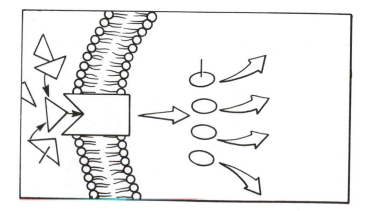

Color the parts of the illustration below as indicated.

RED ☐ pituitary gland
GREEN ☐ pineal gland
YELLOW ☐ thyroid gland
BLUE ☐ parathyroid gland
ORANGE ☐ hypothalamus
BROWN ☐ thymus gland
TAN ☐ pancreas
PINK ☐ testis and ovary
VIOLET ☐ adrenal gland

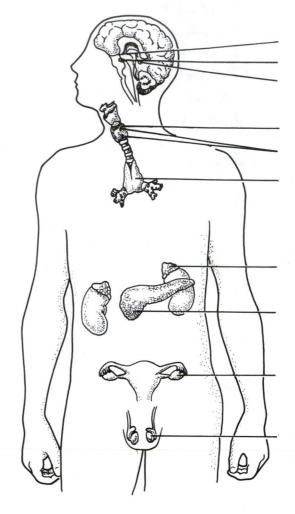

❑

Reproduction

For a species to survive, it is essential that it replace individuals that die. Animal species do this either asexually or, more commonly, sexually. In asexual reproduction, a single parent splits, buds, or fragments, giving rise to two or more offspring that are genetically identical to the parent. This method of reproduction is advantageous for sessile animals that cannot move about to search for mates. In sexual reproduction, sperm contributed by a male parent, and an egg contributed by a female parent, unite, forming a fertilized egg, or zygote, that develops into a new organism. This method of reproduction has the advantage of promoting genetic variety among the offspring, and therefore, giving rise to individuals that may be better able to survive than either parent. Depending on the species, fertilization occurs either outside the body (most aquatic animals) or inside the body (most terrestrial animals). Some animals have both asexual and sexual stages. In some, the unfertilized egg may develop into an adult animal. In still others, both male and female reproductive organs may occur in the same individual. More typically, animals have only a sexual stage, fertilization is required for development to occur, and male and female reproductive organs occur only in separate individuals. The complex structural, functional, and behavioral processes involved in reproduction in vertebrates are regulated by hormones secreted by the brain and gonads. Sexual stimulation in humans results in increased blood flow to reproductive structures and the skin, and increased muscle tension. Sexual response includes sexual desire, excitement, orgasm, and resolution. There are a variety of effective methods of birth control. Next to the common cold, sexually transmitted diseases are the most prevalent communicable diseases in the world.

CHAPTER OUTLINE AND CONCEPT REVIEW
Fill in the blanks.

INTRODUCTION

1 In sexual reproduction, each of two parents contributes a sex cell or _____ containing half of the offspring's genetic endowment.

ASEXUAL REPRODUCTION IS COMMON AMONG SOME ANIMAL GROUPS

2 In asexual reproduction, a single parent endows its offspring with a set of genes identical with its own. Sponges and cnidarians can reproduce by _____, involving the production of a new individual from a group of expendable cells that separate from the "parent's" body.

3 _____ is a form of asexual reproduction whereby a "parent" breaks into several pieces, each piece giving rise to a new individual.

4 Parthenogenesis refers to the production of a complete organism from a(n) _____.

SEXUAL REPRODUCTION IS THE MOST COMMON TYPE OF ANIMAL REPRODUCTION

5 Most hermaphroditic organisms copulate and fertilize each other, however, a few, such as the _____, for example, are capable of self fertilization.

HUMAN REPRODUCTION: THE MALE PROVIDES SPERM

The testes produce sperm

6 The testes, housed in an external sac called the (a)_____, contain the
   (b)_____, where sperm are produced.

7 The initial development of a sperm begins with an undifferentiated stem cell called a
   (a)_____, which enlarges to form the (b)_____
   _____ that goes through meiosis.

8 The _____ at the front of a sperm contains enzymes that assist the sperm in penetrating
   an egg.

9 The _____ are the passage ways that testes take as they descend through the
   abdominal wall into the scrotum.

A series of ducts store and transport sperm

10 Sperm complete their maturation and are stored in the (a)_____, from which
    they are moved up into the body through the (b)_____.

11 During ejaculation, sperm pass through the (a)_____ into the
    (b)_____, which passes through the penis.

The accessory glands produce the fluid portion of semen

12 Human semen contains about (a)_____(#?) sperm suspended in secretions that are
    mostly produced by two glands called the (b)_____.

The penis transfers sperm to the female

13 The penis consists of an elongated (a)_____ that terminates in an expanded portion called the
    (b)_____, which is partially covered by a fold of skin called the
    (c)_____. An operation that removes this "extra" skin is known as
    (c)_____.

14 The penis contains three columns of erectile tissue referred to as (a)_____
    _____. When they become engorged with (b)_____, the penis erects.

Reproductive hormones promote sperm production and maintain masculinity

15 The two gonadotropic hormones (a)_____, stimulate sperm production and secretion of the
    hormone (b)_____, which is responsible for establishing and maintaining primary
    and secondary sex characteristics in the male.

16 The male hormone (a)_____ is produced by the
    (b)_____ cells in the testes.

HUMAN REPRODUCTION: THE FEMALE PRODUCES OVA AND INCUBATES THE EMBRYO

The ovaries produce ova and sex hormones

17 Formation of ova is called (a)_____. In this process, stem cells in the fetus, the
    (b)_____, enlarge, forming the (c)_____, which begin meiosis
    before birth.

18 A developing ovum and the cells immediately around it comprise the (a)_____. At
    puberty, the hormone (b)_____ stimulates development of these cells, causing some of them
    to complete meiosis. One of the resulting cells, the (c)_____, continues
    development to form the ovum.

19 After ovulation, the portion of the follicle that remains in the ovary develops into an endocrine gland
    called the _____.

The oviducts transport the secondary oocyte

20 After ovulation, the ovum enters the _____, where it
may be fertilized.

The uterus incubates the embryo

21 Embryos normally implant in the (a)_____, which is the inner layer of cells of
the uterus. This signals the beginning of pregnancy. If there is no embryo, or the embryo fails to
implant, the bleeding phase known as (b)_____ commences.

22 The lower portion of the uterus, the _____, forms the back wall of the vaginal tube.

The vagina receives sperm
The vulva are external genital structures

23 The structures that comprise the vulva are the _____
_____.

The breasts function in lactation

24 (a)_____ is the production of milk. It begins as a result of stimulation by the
hormone (b)_____ secreted by the (c)_____.

**Reproductive hormones maintain female characteristics and regulate the menstrual
cycle**

25 The first day of menstrual bleeding marks the first day of the menstrual cycle. _____
occurs at about day 14 in an average, typical 28-day menstrual cycle.

26 FSH stimulates follicle development. The developing follicles release (a)_____,
which stimulates development of the endometrium. The two hormones (b)_____
stimulate ovulation, and (c)_____ also promotes development of the corpus luteum.

27 The corpus luteum secretes _____, which stimulates final preparation of the
uterus for possible pregnancy.

SEXUAL RESPONSE INVOLVES PHYSIOLOGICAL CHANGES

28 _____ are two basic
physiological responses to sexual stimulation.

29 The cycle of sexual response includes four phases: the _____
_____.

FERTILIZATION IS THE FUSION OF SPERM AND EGG TO PRODUCE A ZYGOTE

30 Male infertility is usually the result of _____, which means lack of or insufficient
quantities of viable sperm.

BIRTH CONTROL METHODS ALLOW INDIVIDUALS TO CHOOSE

Most hormone contraceptives prevent ovulation

31 Most oral contraceptives are combinations of the synthesized hormones _____
_____.

Use of the intrauterine device (IUD) has declined
Other common contraceptive methods include the diaphragm and condom

32 The condom is the only contraceptive device that affords some protection against _____
_____.

Sterilization renders an individual incapable of producing offspring

33 Male sterilization is by (a)_____ and female sterilization is by (b)_____.

There are three types of abortion

SEXUALLY TRANSMITTED DISEASES ARE SPREAD BY SEXUAL CONTACT

BUILDING WORDS

Use combinations of prefixes and suffixes to build words for the definitions that follow.

Prefixes	The Meaning		Suffixes	The Meaning
acro-	extremity, height		-ectomy	excision
circum-	around, about		-gen(esis)	production of
contra-	against, opposite, opposing		-some	body
endo-	within			
intra-	within			
oo-	egg			
partheno-	virgin			
post-	behind, after			
pre-	before, prior to, in advance of, early			
spermato-	seed, "sperm"			
vaso-	vessel			

Prefix	Suffix	Definition
_____	_____	1. The production of an adult organism from an egg in the absence of fertilization; virgin development.
_____	_____	2. The production of sperm.
_____	_____	3. The organelle (body) at the extreme tip of the sperm head that helps the sperm penetrate the egg.
_____	-cision	4. The removal of all or part of the foreskin by cutting all the way around the penis.
_____	_____	5. The production of eggs or ova.
_____	-metrium	6. The lining within the uterus.
_____	-ovulatory	7. Pertains to a period of time after ovulation.
_____	-menstrual	8. Pertains to a period of time prior to menstruation.
vas-	_____	9. Male sterilization in which the vas deferentia are partially excised.
_____	-congestion	10. The engorgement of vessels or sinusoids with blood.
_____	-uterine	11. Located or occurring within the uterus.
_____	-ception	12. Birth control; techniques or devices opposing conception.

MATCHING

For each of these definitions, select the correct matching term from the list that follows.

_____ 1. Fluid composed of sperm suspended in various glandular secretions that is ejaculated from the penis during orgasm.

_____ 2. A hormone produced by the corpus luteum of the ovary and by the placenta; acts with estradiol to regulate menstrual cycles and to maintain pregnancy.

_____ 3. The largest accessory sex gland of male mammals; secretes a large portion of the seminal fluid.

_____ 4. A small, erectile structure at the anterior part of the vulva in female mammals; homologous to the male penis.

_____ 5. One of the paired female gonads; responsible for producing eggs and sex hormones.

_____ 6. Pocket of endocrine tissue in the ovary that is derived from the follicle cells; secretes the hormone progesterone.

_____ 7. The fusion of the male and female gametes; results in the formation of a zygote.

_____ 8. A sudden expulsion, as in the ejection of semen from the penis.

_____ 9. The release of a mature "egg" from the ovary.

_____ 10. The external sac of skin found in most male mammals that contains the testes and their accessory organs.

Terms:

a. Budding
b. Clitoris
c. Corpus luteum
d. Ejaculation
e. Fertilization

f. Fragmentation
g. Ovary
h. Oviduct
i. Ovulation
j. Progesterone

k. Prostate gland
l. Scrotum
m. Semen
n. Testosterone
o. Zygote

TABLE
Fill in the blanks.

Endocrine Gland	Hormone	Target Tissue	Action in Male	Action in Female
Hypothalamus	GnRH	Anterior pituitary	Stimulates release of FSH and LH	Stimulates release of FSH and LH
#1	FSH	#2	#3	#4
Anterior pituitary	LH	Gonad	#5	#6
#7	#8	General	Establishes and maintains primary & secondary sex characteristics	Not produced
#9	#10	General	Not produced	Establishes and maintains primary & secondary sex characteristics

THOUGHT QUESTIONS
Write your responses to these questions.

1. Define sexual reproduction in broad biological terms. Can you refine your definition to the point of eliminating exceptions?

2. Sketch the vulva of a human and describe the functions of each part.

3. Sketch the reproductive system of a human female and describe the functions of each part.

4. Sketch the external genitalia of a human male and describe the functions of each part.

5. Sketch the reproductive system of a human male and describe the functions of each part.

6. Describe the events of the human menstrual cycle and the hormones that regulate each event.

7. Describe the method of use, the mode of action, and effectiveness of birth control methods.

8. Identify the following for each of the principal sexually transmitted infections: method(s) of contraction and prevention, causative agent, symptoms, short and long term effects, and treatment(s).

MULTIPLE CHOICE

Place your answer(s) in the space provided. Some questions may have more that one correct answer.

____ 1. The tube(s) that transport(s) gametes away from gonads is/are the
   a. vas deferens.
   b. seminal vesicle.
   c. sperm cord.
   d. vagina.
   e. oviduct.

____ 2. The normal, periodic flow of blood from the vagina is associated with
   a. ovulation.
   b. orgasm.
   c. sloughing of the endometrium.
   d. removal of the inner lining of the vagina.
   e. miscarriages.

____ 3. Human female breast size is primarily a function of the amount of
   a. alveoli.
   b. milk glands.
   c. fat.
   d. milk the woman can produce after giving birth.
   e. colostrum.

____ 4. Which of the following is/are true of the human penis?
   a. It has three erectile tissue bodies.
   b. Circumcision involves removal of the prepuce.
   c. The glans is an expanded portion of one of the cavernous bodies.
   d. It has a small bone that articulates with the pelvis.
   e. Erection results from increased blood pressure.

____ 5. A human female at birth already has
   a. primary oocytes in the first prophase.
   b. preovulatory Graafian follicles.
   c. all the formative oogonia she will ever have.
   d. some secondary oocytes in meiosis II.
   e. zona pellucida.

____ 6. The *basic* physiological responses that result from effective sexual stimulation are
   a. vasocongestion.
   b. erection.
   c. vaginal lubrication.
   d. muscle tension.
   e. orgasm.

____ 7. Human chorionic gonadotropin (hCG)
   a. is produced by embryonic membranes.
   b. is secreted by the endometrium.
   c. affects functioning of the corpus luteum.
   d. is originally produced in the pituitary.
   e. causes menstrual flow to begin.

____ 8. The hormone(s) primarily responsible for secondary sexual characteristics is/are
   a. luteinizing hormone.
   b. estrogen.
   c. follicle-stimulating hormone.
   d. testosterone.
   e. human chorionic gonadotropin.

____ 9. The corpus luteum is directly or indirectly involved in
   a. producing estradiol.
   b. producing progesterone.
   c. stimulating glands to produce nutrients for an implanted embryo.
   d. responding to signals from embryonic membrane secretions.
   e. maintaining a newly implanted embryo.

____10. Follicle-stimulating hormone (FSH) stimulates development of
    a. seminiferous tubules.   d. the corpus luteum.
    b. interstitial cells.   e. ovarian follicles.
    c. sperm cells.

____11. The refractory period typically occurs
    a. in men, not women.   d. in the orgasmic phase.
    b. in both men and women.   e. in the resolution phase.
    c. in women, not men.

____12. When used properly, the most effective form of birth control in a fertile person is
    a. condoms.   d. the rhythm method.
    b. douching.   e. spermicidal sponges.
    c. oral contraception.

____13. The first measurable response to effective sexual stimulation
    a. is orgasm.   d. is penile erection.
    b. is ejaculation.   e. is vaginal lubrication.
    c. occurs in the excitement phase.

____14. The form(s) of birth control that can definitely deter contraction of sexually transmitted diseases is/are
    a. condoms.   d. diaphragms.
    b. douching.   e. spermicides.
    c. oral contraception.

____15. A count of 400 million sperm in a single ejaculation would be considered
    a. clinical sterility.   d. about average.
    b. above average.   e. impossible.
    c. below average.

____16. A Papanicolaou test (Pap smear)
    a. can detect early cancer.   d. looks at cells from the menstrual flow.
    b. requires an abdominal incision.   e. examines cells scraped from the cervix.
    c. involves examination of cells taken from the lower portion of the uterus.

____17. Primitive, undifferentiated stem cells that line the seminiferous tubules are called
    a. sperm.   d. secondary spermatocytes.
    b. spermatids.   e. spermatogonia.
    c. primary spermatocytes.

____18. The external genitalia of the human female are collectively known as the
    a. vagina.   d. major and minor lips.
    b. clitoris.   e. vulva.
    c. mons pubis.

____19. Which of the following is/are likely true of PMS?
    a. It causes fatigue.   d. It probably has a psychological basis.
    b. It causes anxiety.   e. Its symptoms usually occur just after the woman's period.
    c. The cause may be due to estradiol-progesterone imbalance.

____20. In gametogenesis in the human male, the cell that develops after the first meiotic division is the
    a. sperm.   d. secondary spermatocyte.
    b. spermatid.   e. spermatogonium.
    c. primary spermatocyte.

____21. The portion of the ovarian follicle that remains behind after ovulation
   a. develops into the corpus luteum.   d. is surrounded by the zona pellucida.
   b. becomes a gland.   e. transforms into the endometrium.
   c. is called the Graafian follicle.

____22. The term gonad refers to the
   a. penis.   d. ovaries.
   b. testes.   e. gametes.
   c. vagina.

____23. Human sperm cell production occurs in the
   a. epididymis.   d. accessory glands.
   b. sperm tube.   e. seminiferous tubules.
   c. testes.

VISUAL FOUNDATIONS
Color the parts of the illustrations below as indicated. Label mitosis, first meiotic division, and second meiotic division.

RED   ☐   sperm
PINK   ☐   large chromosome
YELLOW   ☐   spermatogonia
BLUE   ☐   small chromosome
ORANGE   ☐   primary spermatocyte
BROWN   ☐   secondary spermatocyte
TAN   ☐   spermatids

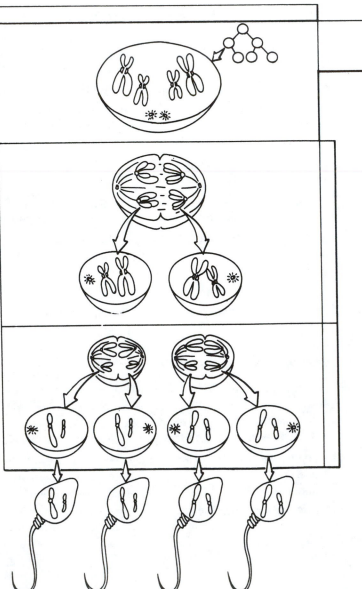

Color the parts of the illustrations below as indicated. Label mitosis, first meiotic division, and second meiotic division.

RED ☐ ovum

GREEN ☐ large chromosome

YELLOW ☐ oogonia

BLUE ☐ small chromosome

ORANGE ☐ primary oocyte

BROWN ☐ secondary oocyte

TAN ☐ polar body

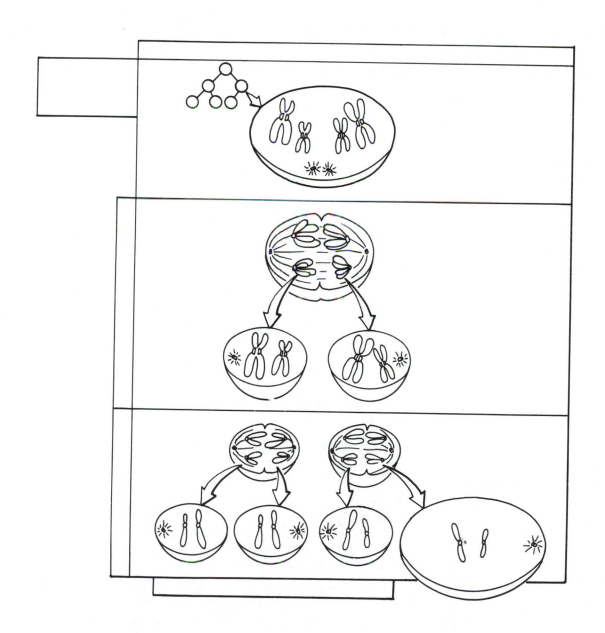

CHAPTER 49

□

Development

Development encompasses all of the changes that take place during the life of an organism from fertilization to death. Guided by instructions encoded in the DNA of the genes, the organism develops from one cell to billions, from a formless mass of cells to an intricate, highly specialized and organized organism. Development is a balanced combination of several interrelated processes: (1) cell division, (2) increase in the number and size of cells, (3) cellular movements that arrange cells into specific structures and appropriate body forms, and (4) biochemical and structural specialization of cells to perform specific tasks. The stages of early development, which are basically similar for all animals, include fertilization, cleavage, gastrulation, and organogenesis. All terrestrial vertebrates have four extraembryonic membranes that function to protect and nourish the embryo. Environmental factors, such as nutrition, vitamin and drug intake, cigarette smoke, and disease-causing organisms, can adversely influence developmental processes. Aging is a developmental process that results in decreased functional capacities of the mature organism.

CHAPTER OUTLINE AND CONCEPT REVIEW
Fill in the blanks.

INTRODUCTION

1 During development, the number of cells (a)_____(increases or decreases?), resulting in growth. Cells undergo (b)_____, a process during which they organize into specific structures. The process by which cells become specialized is called (c)_____.

FERTILIZATION RESTORES THE DIPLOID NUMBER OF CHROMOSOMES

2 Fertilization restores the (a)_____ number of chromosomes, during which time the (b)_____ of the individual is determined.

The first step in fertilization involves contact and recognition

3 The jelly coat, or (a)_____, in mammals is contacted by the sperm, causing the (b)_____ to release enzymes that hydrolyze the jelly coat down to the (c)_____ membrane.

4 _____, a species-specific protein of the acrosome, adheres to a receptor on the vitelline membrane.

Sperm entry is regulated

5 Microvilli on the egg elongate, forming the (a)_____, which contracts, thereby drawing the (b)_____ into the egg.

Sperm and egg pronuclei fuse

6 _____ draw the sperm nucleus toward the egg nucleus.

Fertilization activates the egg

7 The increase in (a)_____ ions within egg cytoplasm is associated with the (b)_____ reaction. It blocks polyspermy and triggers metabolic activity in the egg.

DURING CLEAVAGE THE ZYGOTE DIVIDES, GIVING RISE TO MANY CELLS

8 The zygote undergoes _____, forming a morula and then a blastula.

Cleavage provides building blocks for development
The amount of yolk determines the pattern of cleavage

9 The isolecithal eggs of (a)_____ undergo
(b)_____ cleavage, that is, the entire egg divides into cells of about the same size.
(c)_____ cleavage of isolecithal eggs is common in deuterostomes, while some protostomes
undergo (d)_____ cleavage.

10 The (a)_____ cleavage that occurs in the telolecithal eggs of reptiles and birds is
restricted to the (b)_____.

THE GERM LAYERS FORM DURING GASTRULATION

11 The blastula becomes a three-layered embryo during (a)_____; these layers,
collectively called (b)_____, include the _____.

Each germ layer has a specific fate

12 (a)_____ gives rise to the lining of the gut, the lungs, and the liver. (b)_____
forms epidermis and the nervous system. The (c)_____ gives rise to the skeleton, muscle,
circulatory system, excretory system, and reproductive system.

The pattern of gastrulation is affected by the amount of yolk

13 In the sea star or *Amphioxus*, cells from the blastula wall invaginate and form a cavity called the
_____.

14 In the amphibian, cells from the animal pole move down over the yolk-rich cells and invaginate,
forming the dorsal lip of the _____.

15 In the bird, invagination occurs at the _____ and no archenteron forms.

ORGANOGENESIS BEGINS WITH THE DEVELOPMENT OF THE NERVOUS SYSTEM

16 The (a)_____ induces the formation of the neural plate, from which cells migrate
downward thus forming the (b)_____, flanked on each side by neural folds.
When neural folds fuse, the (c)_____ is formed. This structure gives rise to the
(d)_____.

17 (a)_____ meet the (b)_____ thereby
forming branchial arches that give rise to elements of the face, jaws, and neck.

EXTRAEMBRYONIC MEMBRANES PROTECT AND NOURISH THE EMBRYO

18 Terrestrial vertebrates have four extraembryonic membranes: _____
_____.

19 The _____ is a fluid-filled sac that surrounds the embryo and keeps it moist.

20 The _____, an outgrowth of the digestive tract, fuses with the chorion, forming the
highly vascular chorioallantoic membrane.

The placenta is an organ of exchange

HUMAN PRENATAL DEVELOPMENT REQUIRES ABOUT 266 DAYS

Development begins in the oviduct

21 The (a)_____ dissolves when the embryo enters the uterus. At this time, the
embryo is in the (b)_____ stage, but now begins to differentiate into a
(c)_____.

The embryo implants in the wall of the uterus

22 The trophoblast cells of the blastocyst secrete enzymes that digest an opening into the uterine wall in a process called (a)_____. The process begins at about the (b)_____ day of development.

The placenta is an organ of exchange

23 In placental mammals, the _____ tissues give rise to the placenta, the organ of exchange between mother and developing child.

24 The (a)_____ connects the embryo with the placenta. It contains the (b)_____.

Organ development begins during the first trimester

25 Gastrulation occurs during the _____ weeks of human development.

26 After the first two months of development, the embryo is referred to as a _____.

27 Sex (gender) can be determined by external observation by _____.

Development continues during the second and third trimesters
The birth process may be divided into three stages

28 The normal duration of a human pregnancy, usually called the (a)_____, is (b)_____ days. It culminates in (c)_____, the process of birth.

29 During the first stage of labor the (a)_____ becomes dilated and effaced; during the second stage the (b)_____; and during the third stage the (c)_____.

The neonate must adapt to its new environment
Environmental factors affect the embryo

30 A sonogram is a picture taken by the technique known as (a)_____. This technique and (b)_____, the analysis of intrauterine fluids, are helpful in anticipating potential birth defects.

THE HUMAN LIFE CYCLE EXTENDS FROM FERTILIZATION TO DEATH

31 The human life cycle can be divided into the following stages: _____ _____.

Homeostatic response to stress decreases during aging

BUILDING WORDS
Use combinations of prefixes and suffixes to build words for the definitions that follow.

Prefixes	The Meaning	Suffixes	The Meaning
arch-	primitive	-blast	embryo
blasto-	embryo	-mere	part
holo-	whole, entire		
iso-	equal		
mero-	part, partial		
neo-	new, recent		
telo-	end		
tri-	three		
tropho-	nourishment		

Prefix	Suffix	Definition
_____	-lecithal	1. Having an accumulation of yolk at one end (vegetal pole) of the egg.
_____	-lecithal	2. Having a fairly equal distribution of yolk in the egg.
_____	_____	3. Any cell that is a part of the early embryo (specifically, during cleavage).
_____	-cyst	4. The blastula of the mammalian embryo.
_____	-pore	5. The opening into the archenteron of an early embryo.
_____	-blastic	6. A type of cleavage in which the entire egg divides.
_____	-blastic	7. A type of cleavage in which only the blastodisc divides; partial cleavage.
_____	-enteron	8. The primitive digestive cavity of a gastrula.
_____	_____	9. The extraembryonic part of the blastocyst that chiefly nourishes the embryo or that develops into fetal membranes with nutritive functions.
_____	-mester	10. A term or period of three months.
_____	-nate	11. A newborn child.

MATCHING

For each of these definitions, select the correct matching term from the list that follows.

_____ 1. An norepinephrine membrane that forms a fluid-filled sac for the protection of the developing embryo.

_____ 2. The first of several cell divisions in early embryonic development that converts the zygote into a multicellular blastula.

_____ 3. Any of the three embryonic tissue layers.

_____ 4. The yolk pole of a vertebrate or echinoderm egg.

_____ 5. Usually a spherical structure produced by cleavage of a fertilized ovum; consists of a single layer of cells surrounding a fluid-filled cavity.

_____ 6. The outer germ layer that gives rise to the skin and nervous system.

_____ 7. The attachment of the developing embryo to the uterus of the mother.

_____ 8. The development of the form and structures of an organism and its parts.

_____ 9. Early stage of embryonic development during which the embryo has two layers and is cup-shaped.

_____ 10. The fusion of the egg and the sperm, resulting in the formation of a zygote.

Terms:

a. Amnion
b. Animal pole
c. Blastula
d. Chorion
e. Cleavage
f. Ectoderm
g. Endoderm
h. Fertilization
i. Gastrula
j. Germ layer
k. Implantation
l. Morphogenesis
m. Placenta
n. Vegetal pole

TABLE
Fill in the blanks.

Human Developmental Process	Approximate Time	Event
Fertilization	0 hour	Restores the diploid number of chromosomes
#1	2, 3 weeks	Blastocyst becomes a three-layered embryo
Cleavage	#2	Formation of the embryo
#3	#4	Activates the egg
Implantation	#5	#6
#7	0 hour	Establishes the sex of the offspring
#8	2.5 weeks	Neural plate begins to form
#9	#10	Morula is formed, reaches uterus
Parturition	#11	#12

THOUGHT QUESTIONS
Write your responses to these questions.

1. What is fertilization? What are the processes by which it occurs?
2. Define each of the following terms and explain how each arises during development: zygote, morula, blastula, gastrula, germ layers, and organs.
3. Describe the major events in each trimester of human development, including the development and functions of extraembryonic membranes and the placenta.
4. Describe the three stages of labor.
5. How is the human fetus affected by the pregnant woman's lifestyle? Is it affected by ambient external stimuli? Explain your answers.
6. In your opinion, when does human life begin? Create both pro and con arguments for the assertion that human life begins at each of the following "stages:" gamete formation, fertilization, implantation, first trimester, second trimester, third trimester, "quickening" (i.e., first detected fetal movement), birth, postnatal completion of the nervous system, when he/she graduates from college.

MULTIPLE CHOICE
Place your answer(s) in the space provided. Some questions may have more that one correct answer.

_____ 1. The liver, pancreas, trachea, and pharynx
   a. arise from Hensen's node.
   b. are formed prior to the primitive streak.
   c. are derived from endoderm.
   d. are derived from ectoderm.
   e. are derived from mesoderm.

_____ 2. A hollow ball of several hundred cells
   a. is a morula.
   b. is a blastula.
   c. surrounds a fluid-filled cavity.
   d. is the gastrula.
   e. contains a blastocoel.

_____ 3. The first organ system that begins to form during organogenesis is the
   a. circulatory system.
   b. reproductive system.
   c. excretory system.
   d. nervous system.
   e. muscular system.

_____ 4. The release of calcium ions from cortical granules in the egg
   a. follows entry into the egg by a sperm.
   b. is part of the cortical reaction.
   c. is followed by a burst of protein synthesis in the egg.
   d. triggers metabolic changes within the egg.
   e. is part of the mechanism that prevents polyspermy.

_____ 5. The duration of pregnancy is referred to as the
   a. parturition.
   b. induction cycle.
   c. gestation period.
   d. neonatal period.
   e. prenatal period.

_____ 6. The fertilization membrane
   a. promotes binding of sperm to egg.
   b. facilitates egg-sperm recognition.
   c. prevents polyspermy.
   d. expedites entry of sperm into egg.
   e. blocks entrance of sperm.

_____ 7. The cavity within the embryo that is formed during gastrulation is the
   a. archenteron.
   b. gastrulacoel.
   c. blastocoel.
   d. blastopore.
   e. neural tube.

_____ 8. An embryo is called a fetus
   a. when the brain forms.
   b. when the limbs forms.
   c. when its heart begins to beat.
   d. after two months of development.
   e. once it is firmly implanted in the endometrium.

_____ 9. Embryonic cells are arranged in three distinct germ layers (endoderm, mesoderm, ectoderm) during
   a. cleavage.
   b. gastrulation.
   c. blastulation.
   d. organogenesis.
   e. blastocoel formation.

_____10. Amniotic fluid
   a. replaces the yolk sac in mammals.
   b. is secreted by the allantois.
   c. is secreted by an embryonic membrane.
   d. fills the space between the amnion and chorion.
   e. fills the space between the amnion and embryo.

_____11. Recognition of species compatibility between a sperm and an egg is due to the recognition protein
   a. acrosin.
   b. actin.
   c. vitelline.
   d. bindin.
   e. pellucidin.

_____12. The organ of exchange between a placental mammalian embryo and mother develops from
   a. chorion and amnion.
   b. chorion and allantois.
   c. chorion and uterine tissue.
   d. amnion and uterine tissue.
   e. amnion and allantois.

_____13. Eggs that have a large amount of yolk concentrated at one pole
   a. are isolecithal.
   b. are telolecithal.
   c. have a metabolically active animal pole.
   d. have a vegetal pole that is metabolically active.
   e. do not have food for the embryo in the egg.

_____14. _____ gives rise to skeletal tissues, muscle, and the circulatory system.
   a. ectoderm
   b. mesoderm
   c. endoderm
   d. one of the germ layers
   e. the blastocoel

VISUAL FOUNDATIONS
Color the parts of the illustration below as indicated.

RED   ☐ umbilical artery
GREEN   ☐ inner cell mass
YELLOW   ☐ yolk sac
BLUE   ☐ amniotic cavity and amnion
ORANGE   ☐ chorionic cavity and chorion
BROWN   ☐ trophoblast
TAN   ☐ uterine epithelium
PINK   ☐ embryo
VIOLET   ☐ placenta

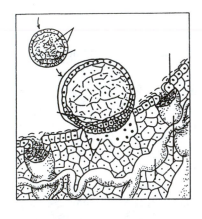
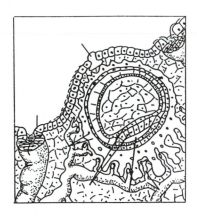
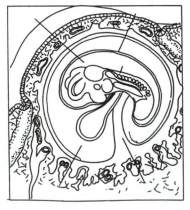
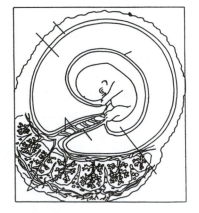

CHAPTER 50

Animal Behavior

Animal behavior consists of those movements or responses an animal makes in response to signals from its environment; it tends to be adaptive and homeostatic. The behavior of an organism is as unique and characteristic as its structure and biochemistry. An organism adapts to its environment by synchronizing its behavior with cyclic change in its environment. Although behaviors are inherited, they can still be modified by experience. Behavioral ecology focuses on the interactions between animals and their environments and on the survival value of their behavior; migratory and foraging behaviors are studied by behavioral ecologists. Social behaviors, that is, interactions between two or more animals of the same species, have advantages and disadvantages. Advantages include confusing predators, repelling predators, and finding food. Disadvantages include competition for food and habitats, and increased ease of disease transmission. Many species that engage in social behaviors form societies. Some societies are loosely organized, whereas others have a complex structure. A system of communication reinforces the organization of the society. Animals communicate in a wide variety of ways; some use sound, some use scent, and some use pheromones. Examples of various types of social behaviors include the formation of dominance hierarchies, territoriality, courtship, pair bonding, parenting, playing, and altruistic behaviors. Some invertebrate societies exhibit elaborate and complex patterns of social interactions, such as the bees, ants, wasps, and termites. Vertebrate societies are far less rigid.

CHAPTER OUTLINE AND CONCEPT REVIEW
Fill in the blanks.

INTRODUCTION
1 Behavior refers to the responses an organism makes to _____ from its environment.

BEHAVIOR IS ADAPTIVE
2 _____ is the study of the characteristics, adaptive value, and evolutionary history of innate, species-typical behaviors. Behavioral genetics is concerned with how behavioral traits are inherited.

BIOLOGICAL RHYTHMS ANTICIPATE ENVIRONMENTAL CHANGES
A variety of behavioral cycles occur among organisms
3 The behavioral cycle of activity for animals depends on when they are most active; they are categorized as (a)_____ when they are active during the day, (b)_____ when active at night, and (c)_____ when active in the twilight hours.

Biological rhythms are controlled by an internal clock
4 Most organisms have a number of internal timing mechanisms that regulate their _____.

BEHAVIOR CAPACITY IS INHERITED AND IS MODIFIED BY ENVIRONMENTAL FACTORS

Behavior develops

5 Behavior is influenced primarily by two systems of the body, namely, the _____ systems.

Some behavior patterns have strong genetic components

6 _____ refers to inherited (innate) behaviors.

7 Fixed action pattern (FAP) behaviors are elicited by a _____.

Behavior is modified by learning

8 Learning is defined as a change in behavior due to _____.

In classical conditioning, a reflex becomes associated with a new stimulus

9 In classical conditioning, an animal makes an association between an irrelevant stimulus and a normal body response. It occurs when a (a)_____ becomes a substitute for (b)_____ stimulus.

In operant conditioning, spontaneous behavior is reinforced

10 In operant conditioning, the behavior of the animal is rewarded or punished after it performs a behavior discovered by chance. Repetitive rewards bring about
(a)_____, and consistent punishment induces
(b)_____.

Imprinting is a form of learning that occurs during a critical period

11 In imprinting, an animal forms a _____ to an object during a critical period of its life.

Habituation enables an animal to ignore irrelevant stimuli
Insight learning uses recalled events to solve new problems
Learning abilities are biased

12 The _____ things appear to be most easily learned.

BEHAVIORAL ECOLOGY EXAMINES INTERACTION OF ANIMALS WITH THEIR ENVIRONMENTS

Migration is triggered by environmental changes

13 The components of migration include the (a)_____ and the mechanisms that allow organisms to (b)_____.

Efficient foraging behavior contributes to survival

14 _____ refers to an animal's ability to obtain food in the most efficient manner.

15 The method that an animal uses to obtain food is called its _____.

16 The most efficient method of obtaining food reduces three important variables to their minimums. These variables are _____ _____.

SOCIAL BEHAVIOR HAS BOTH BENEFITS AND COSTS

17 A group of cooperating individuals in the same species comprises a _____.

Communication is necessary for social behavior
Animals communicate in a wide variety of ways
Pheromones are chemical signals used in communication

18 _____ are chemical signals that convey information between members of a species.

Animals often form dominance hierarchies

19 In some animals, the hormone _____ is the apparent cause of their aggressive behaviors.

Many animals defend territory

20 An animal's territory is the portion of its _____ that it defends from members of the same species, and occasionally different species.

Sexual behavior is generally social

21 _____ behavior ensures that the male is a member of the same species, and it provides the female with a means of evaluating the male.

22 For some animals, mating is the only social contact they experience; it requires three basic elements of social conduct, namely, _____

_____.

Pair bonds establish reproductive cooperation

23 A pair bond is a stable relationship between a male and a female that ensures cooperation in two behaviors that are essential to continuation of the species. These are

_____.

Many organisms care for their young
Play is often practice behavior
Elaborate societies are found among the social insects

24 Elaborate societies with well defined division of labor occurs in the two insect groups

_____.

Vertebrate societies tend to be relatively flexible
Kin selection could produce altruistic behavior

25 More complex social groups tend to exhibit _____, wherein an individual seemingly acts for the benefit of others rather than for itself.

Sociobiology explains altruism by kin selection

BUILDING WORDS
Use combinations of prefixes and suffixes to build words for the definitions that follow.

Prefixes	The Meaning
socio-	social, sociological, society

Prefix	Suffix	Definition
_____	-biology	1. The school of ethology that focuses on the evolution of social behavior through natural selection.

MATCHING
For each of these definitions, select the correct matching term from the list that follows.

_____ 1. Means by which activities of plants or animals are adapted to regularly-recurring changes.

_____ 2. An animal that is most active during the day.

_____ 3. Behaviors that are inherited and typical of the species.

_____ 4. A substance secreted by organisms into the external environment that influences the development or behavior of other members of the same species.

_____ 5. The theory that animals feed in a manner that maximizes benefits and/or minimizes costs.

_____ 6. The process by which organisms become accustomed to a stimulus and cease to respond to it.

_____ 7. A form of rapid learning by which a young bird or mammal forms a strong social attachment to an individual or object within a few hours after hatching or birth.

_____ 8. Behavior in which one individual appears to act in such a way as to benefit others rather than itself.

_____ 9. An internal rhythm that approximates the 24-hour day.

_____10. The study of animal behavior under natural conditions.

Terms:

a. Altruistic behavior
b. Biological clock
c. Circadian rhythm
d. Crepuscular animal
e. Diurnal animal

f. Ethology
g. Habituation
h. Imprinting
i. Innate behaviors
j. Migration

k. Nocturnal animal
l. Optimal foraging
m. Pheromone
n. Sign stimulus
o. Territoriality

TABLE
Fill in the blanks.

Classification of Behavior	Example of the Behavior
Innate	Egg rolling in graylag goose
#1	Wasp responding to cone arrangement to locate nest
#2	Dog salivating at sound of bell
#3	Child sitting quietly for praise
#4	Birds tolerant of human presence
#5	Chicks learning the appearance of the parent
#6	Primate stacking boxes to reach food

THOUGHT QUESTIONS
Write your responses to these questions.

1. What is meant by a biological rhythm? Give examples and a possible cause for each.
2. Are animal behaviors caused mostly by inherited factors, environmental imperatives, or learning, or a combination of these factors? Explain your answer. Would you modify your position or stick by your answer when considering human behavior? Why?
3. Explain each of the following forms of learned behavior: classical conditioning, operant conditioning, habituation, and insight learning. How do conditioning and habituation differ? Which form of learning is most applicable to the biology student?
4. What is imprinting? Explain how imprinting provides an adaptive advantage?
5. How do nonhuman animals communicate? How do humans communicate? Are there any basic differences in the methods of communication employed by humans and other primates? How about differences between humans and an earthworm? Are there any basic differences between any nonhuman animals and humans? If so, at what point in phylogeny do they appear?
6. Explain dominance hierarchy. Does it apply to a college campus?

7. What is pair bonding? Can pair bonding occur between humans? If so, are all pair bonded humans married? Are all married humans pair bonded? Can two humans of the same gender become pair bonded? Explain your answers in both social and biological terms.

MULTIPLE CHOICE
Place your answer(s) in the space provided. Some questions may have more that one correct answer.

_____ 1. A behavior that resembles a simple reflex in some ways and a volitional behavior in others, and is triggered by a sign stimulus is best described as a/an

a. releaser.   d. FAP.

b. conditioned reflex.   e. unconditioned reflex.

c. conditioned stimulus.

_____ 2. If your dog perks up its ears when you clap your hands, but stops perking its ears after you have repeatedly clapped your hands for a period of time, your dog has displayed a form of

a. learning.   d. sensory adaptation.

b. operant conditioning.   e. habituation.

c. classical conditioning.

_____ 3. A change in behavior derived from experience in the environment is the result of

a. a FAP.   d. redirected behavior.

b. learning.   e. habituation and/or sensitization.

c. inherited behavioral characteristics.

_____ 4. Suppose you decide to perform an experiment with a type of lizard that instantly attacks any lizard with green scales on its sides. You place a live lizard with green scales on its sides in a cage next to a styrofoam block with green paint on its sides. Your experimental lizard ignores the spotted lizard, preferentially attacking the styrofoam block. The styrofoam-attacking lizard is responding to a

a. releaser.   d. displacement syndrome.

b. sign stimulus.   e. redirected behavioral syndrome.

c. conditioned reflex.

_____ 5. Which of the following is/are used by a worker honey bee to indicate the distance of a food source?

a. dance speed   d. angle relative to gravity

b. frequency of abdominal waggles   e. angle relative to sun

c. angle relative to a north–south axis

Use the following list to answer questions 6-10, which are concerned with the different approaches to the study of animal behavior.

a. ethology   d. behavioral ecology

b. behavioral genetics   e. neurobiology

c. social behaviorist

_____ 6. A field of study that would likely create a model about the behaviors of individuals in the population; a model that is designed to explain how a population is adapted to its environment.

_____ 7. A person in this field would probably want to know about the behavior of a hybrid produced from two strains with opposing behavioral characteristics.

_____ 8. The study of the behavior of animals in their natural habitats.

____ 9. A person in this field would be interested in behavioral changes that result from destruction of portions of the brain.

____10. A person in this field would be interested in interactions between members of the same species.

VISUAL FOUNDATIONS

Color the parts of the illustration below as indicated. Label the behavioral cycle and the physiological cycle.

RED	❑	molting, gonad development, and physiological cycle
GREEN	❑	spring
YELLOW	❑	summer
BLUE	❑	winter
ORANGE	❑	autumn
VIOLET	❑	migration, mating, and behavioral cycle

PART VIII

Ecology

CHAPTER 51

Ecology and the Geography of Life

Ecology is the study of the relationships among organisms and between organisms and their nonliving environment. Ecologists study populations, communities, ecosystems, the biosphere, and the ecosphere. Living things are restricted to areas where the environment and potential lifestyles fit their adaptations. The greater the differences among habitats, the greater the differences among the organisms that inhabit them. Major terrestrial life zones (biomes) are large relatively distinct ecosystems characterized by similar climate, soil, plants, and animals, regardless of where they occur on earth. Their boundaries are determined primarily by climate. In terrestrial life zones, temperature and precipitation are the major determinants of plant and animal inhabitants. Examples of biomes include tundra; the taiga, or boreal forests; temperate forests; temperate grasslands; chaparral; deserts; tropical grasslands, or savanna; and tropical rain forests. In aquatic life zones, salinity, dissolved oxygen, light, and mineral nutrients are the major determinants of plant and animal inhabitants. Freshwater ecosystems include flowing water (rivers and streams), standing water (lakes and ponds), and freshwater wetlands. Marine ecosystems include estuaries, the neritic province, and the oceanic province. All terrestrial and aquatic life zones interact with one another.

CHAPTER OUTLINE AND CONCEPT REVIEW
Fill in the blanks.

INTRODUCTION

1 A group of individuals in the same species living together in the same area is a
   (a)_____. All of these "groups" interacting in a given area comprise a
   (b)_____, and this entity together with its abiotic environmental elements is an
   (c)_____.

BIOMES ARE LARGELY DETERMINED BY CLIMATE

2 Biomes are terrestrial life zones containing characteristic geographic assemblages. _____ is the most influential factor in determining the characteristics and boundaries of a biome.

   Cold, boggy plains of the far north are called tundra

3 Tundra is the northernmost biome. It is characterized by low-growing vegetation, a short growing season, and a permanently frozen layer of subsoil called _____.

   Taiga is an evergreen forest of the north
   Temperate rain forest is characterized by cool weather, dense fog, and high precipitation

4 Coniferous temperate-zones that receive more than 80 inches (over 200 cm) of precipitation annually, like those in northwest North America, are specifically referred to as temperate _____
   _____.

Temperate deciduous forest has a dense canopy of broad-leaf trees that overlie saplings and shrubs

5 The temperate deciduous forest receives approximately _____ inches (cm) of rain and is dominated by broad-leaved trees.

Grasslands occur in temperate areas of moderate precipitation

Chaparral is a thicket of evergreen shrubs and small trees

6 Chaparral is an assemblage of small-leaved shrubs and trees in areas of modest rainfall and mild winters, climatic conditions referred to as _____ climate.

Deserts are arid ecosystems

7 Deserts are produced by low rates of precipitation and contain organisms with water-conserving adaptations. Many desert plants secrete toxic chemicals that inhibit growth of nearby competitors, an adaptation known as _____.

Savanna is a tropical grassland with scattered trees

8 The best known savanna, with its large herds of grazing, hoofed animals and predators, is the (a)_____ savanna. Smaller savannas also occur on the continents of (b)_____.

Tropical rain forests are lush equatorial forests

9 Tropical rain forests have about (a)_____ inches (cm) of rainfall annually. They have many diverse species and are noted for tall foliage, many epiphytes, and leached soil. Most of the nutrients in tropical rain forests are tied up in the (b)_____, not in the soil.

AQUATIC ECOSYSTEMS OCCUPY MOST OF EARTH'S SURFACE

10 Aquatic life zones differ from terrestrial largely due to the different properties of water and air. The three principal limiting factors in water habitats include
_____.

11 Aquatic life is ecologically divided into the free-floating (a)_____, the strong swimming (b)_____, and the bottom dwelling (c)_____ organisms.

12 The base of the aquatic food chain is composed of the free-floating (a)_____, while the free-floating heterotrophic protozoa and small animals comprise the (b)_____.

FRESHWATER ECOSYSTEMS ARE IMPORTANT ECOLOGICALLY

Rivers and smaller streams are flowing-water ecosystems

13 Rivers and streams are flowing-water ecosystems. The kinds of organisms they contain depends primarily on the _____.

Lakes and ponds are standing-water ecosystems

14 Lakes and ponds are standing-water ecosystems. Large lakes have three zones: the (a)_____ zone is the shallow water area along the shore; the (b)_____ zone is the open water area away from shore that is illuminated by sunlight; and the (c)_____ zone, which is the deeper, open water area beneath the illuminated portion. Of these, the (d)_____ zone is the most productive.

15 Vertical thermal layering of standing water is called (a)_____. It is usually marked in the summer by an abrupt temperature transition at some depth called the (b)_____. When atmospheric temperatures drop in fall, vertical layers mix, a phenomenon known as (c)_____. In spring, surface waters sink and bottom waters return to the surface. This is called (d)_____.

Freshwater wetlands are transitional between aquatic and terrestrial ecosystems

16 Freshwater wetlands may be (a)_____, where grasslike plants dominate, or
(b)_____, where trees and shrubs dominate, or they may be small, shallow ponds called
(c)_____, or (d)_____ that are dominated by mosses.

ESTUARIES OCCUR WHERE FRESH WATER AND SALTWATER MEET

17 The estuary community is very productive and serves as a nursery for young stages of many aquatic
organisms. It often contains _____, shallow, swamplike areas dominated by
grasses where salinity fluctuates considerably.

MARINE ECOSYSTEMS DOMINATE EARTH'S SURFACE

The intertidal zone is transitional between land and ocean

18 The main marine habitats are similar to those of fresh water but are modified by tides and currents.
The marine habitat near the shoreline, where oxygen, light, and food are most abundant is the
_____. Organisms living there have adaptations enabling them to resist
wave action.

Coral reefs are an important part of the benthic environment

The neritic province consists of shallow waters close to shore

19 Sharks, porpoises, and the larger benthic organisms are mostly confined to the shallower waters of
the region known as the (a)_____. The portion of this region that contains
sufficient light to support photosynthesis is called the (b)_____ region.

The oceanic province comprises most of the ocean

20 The deeper region of the open ocean, beyond the reach of light, is known as the
_____.

BUILDING WORDS

Use combinations of prefixes and suffixes to build words for the definitions that follow.

Prefixes	The Meaning
bio-	life
hydro-	water
inter-	between, among
litho-	stone
phyto-	plant
thermo-	heat, hot
zoo-	animal

Prefix	Suffix	Definition
_____	-plankton	1. Planktonic plants; primary producers.
_____	-plankton	2. Planktonic protists and animals.
_____	-tidal	3. Pertains to the zone of a shoreline between high tide and low tide.
_____	-cline	4. A layer of water marked by an abrupt temperature transition.
_____	-sphere	5. The entire zone of air, land, and water at the surface of the earth that is occupied by living things.

_____ -sphere   6. The water on or surrounding the earth, including the oceans and the water in the atmosphere.

_____ -sphere   7. The portion of the earth that is composed of rock.

MATCHING
For each of these definitions, select the correct matching term from the list that follows.

____ 1. The study of the interrelations between living things and their environment, both physical and biotic.

____ 2. Northern coniferous forest biome found primarily in Canada, northern Europe, and Siberia.

____ 3. The community of organisms and its nonliving, physical environment.

____ 4. An assemblage of different species of organisms that live in a defined area or habitat.

____ 5. Region of aquatic habitats lying near enough the surface that sufficient light is present for photosynthesis to take place.

____ 6. The open water area away from the shore of a lake or pond, extending down as far as sun light penetrates.

____ 7. Organisms that live on the bottom of oceans or lakes.

____ 8. The interactions among and between all of the Earth's living organisms in air, land and water.

____ 9. Distinctive vegetation type characteristic of certain areas with cool moist winters and long dry summers.

____10. A group of organisms of the same species that live in the same geographical area at the same time.

Terms:

a. Benthos
b. Biome
c. Chaparral
d. Community
e. Ecology

f. Ecosphere
g. Ecosystem
h. Ecotone
i. Euphotic region
j. Limnetic zone

k. Neritic province
l. Population
m. Savanna
n. Taiga
o. Tundra

TABLE
Fill in the blanks.

Biome	Climate	Soil	Characteristic Plants and Animals
Temperate rain forest	Cool weather, dense fog, high rainfall	Relatively nutrient-poor	Large conifers, epiphytes, squirrels, deer
Tundra	#1	#2	#3
#4	Not as harsh as tundra	Acidic, mineral poor, decomposed needles	Coniferous trees, medium to small animals
Temperate forest	#5	#6	#7
#8	Moderate, uncertain rainfall	#9	Grasses, grazing animals and predators such as wolves

Biome	Climate	Soil	Characteristic Plants and Animals
Chaparral	#10	Thin, infertile	#11
#12	Temperate and tropical, low rainfall	#13	Organisms adapted to water conservation, small animals
#14	Tropical areas with low or seasonal rainfall	#15	Grassy areas with scattered trees; hoofed mammals, large predators
Tropical rain forest	#16	#17	#18

THOUGHT QUESTIONS
Write your responses to these questions.

1. Define each of the following terms: ecology, population, community, ecosystem, biosphere, ecosphere.
2. What are the principal abiotic and biotic factors that distinguish each of the principal biomes?
3. What are the principal abiotic and biotic factors that distinguish each of the aquatic ecosystems?
4. Explain how human activities are impacting each of the biomes and each of the aquatic ecosystems.

MULTIPLE CHOICE
Place your answer(s) in the space provided. Some questions may have more that one correct answer.

_____ 1. Water lilies, water striders, and crayfish are found mainly in
   a. the littoral zone.
   b. the limnetic zone.
   c. the profundal zone.
   d. a shallow lake.
   e. the area just below the compensation point.

_____ 2. Small or microscopic organisms carried about by currents are
   a. littoral.
   b. all producers.
   c. nektonic.
   d. benthic.
   e. planktonic.

_____ 3. The area(s) essentially corresponding to the waters above the continental shelf is/are the
   a. littoral zone.
   b. limnetic zone.
   c. profundal zone.
   d. neritic zone.
   e. benthic zone.

_____ 4. Communities dominated by evergreens and containing acid, mineral-poor soil is/are
   a. in South America.
   b. called taiga.
   c. temperate coniferous forests.
   d. typically impregnated with a layer of permafrost.
   e. boreal forest.

_____ 5. The two most productive and diverse of communities are the _____ and _____
   a. deciduous forest.
   b. tropical rain forest .
   c. coniferous forest.
   d. coral reef.
   e. grassland.

____ 6. One expects to find organisms with adaptations for retaining moisture and clinging securely to the substrate in the
   a. salt marsh.   d. littoral zone.
   b. intertidal zone.   e. estuary.
   c. profundal zone.

____ 7. An aquatic zone at the juncture of a river and the sea is a/an
   a. salt marsh.   d. littoral zone.
   b. intertidal zone.   e. estuary.
   c. profundal zone.

____ 8. The factor that determines the nature of the biogeographic realms more than any other single factor is
   a. soil.   d. types of animals.
   b. climate.   e. altitude.
   c. latitude.

____ 9. A habitat with short grass and somewhat greater precipitation than moist deserts is a
   a. tropical habitat.   d. deciduous forest.
   b. steppes.   e. muskeg.
   c. boreal forest.

____ 10. The relatively moist forests on the western slopes of North America are
   a. boreal forests.   d. steppes.
   b. deciduous forests.   e. typified by Mediterranean climates.
   c. temperate communities.

____ 11. The communities containing the most comprehensive coverage of permafrost are called
   a. polar.   d. tundra.
   b. steppes.   e. muskegs.
   c. boreal.

____ 12. Most of midwestern North America is
   a. grassland.   d. temperate.
   b. deciduous forest.   e. chaparral.
   c. temperate rain forest.

____ 13. A community with two distinct soil layers abounding in trees, reptiles, and amphibians is
   a. grassland.   d. temperate.
   b. deciduous forest.   e. chaparral.
   c. temperate rain forest.

____ 14. An adaptation by which plants discourage competition through secretion of toxic substances is
   a. allelopathy.   d. typical of tropic plants.
   b. typical of vines.   e. found in deserts.
   c. facilitated by secretions of roots or leaves.

VISUAL FOUNDATIONS
Color the parts of the illustration below as indicated.

RED   ❑   pelagic

GREEN   ❑   intertidal zone

YELLOW   ❑   neritic province

BLUE   ❑   abyssal

ORANGE   ❑   benthic

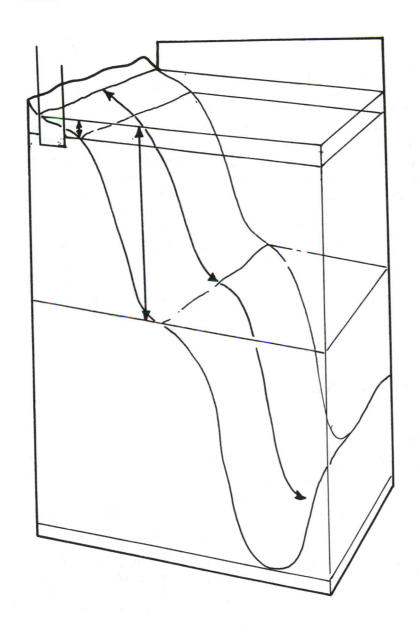

Population Ecology

A population is a group of organisms of a single species living in a particular geographic area at a given time. Populations have properties that do not exist in their component individuals or in the community of which the population is a part. These properties include birth rates, death rates, growth rates, population density, sex ratios, age structure, and gene pools. Population ecology deals with the numbers of particular organisms that are found in an area and how and why those numbers change or remain fixed over time. Populations must be understood in order to manage forests, field crops, game, fishes, and other populations of economic importance. Population growth is determined, in part, by the rate of arrival of organisms through immigration and births, and the rate of departure through emigration and deaths. Population growth curves take a variety of forms. Their shapes are primarily a function of limiting factors and the theoretical maximum rate a population can grow. Population growth is limited by density-dependent factors such as predation, disease, and competition, and density-independent factors, such as hurricanes and fires. Each species has its own survival strategy, that is, its own reproductive characteristics, body size, habitat requirements, migration patterns, and other characteristics that enable it to survive. Most organisms fall into one of two survival strategies: One emphasizes a high rate of natural increase; these organisms often have small body sizes, bear large numbers of young once per lifetime, and inhabit variable environments. The second strategy emphasizes maintenance of a population near the carrying capacity of the environment; these organisms often have large body sizes, repetitive reproductive cycles, long life spans, and inhabit stable environments. The principles of population ecology apply to humans as well as other organisms.

CHAPTER OUTLINE AND CONCEPT REVIEW
Fill in the blanks.

INTRODUCTION
1 Populations of organisms have properties that individual organisms do not have. Some properties of importance are _____
_____.

DENSITY AND DISPERSION ARE IMPORTANT FEATURES OF POPULATIONS
2 _____ is the number of individuals of a species per unit of habitat area.

3 Individuals within a population may exhibit a characteristic spacing. For example, (a)_____ exists when individuals are not spaced in a definable, predictable pattern; (b)_____ is even, regular spacing of individuals; and (c)_____ applies when individuals are more-or-less confined to a limited portion of their potential range.

MATHEMATICAL MODELS DESCRIBE POPULATION GROWTH
4 Changes in the size of a population result from the difference between the _____
_____.

5 The rate of growth in a population is expressed as.

$$\frac{\Delta N}{\Delta t} = b - d \quad , \text{where}$$

ΔN is the (a)_____, Δt is the
(b)_____, b is the birth rate or (c)_____, and d is the death
rate or (d)_____.

Migration affects the growth rate in some populations
Each population has a characteristic biotic potential
No population can increase exponentially indefinitely

6 Population growth curves take a variety of forms, their shapes primarily a function of biotic potential
and limiting factors. The (a)_____ growth curve shows an initial lag phase of population
growth, followed by a logarithmic phase, followed by a static or exhaustion phase. The
(b)_____ growth curve resembles the logarithmic portion of the s-curve and occurs in the
virtual absence of limiting factors.

7 (a)_____ is the theoretical maximum rate at which a population can grow. The
maximum potential rate of increase cannot be sustained due to the constraints upon growth
collectively known as (b)_____. The largest population
that can be sustained by a particular environment is that environment's (c)_____.

POPULATION SIZE VARIES OVER TIME
Density-dependent factors regulate population size

8 Density-dependent limiting factors are most effective at _____ (high or low?) population
densities, and therefore they tend to stabilize populations.

9 _____ is an effective density-dependent limiting factor because
transmission is more likely among members of dense populations.

Density-independent factors influence populations
Natural populations may fluctuate chaotically

DIFFERENT SPECIES HAVE DIFFERENT REPRODUCTIVE STRATEGIES

10 Each species has a life history strategy, The (a)_____ strategy is employed mainly
by pioneer organisms. It emphasizes wide and quick dispersal, large numbers of seeds or offspring,
and genetic uniformity. A (b)_____ strategy is characteristic of inhabitants of
mature communities where replacement is slow.

Survivorship is related to *r* and *K* selection

THE PRINCIPLES OF POPULATION ECOLOGY APPLY TO HUMAN POPULATIONS

11 ZPG stands for _____, which is the point where natality
equals mortality.

Not all countries have the same growth rates

12 The amount of time that it takes for a population to double in size is its (a)_____.
In general, the shorter this time, the (b)_____(more or less?) developed the country.

13 The number of children that a couple must produce to replace themselves is known as the
(a)_____, a value that is generally relatively

(b)_____(higher or lower?) in developed countries and (c)_____(higher or lower?) in undeveloped and developing countries.

14 The average number of children born to a group of women in their lifetimes is called the (a)_____. On a worldwide basis, this value is currently (b)_____(higher or lower?) than the replacement value.

The age structure of a country can be used to predict future population growth

BUILDING WORDS

Use combinations of prefixes and suffixes to build words for the definitions that follow.

Prefixes	The Meaning
inter-	between, among
intra-	within

Prefix	Suffix	Definition
_____	-specific competition	1. Competition for resources within a population.
_____	-specific competition	2. Competition for resources among populations.

MATCHING

For each of these definitions, select the correct matching term from the list that follows.

_____ 1. A pattern of spacing in which individuals in a population are spaced unpredictably.

_____ 2. Growth that occurs at a constant rate of increase over a period of time.

_____ 3. The rate at which organisms produce offspring.

_____ 4. The maximum number of organisms that a habitat can support.

_____ 5. Any factor, such as climate, that does not depend on the density of populations.

_____ 6. The number of offspring a couple must produce in order to "replace" themselves.

_____ 7. The movement of individuals out of a population.

_____ 8. The maximum rate of increase of a species that occurs when all environmental conditions are optimal.

_____ 9. The amount of time it takes for a population to double in size, assuming that its current rate of increase does not change.

_____10. An abrupt decline in a population from a high to very low population density.

Terms:

a. Biotic potential	f. Doubling time	l. Population crash
b. Carrying capacity	g. Emigration	m. Random dispersion
c. Contest competition	h. Exponential growth	n. Replacement-level
d. Density-dependent factor	i. Immigration	fertility
e. Density-independent factor	j. Mortality	o. Uniform dispersion
	k. Natality	

TABLE
Fill in the blanks.

Examples of Environmental Factors That Affect Population Size	Category of Factors That Affect Population Size	Description
Predation	Density-dependent	As density increases, predators are more likely to encounter a prey individual
Fire	#1	#2
Contest competition	#3	#4
Disease	#5	#6
Hurricane	#7	#8
#9	#10	Population members share limited resources equally

THOUGHT QUESTIONS
Write your responses to these questions.

1. Describe population density and dispersion and explain how they are related.

2. Define biotic potential and carrying capacity and how each relates to the J-curve and the S-curve.

3. What is meant by a K strategy? Give two examples. What is meant by an r strategy? Give two examples. Can any given species or population utilize both strategies?

4. What is a survivorship curve? Distinguish between Types I, II, and III curves and identify which of these curves corresponds to "r selection" and which applies to "K selection."

5. Using the information provided in the textbook, and assuming current trends in infant mortality, fertility rates, and birth rates and death rates prevail, what would you predict will happen to human population distributions on earth over the next few decades? What would you predict regarding distribution of essential human resources such as space, food, and clean water? What would you predict regarding distribution of wealth? Do any of your predictions have any bearing on survival of the human species? Are politics involved? Is education involved? Explain your answers.

MULTIPLE CHOICE
Place your answer(s) in the space provided. Some questions may have more that one correct answer.

Questions 1-5 pertain to the following equation. Use the list of choices below to answer them.

$$\frac{\Delta N}{\Delta t} = b - d$$

a. growth only   d. ΔN g. value "b"
b. speed of growth   e. Δt h. value "d"
c. change only   f. $b - d$ i. $\Delta N / \Delta t$

_____ 1. Birth rate.

_____ 2. Death rate.

_____ 3. Population change independent of time.

_____ 4. Rate of population change.

_____ 5. The parameter derived from the equation.

_____ 6. The equation $dN/dt = rN$ is used to derive the
    a. biotic potential (r_m).   d. per capita growth rate (r).
    b. instantaneous growth rate.   e. growth rate at carrying capacity.
    c. decline in growth due to limiting factors.

_____ 7. During the exponential growth phase of bacterial growth, the population
    a. increases moderately.   d. decreases dramatically.
    b. increases dramatically.   e. does not change substantially.
    c. decreased moderately.

_____ 8. A cultivated field of wheat would most likely display
    a. uniform distribution.   d. no distribution pattern.
    b. random distribution.   e. a form of distribution not found in nature.
    c. clumped distribution.

_____ 9. The difference in the biotic potentials of a cow and mouse is largely due to their respective
    a. litter sizes.   d. limiting factors.
    b. habitats.   e. choices of food.
    c. sizes.

_____10. In the logistic equation, if environmental resistance is sustained, "K" is the
    a. growth rate.   d. carrying capacity.
    b. birth rate.   e. equivalent of the lag phase.
    c. mortality rate.

_____11. A grove of trees that originated from one seed displays
    a. uniform distribution.   d. no distribution pattern.
    b. random distribution.   e. a form of distribution not found in nature.
    c. clumped distribution.

_____12. Population growth is most frequently checked by which *one* of the following?
    a. Environmental resistance.   d. Behavioral modifications.
    b. Predation.   e. Resource depletion.
    c. Disease.

_____13. The inverse of biotic potential is
    a. natality.   d. birth rate minus death rate.
    b. mortality.   e. a population crash.
    c. b − d.

_____14. r-strategists usually
    a. are mobile animals.   d. survive well in the long term.
    b. have a high "r" value.   e. develop slowly.
    c. inhabit variable environments.

_____15. Examples of density-dependent limiting factors include
    a. competition.   d. predation.
    b. disease.   e. behavioral mechanisms.
    c. resource depletion.

_____16. The redwood stands in California are examples of organisms that
    a. use the K-strategy.   d. live in a climax community.
    b. use the r-strategy.   e. pioneer new habitats.
    c. often experience local extinction.

_____17. The number of individuals of a species per unit of habitat area is the _____ for that species.
    a. distribution curve   d. ΔN
    b. distribution pattern   e. b − d
    c. population density

VISUAL FOUNDATIONS

Color the parts of the illustration below as indicated. Label stable, declining, and expanding populations.

    RED   ❑   post-reproductive

    GREEN   ❑   reproductive

    YELLOW   ❑   pre-reproductive

CHAPTER 53

❏

Community Ecology

A community is an association of organisms of different species living and interacting together. Communities interact with and influence one another in numerous ways. Communities exist within communities. For example, a community of microorganisms lives within the gut of a termite that itself is a member of a community of organisms within a rotten log that is part of a forest community, and so on. A community and its physical environment is inseparably linked in what is called an ecosystem. Communities contain producers, consumers, and decomposers. Producers are the photosynthetic organisms that are the basis of most food chains. Consumers feed on other organisms. Decomposers recycle the components of corpses and wastes by feeding on them. The primary producers, consumers, and decomposers of a community interact with one another in complex ways. These interactions include predation, symbiosis, and competition. Symbiosis is an intimate long-term association between two or more different species. In one form of symbiosis, both partners benefit. In another, one partner benefits and the other neither benefits nor is harmed. In a third, one partner benefits and the other is adversely affected. Every organism has its own ecological niche, that is, its own role within the structure and function of the community. Its niche reflects the totality of its adaptations, its use of resources, and its lifestyle. Communities vary greatly in the number of species they contain. The number is often high where the number of potential ecological niches is great, where a community is not isolated or severely stressed, at the edges of adjacent communities, and in communities with long histories. A community develops gradually over time, through a series of stages, until it reaches a state of maturity; species in one stage are replaced by different species in the next stage. Most ecologists believe that the species composition of a community is due primarily to factors other than biological interactions.

CHAPTER OUTLINE AND CONCEPT REVIEW
Fill in the blanks.

INTRODUCTION

1 A biological community consists of a group of organisms that interact and live together. A complete and independent community, interacting with the nonliving environment, makes up an

_____.

COMMUNITIES CONTAIN AUTOTROPHS AND HETEROTROPHS

2 The major community roles are those of producer, consumer, and decomposer. Producers, which are also called (a)_____, are the photosynthetic organisms that are the basis of most food chains. Consumers, also called (b)_____, feed on other organisms. Consumers are further divided into the (c)_____ consumers that eat producers, and (d)_____ consumers that eat other consumers. (e)_____ eat both plants and animals. Decomposers, which are also known as (f)_____, recycle the components of corpses and wastes by feeding on them.

3 _____ work together with decomposers by consuming the organic matter of plant and animal remains.

LIVING ORGANISMS INTERACT IN A VARIETY OF WAYS

Natural selection shapes both prey and predator

Symbiosis involves close associations in which one species usually lives on or in the other species

THE NICHE DESCRIBES AN ORGANISM'S ROLE IN THE COMMUNITY

4 The role an organism plays in an ecosystem is its niche. A niche reflects the organism's total adaptational package. The role an organism *could* potentially play in a community is its (a)_____ niche, while the role it *actually* plays is its (b)_____ niche.

Competition between two species with identical or similar niches leads to competitive exclusion

5 Competitive exclusion means that one species is excluded from a niche as a result of _____, most often for a limited resource.

Limiting factors restrict an organism's ecological niche

COMMUNITIES VARY IN SPECIES RICHNESS

6 Species diversity is inversely related to (a)_____ of a habitat. For example, island communities are generally less diverse than continental communities. The (b)_____ is a phrase used to describe the relatively greater diversity near the borders of a community compared to the interior portions of the same community.

Species richness causes community stability — or does it?

SUCCESSION IS COMMUNITY CHANGE OVER TIME

7 The relatively stable, often long term stage at the end of ecological succession is the _____.

Sometimes a community develops in a "lifeless" environment

8 (a)_____ occurs in a habitat that is initially devoid of life. For example, lichens may invade a bare rock, altering its composition enough for other organisms to "take hold." As the first to colonize the "new lifeless" habitat, initial invaders like the lichen, are referred to as (b)_____ communities. In general, the order in which plants are established in a new community is: lichens —> (c)_____ —> grasses —> (d)_____ —> and finally (e)_____.

Sometimes communities develop in an environment where a previous community existed

9 _____ begins with an existing community. Abandoned farmlands are examples.

ECOLOGISTS CONTINUE TO STUDY THE NATURE OF COMMUNITIES

10 The view that the species in a community cooperate like organs in the body of a complex organism is known as the (a)_____ model of a community. The opposing view, namely that biological interactions are less important than abiotic factors in producing communities, and that, in fact, communities may not be biological entities at all, is the (b)_____ model of communities.

BUILDING WORDS

Use combinations of prefixes and suffixes to build words for the definitions that follow.

Prefixes	The Meaning	Suffixes	The Meaning
auto-	self, same	-troph	nutrition, growth, "eat"
hetero-	different, other	-vore	eating
omni-	all		

Prefix	Suffix	Definition
carni-	_____	1. Any animal that eats flesh.
_____	_____	2. Any animal that eats all kinds of foods indiscriminately.
_____	_____	3. An organism that produces its own food.
_____	_____	4. An organism that is dependent upon other organisms for food, energy, and oxygen.
herbi-	_____	5. Any animal that eats plants.

MATCHING

For each of these definitions, select the correct matching term from the list that follows.

____ 1. The sequence of changes in a plant community over time.

____ 2. An ecological succession that occurs after some disturbance destroys the existing vegetation.

____ 3. An organism that is capable of producing disease.

____ 4. Organic debris.

____ 5. Resemblance of different species, each of which is equally obnoxious to predators.

____ 6. A consumer that eats producers.

____ 7. An intimate relationship between two or more organisms of different species.

____ 8. A symbiotic relationship between individuals of different species in which one benefits from the relationship and one is harmed.

____ 9. A fairly broad transition region between adjacent biomes.

____10. A type of symbiosis in which one organism benefits and the other one is neither harmed nor helped.

Terms:

a. Batesian mimicry
b. Commensalism
c. Detritus
d. Ecological niche
e. Ecotone

f. Edge effect
g. Mullerian mimicry
h. Mutualism
i. Parasitism
j. Pathogen

k. Primary consumer
l. Secondary succession
m. Succession
n. Symbiosis

TABLE
Fill in the blanks.

Kind of Consumer	Category of Food Preference	Example	What They Eat
Primary consumer	Exclusively herbivores	Deer, grasshoppers	Producers
#1	Carnivores	Lions, spiders	#2
#3	Omnivores	Bears, pigs, humans	#4
Detritus feeder	#5	Snail, crabs, worms, termites, earthworms	Dead organic matter

THOUGHT QUESTIONS
Write your responses to these questions.

1. What is an ecosystem? What is a community? How do they differ?
2. What is symbiosis? Is predation a form of symbiosis? Explain mutualism, commensalism, and parasitism. Would a "typical marriage" between two healthy humans qualify as a form of symbiosis? If so, would you call it mutualism, commensalism, or parasitism? How would you classify the relationship between a pregnant human and a developing fetus? How about the relationship between supportive parents and a dependent college student? Justify your answers in both biological and social terms.
3. Explain the nature of and differences between an organism's habitat and niche.
4. Describe the effects of competitive exclusion and limiting factors.
5. Using realistic examples, describe primary succession, secondary succession, and the climax community.

MULTIPLE CHOICE
Place your answer(s) in the space provided. Some questions may have more that one correct answer.

____ 1. Critically important minerals and other essential substances would be permanently lost to organisms if it were not for the
    a. producers.   d. consumers.
    b. nitrogen fixers.   e. decomposers.
    c. phosphorus fixers.

____ 2. The totality of an organism's adaptations, its use of resources, and the lifestyle to which it is fitted are all embodied in the term
    a. ecosystem.   d. pyramid.
    b. community.   e. niche.
    c. food web.

____ 3. The nonliving environment together with different interacting species is a/an
    a. ecosystem.   d. pyramid.
    b. community.   e. niche.
    c. food web.

____ 4. An association of different species of organisms interacting and living together is a/an
    a. ecosystem.   d. pyramid.
    b. community.   e. niche.
    c. food web.

____ 5. The major roles of organisms in a community are
    a. producers.   d. consumers.
    b. nitrogen fixers.   e. decomposers.
    c. phosphorus fixers.

____ 6. The difference between an organism's potential niche and the niche it comes to occupy may result from
    a. competition.   d. overabundance of resources.
    b. an inverted pyramid.   e. limited space or resources.
    c. factors that exclude it from part of its fundamental niche.

____ 7. All food energy in the biosphere is ultimately provided by
    a. producers.   d. consumers.
    b. nitrogen fixers.   e. decomposers.
    c. phosphorus fixers.

CHAPTER 54

Ecosystems and the Ecosphere

Matter, the material of which living things are composed, cycles from the living world to the abiotic physical environment and back again. All materials vital to life are continually recycled through ecosystems and so become available to new generations of organisms. Key cycles include the carbon cycle, the nitrogen cycle, the phosphorus cycle, and the water, or hydrologic, cycle. Although matter is cyclic in ecosystems and the ecosphere, energy flow is not. Energy cannot be recycled and reused. Energy moves through ecosystems in a linear, one-way direction from the sun to producer to consumer to decomposer. Much of this energy is converted to less useful heat as the energy moves from one organism to another. Life is possible on earth because of its unique environment.

CHAPTER OUTLINE AND CONCEPT REVIEW
Fill in the blanks.

INTRODUCTION

1 A community *and* its abiotic components together make up an (a)_____, all of these ecological units combined comprise the Earth's (b)_____.

MATTER CYCLES THROUGH ECOSYSTEMS

2 The cycling of matter through the biota to the abiota and back again is referred to as
(a)_____. Four of these cycles that are of particular importance
are the cycles for (b)_____
_____.

Carbon dioxide is the pivotal molecule of the carbon cycle

3 Carbon is incorporated into the biota, or "fixed," primarily through the process of
(a)_____, and in the short term much of it is returned to the abiotic
environment by (b)_____. Some carbon is tied up for long periods of time in wood
and fossil fuels, which may eventually be released through combustion.

Bacteria are essential to the nitrogen cycle

4 The nitrogen cycle has five steps: (a)_____, which is the conversion of
nitrogen gas to ammonia or nitrate; the conversion of ammonia to nitrate, called
(b)_____; absorption of nitrates and ammonia by plant roots, called
(c)_____; the release of organic nitrogen back into the abiotic environment in the
form of ammonia in waste products and by decomposition, a process known as
(d)_____; and finally, to complete the cycle, (e)_____
reduces nitrates to gaseous nitrogen.

The phosphorus cycle lacks a gaseous component

5 Erosion of (a)_____ releases inorganic phosphorus to the (b)_____ where it is taken up by
plant roots.

Water circulates in the hydrologic cycle

6 The hydrological cycle results from the evaporation of water from both land and sea and its subsequent precipitation when air is cooled. The movement of water from land to seas is called (a)_____. Water also seeps down through the soil to become (b)_____, which supplies much of the water to streams, soil, and plants.

THE FLOW OF ENERGY THROUGH ECOSYSTEMS IS LINEAR

7 Energy flows through ecosystems. It enters in the form of (a)_____ energy, which plants can convert to useful (b)_____ energy that is stored in the bonds of molecules. Since energy conversions are never 100% efficient, when it is converted from one form to another, or moved from the abiota to the biota and back again, some of it is converted to (c)_____ energy and returned permanently to the abiotic environment where organisms cannot use it.

Energy flow describes who eats whom in ecosystems

8 Each "link" in a food chain is a trophic level, the first and most basic of which is comprised of the photosynthesizers, or (a)_____. Once "fixed," materials and energy pass linearly through the chain from photosynthesizers to the (b)_____, and then to the (c)_____.

9 A complex of interconnected food chains is a _____.

Ecological pyramids illustrate how ecosystems work

10 The three main types of ecological pyramids are the pyramid of (a)_____, which illustrates how many organisms there are in each trophic level; the pyramid of (b)_____, which is a quantitative estimate of the amount of living material at each level; and the pyramid of (c)_____, a somewhat unique representation since this pyramid, unlike the others, and because of the second law of thermodynamics, can never be inverted.

Ecosystems vary in productivity

11 The rate at which carbon compounds are synthesized by photosynthesis is known as the (a)_____ of an ecosystem, and the energy that remains as biomass after energy for respiration is "consumed" is the (b)_____, which basically represents plant growth.

ABIOTIC FACTORS INFLUENCE WHERE AND HOW SUCCESSFULLY AN ORGANISM CAN SURVIVE

THE SUN WARMS EARTH

Temperature changes with latitude
Temperature changes with season

12 Seasons result from two factors: (a)_____, which varies throughout the year, and most importantly, from (b)_____, the tilt of the Earth's axis relative to the sun.

ORGANISMS DEPEND ON THE ATMOSPHERE

Atmospheric circulation is driven by the sun

13 Winds generally blow from (a)_____(high or low?) pressure areas to (b)_____(high or low?) pressure areas, but they are deflected from these paths by Earth's rotation, a phenomenon known as the (c)_____.

THE GLOBAL OCEAN COVERS MOST OF THE EARTH'S SURFACE

Surface ocean currents are driven by winds and by the Coriolis effect

14 Prevailing winds push masses of water along the surfaces of oceans. These movements, or (a)_____, move in large circular patterns called (b)_____.

The ocean interacts with the atmosphere

TEMPERATURE AND PRECIPITATION ARE IMPORTANT ASPECTS OF CLIMATE

15 _____ is the average weather conditions, including temperature and rainfall, that prevail in a given location over an extended period of time.

Precipitation patterns are affected by air and water movements, and by surface features of the land

16 Local climate is determined largely by the angle at which the solar input falls to earth's surface. Angles are usually greatest and least variable in (a)_____ regions, producing hotter, less variable climates in these regions. Air in this region is heated, causing it to expand and rise, which in turn causes it to cool and sink again. Most of the cool air sinks back into the region from which it came, but some of it flows toward the (b)_____ where it cools and sinks.

17 _____ is greatest where air heavily saturated with water vapor cools to the dew point.

Some mountains cause rain shadows

18 _____ develop in the rain shadows of mountain ranges and continental interiors.

Microclimates are local variations in climate

19 _____ refers to the relatively long term average weather conditions in a small locale, these conditions usually attributable to the unique topological features in that location.

BUILDING WORDS

Use combinations of prefixes and suffixes to build words for the definitions that follow.

Prefixes	The Meaning
atmo-	air
bio-	life
hydro-	water
micro-	small

Prefix	Suffix	Definition
_____	-mass	1. The total weight of all living things in a particular habitat.
_____	-sphere	2. The gaseous envelope surrounding the earth; the air.
_____	-logic	3. The water cycle which includes evaporation, precipitation and flow to the seas.
_____	-climate	4. A localized variation in climate; the climate of a small area.

MATCHING
For each of these definitions, select the correct matching term from the list that follows.

_____ 1. The distance of an organism in a food chain from the primary producers of a community.

_____ 2. Large, thick-walled cells of certain cyanobacteria that are the site of nitrogen fixation.

_____ 3. A community of organisms and its nonliving, physical environment.

_____ 4. The system of interconnected food chains in a community.

_____ 5. The ability of certain microorganisms to bring atmospheric nitrogen into chemical combination.

_____ 6. An area on the downwind side of a mountain range with very little precipitation.

_____ 7. The interactions among and between all of the Earth's living organisms in air, land and water.

_____ 8. A sequence of organisms through which energy is transferred from its ultimate source in a plant; each organism eats the preceding member and is eaten by the following member of the sequence.

_____ 9. A graphical representation of the relative energy value at each trophic level.

_____10. The oxidation of ammonium salts into nitrites and nitrates by soil bacteria.

Terms:

a.	Assimilation	f.	Food chain	k.	Nitrogen fixation
b.	Coriolis effect	g.	Food web	l.	Rain shadow
c.	Ecological pyramid	h.	Fossil fuel	m.	Trophic level
d.	Ecosphere	i.	Heterocyst	n.	Upwelling
e.	Ecosystem	j.	Nitrification		

TABLE
Fill in the blanks.

Biogeochemical Cycle	Material Cycled	Description
Phosphorus	Phosphorus	No gas formed; dissolves in water, is taken into plants
#1	#2	As a gas, is taken into plants, converted into glucose; returned to atmosphere by respiration
#3	#4	Converted from a gas into form usable by plants, then fixed in plant tissue, then consumed by animals
#5	Water	#6

THOUGHT QUESTIONS
Write your responses to these questions.

1. Describe the flow of energy through a terrestrial ecosystem. Do the same for any aquatic ecosystem.
2. Produce diagrams for each of the following cycles: carbon, nitrogen, phosphorus, hydrogen. Do any of these cycles affect one another? Which ones and how?
3. Sketch and explain the meaning of a pyramid of numbers, biomass, and energy. Place the principal "players" in each level of the pyramid for both terrestrial and aquatic environments. (For example, phytoplankton are principal aquatic producers.) Where would you place humans on the pyramid?

MULTIPLE CHOICE
Place your answer(s) in the space provided. Some questions may have more that one correct answer.

_____ 1. The rate at which organic matter is incorporated into plant bodies is
    a. significant only in a changing ecosystem.
    b. called gross primary productivity.
    c. called net primary productivity.
    d. equal to the primary producer's metabolic rate.
    e. evident only in a climax community.

_____ 2. Which of the following would be positioned at the top of a typical pyramid?
    a. Producers.
    b. Consumers.
    c. The trophic level with the greatest biomass.
    d. Trophic level with the least numbers.
    e. Trophic level with the least energy.

_____ 3. Most carbon is fixed _____ and liberated _____
    a. in proteins/as organic compounds.
    b. in CO_2/as complex compounds.
    c. in complex compounds/as CO_2.
    d. in humus/by cyanobacteria.
    e. by plants/by plants.

_____ 4. Activities performed by microorganisms involved in the nitrogen cycle include
    a. liberation of nitrogen from nitrate.
    b. oxidization of ammonia to nitrates.
    c. production of ammonia from proteins, urea, or uric acid.
    d. fixation of molecular nitrogen.
    e. continuous cycling of nitrogen.

_____ 5. Phosphorus enters aquatic communities by means of
    a. decomposers.
    b. producers.
    c. primary consumers.
    d. secondary consumers.
    e. erosion.

_____ 6. The trophic levels found at the two extremes (top and bottom) of a pyramid are
    a. producers.
    b. nitrogen fixers.
    c. phosphorus fixers.
    d. consumers.
    e. decomposers.

_____ 7. Compared to the overall regional climate, a microclimate
    a. may be significantly different.
    b. is insignificant to organisms.
    c. is more important to resident organisms.
    d. covers a larger area.
    e. exists for a brief period.

_____ 8. A major ocean current that moves northward along the west coast of South America
    a. warms coastal Europe.
    b. warms the U.S.
    c. cools South America.
    d. is called the Gulf Stream.
    e. reaches Alaska.

_____ 9. The totality of the Earth's living inhabitants is the
    a. ecosphere.
    b. lithosphere.
    c. biosphere.
    d. biome.
    e. hydrosphere.

_____10. The least pronounced seasonal temperature differences are found in the
    a. temperate zones.
    b. equatorial areas.
    c. United States.
    d. polar regions.
    e. subpolar areas.

_____11. When it is warmest in the United States, it is coolest in
    a. the Southern Hemisphere.
    b. the Northern Hemisphere.
    c. Rio de Janiero.
    d. Europe.
    e. Mexico.

_____12. A burrow in a hostile desert environment is an example of a/an
    a. subterranean biome.   d. microclimate.
    b. O-horizon.   e. ecosystem.
    c. macroclimate.

_____13. All of Earth's living inhabitants and the abiotic environmental factors comprise the
    a. ecosphere.   d. biome.
    b. lithosphere.   e. hydrosphere.
    c. biosphere.

_____14. The main collector(s) of solar energy used to power life processes is/are the
    a. producers.   d. green leaves.
    b. atmosphere.   e. photosynthetic organisms.
    c. hydrosphere.

VISUAL FOUNDATIONS

Color the parts of the illustration below as indicated. Label N_2 in the atmosphere, NO_3^- in the soil, and NH_3 in the soil.

RED   ☐   nitrogen-fixation
GREEN   ☐   plants
YELLOW   ☐   ammonification
BLUE   ☐   atmosphere
ORANGE   ☐   assimilation
BROWN   ☐   soil
TAN   ☐   animal
PINK   ☐   nitrification
VIOLET   ☐   denitrification

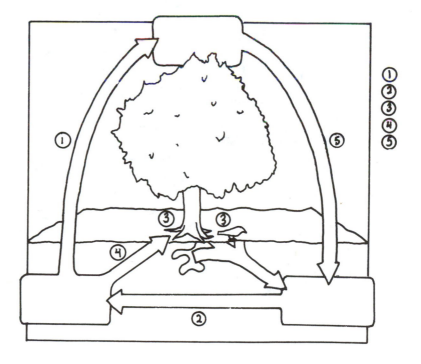

Color the parts of the illustration below as indicated.

RED ☐ equator

GREEN ☐ trade winds

YELLOW ☐ warm air

BLUE ☐ cool air

ORANGE ☐ westerlies

VIOLET ☐ polar easterlies

Humans in the Environment

Although humans have been on earth a comparatively brief span of time, our biological success has been unparalleled. We have expanded our biological range into almost every habitat on earth. In the process, we have placed great strain on the Earth's resources and resilience, and profoundly affected other life forms. As a direct result of our actions, many environmental concerns exist today. Species are disappearing from earth at an alarming rate. Conservation biologists estimate that at least one species becomes extinct every day. Although forests provide us with many ecological benefits, including watershed protection, soil erosion prevention, climate moderation, and wildlife habitat, deforestation is occurring at an unprecedented rate. The production of atmospheric pollutants that trap solar heat in the atmosphere threaten to alter the earth's climate. Global warming may cause a rise in sea level, changes in precipitation patterns, death of forests, extinction of animals and plants, and problems for agriculture. It could result in the displacement of thousands or even millions of people. Ozone is disappearing from the stratosphere at an alarming rate; during the 1980s, its rate of depletion was roughly three times the rate in the 1970s. The ozone layer helps to shield Earth from damaging ultraviolet radiation. If the ozone were to disappear, Earth would become uninhabitable for most forms of life.

CHAPTER OUTLINE AND CONCEPT REVIEW
Fill in the blanks.

SPECIES ARE DISAPPEARING AT AN ALARMING RATE

1 When the number of individuals in a species is reduced to the point where extinction seems imminent, the species is said to be (a)_____, and when numbers are seriously reduced so that extinction is feared but thought to be less imminent, the species is said to be (b)_____.

2 Although a variety of human activities threaten the continued existence of many organisms, most species facing extinction reach this perilous state as the result of
_____.

Human activities contribute to declining biological diversity
There are two types of conservation efforts to save wildlife

3 *In situ* preservation attempts to preserve species in (a)_____ by establishing reserves (e.g., parks), while *ex situ* conservation tries to preserve species in (b)_____ such as zoos and seed banks.

DEFORESTATION IS OCCURRING AT AN UNPRECEDENTED RATE
Where and why are forests disappearing?

4 Agriculture replaces diverse natural communities with communities consisting of one or two specially bred species. These artificial communities are unstable, simple, and lack some important mineral cycles, and modern agriculture requires substantial energy and has great environmental impact. For example, most of the surviving tropical forests, which consists mostly of the (a)_____ in South America and the (b)_____ forest in Africa, are being destroyed by

individual farmers who clear forest areas by cutting down and burning forest plants. This practice, known as (c)_____ agriculture, ultimately leads to barren, unproductive land.

5 In addition to agriculture, which may be the main culprit, a significant amount of deforestation also results from (a)_____ when it proceeds at a faster rate than is sustainable, as is occurring in the forests of Malaysia, and from the destruction of trees to provide open (b)_____.

CERTAIN ATMOSPHERIC POLLUTANTS MAY AFFECT EARTH'S CLIMATE
Greenhouse gases cause global warming

6 Gases that are accumulating in the atmosphere as a result of human activities include (a)_____
_____. Gas powered vehicles cause three gases to accumulate, namely, (b)_____. Landfills produce (c)_____, and leaking cooling units release (d)_____.

7 Since CO_2 and other gases that are being added to the atmosphere trap heat, global warming is occurring, a phenomenon known as the _____.

Global warming could profoundly affect all life

8 Models developed through computer simulations predict that a doubling of atmospheric CO_2 will raise the average temperature on Earth by (a)_____ degrees Centigrade within (b)_____ years, which will profoundly affect the entire Earth.

Many actions have been suggested to deal with global warming

THE AMOUNT OF STRATOSPHERIC OZONE IS DROPPING
Certain chemicals destroy stratospheric ozone

9 CFCs and other human-produced compounds move upward into the stratosphere where ultraviolet radiation breaks them down. For the most part, three elements result, namely,
(a)_____. Of these, (b)_____ is particularly harmful since a single atom can break down thousands of ozone molecules.

Ozone depletion harms living organisms

10 Ozone shields Earth from much of the harmful _____ that comes from the sun. Ozone depletion is occurring at an alarming rate.

International cooperation will prevent significant depletion of the ozone layer

ENVIRONMENTAL PROBLEMS ARE INTERRELATED

BUILDING WORDS
Use combinations of prefixes and suffixes to build words for the definitions that follow.

Prefixes	The Meaning
de-	to remove
strat(o)-	layer

Prefix	Suffix	Definition
_____	-forestation	1. The removal or destruction of all tree cover in an area.
_____	-sphere	2. The layer of the atmosphere between the troposphere and the mesosphere, containing a layer of ozone that protects life by filtering out much of the sun's ultraviolet radiation.

MATCHING

For each of these definitions, select the correct matching term from the list that follows.

____ 1. The warming of the Earth resulting from the retention of atmospheric heat caused by the build-up of certain gases, especially carbon dioxide.

____ 2. A species whose population is low enough for it to be at risk of becoming extinct.

____ 3. The number and variety of living organisms.

____ 4. The disappearance of a species from a given habitat.

____ 5. Carbon dioxide, methane, nitrogen oxide, CFCs and ozone.

____ 6. A method of controlled breeding in zoos.

____ 7. Efforts to preserve biological diversity in the wild.

____ 8. The production of enough food to feed oneself and one's family with little left over to sell or reserve for bad times.

____ 9. A species whose numbers are so severely reduced that it is in imminent danger of becoming extinct.

____10. The layer in the upper atmosphere that helps shield the Earth from damaging ultraviolet radiation.

Terms:

a.	Artificial insemination	g.	Greenhouse effect	m.	Slash-and-burn agriculture
b.	Biological diversity	h.	Greenhouse gases	n.	Subsistence agriculture
c.	Endangered species	i.	Host mothering	o.	Threatened species
d.	Ex situ conservation	j.	In situ conservation		
e.	Extinction	k.	Keystone species		
f.	Genetic diversity	l.	Ozone		

TABLE

Fill in the blanks.

Environmental Issue	Why It Is An Issue	Cause of Problem	Solution
Declining biological diversity	Diversity contributes to a sustainable environment; lost opportunities, lost solutions to future problems	Commercial hunting, habitat destruction, efforts to eradicate or control species, commercial harvesting	*In situ* conservation, *ex situ* conservation, habitat protection
#1	Forests are needed for habitat, watershed protection, prevention of erosion	#2	#3
#4	Threatens food production, forests, biological diversity; could cause submergence of coastal areas, change precipitation patterns	#5	#6
Ozone depletion	Ozone layer protects living organisms from exposure to harmful amounts of UV radiation	#7	#8

THOUGHT QUESTIONS
Write your responses to these questions.

1. What is a threatened species? What causes a threatened species to be called an endangered species? When is a species considered extinct?
2. Why is biological diversity important? Why is diversity declining?
3. What are the "greenhouse" gases and how are they causing global warming?
4. How do you think life would be affected by the pollution of phytoplankton and the destruction of the rain forests? Explain your answer in terms of the food supply, oxygen and carbon concentrations, global warming and the "greenhouse effect," and changes in sea level and weather patterns on earth.

REFLECTIONS
The subject matter covered in this chapter is of critical importance to the continuance of life on earth as we know it. This chapter points out some of the dangers humanity faces if we continue some of our habits. Biologists and other responsible citizens can make a difference. One might even say that we have a responsibility to make a difference. A few genuinely concerned individuals will assume leadership roles in bringing about the social, cultural, and political changes required to protect the environment. All of us can participate in assuring that our children and grandchildren will inherit a clean, healthy, and productive environment.

We encourage you to reflect on what your personal role could be to bring about needed reforms. What might you contribute to the welfare of the planet? What changes are needed to ensure that future generations can enjoy a reasonably good quality of life? Think about your lifestyle and aspirations. Are they compatible with a long term investment in the future of the planet?

VISUAL FOUNDATIONS
Color the parts of the illustration below as indicated.

RED ☐ heat re-radiated back to Earth   ORANGE ☐ heat escaping to space

GREEN ☐ Earth   PINK ☐ stratosphere

YELLOW ☐ solar energy   TAN ☐ CO_2 in stratosphere

BLUE ☐ troposphere

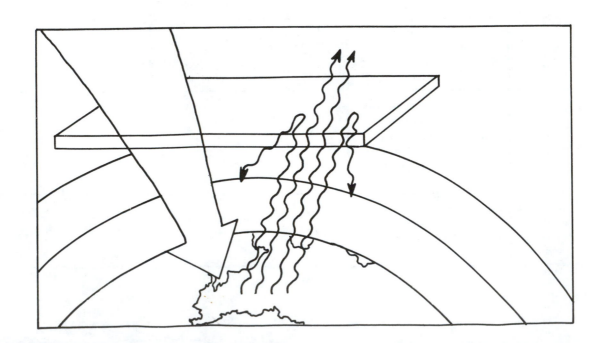

Color the parts of the illustration below as indicated. Label the southern and northern hemispheres.

RED ☐ antarctic ozone hole

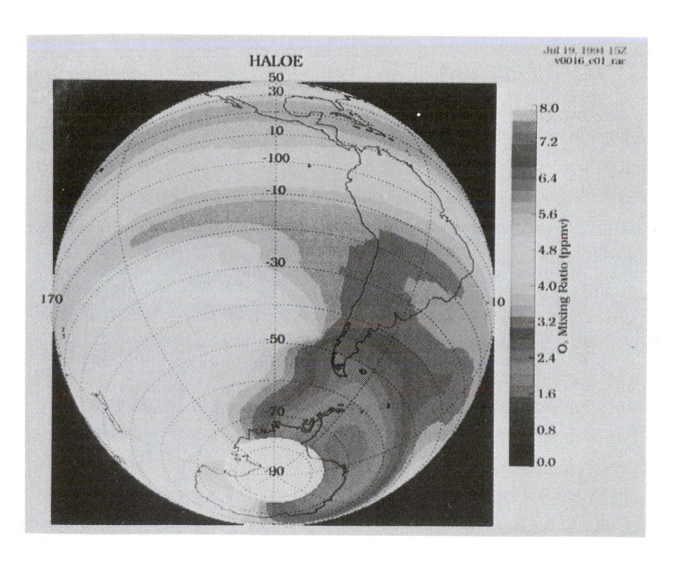

ANSWERS

❏

Chapter 1

Concept Questions: 1. composed of cells, organized, maintain homeostasis, metabolize, move, grow, respond to stimuli, adapt, reproduce   2. cell theory   3. size and/or number   4. development   5. chemical activities   6. homeostasis   7. sexually or asexually   8. evolution   9. genes   10. hormones and neurotransmitters   11. *On the Origin of Species By Means of Natural Selection*   12. (a)vary, (b)survive, (c)survive   13. environmental changes (or pressures)   14. evolution   15. cell   16. biosphere   17. taxonomy   18. genus and species   19. kingdom   20. (a)Prokaryotae, (b)Protista, (c)Fungi, (d)Plantae, (e)Animalia   21. cellular respiration   22. producers, consumers, and decomposers   23. scientific method   24. deductive and inductive   25. hypothesis   26. observations and experiments

Building Words: 1. biology   2. ecology   3. asexual   4. homeostasis   5. ecosystem   6. biosphere   7. autotroph   8. heterotroph   9. protozoa   10. photosynthesis

Matching: 1. j   2. g   3. m   4. c   5. l   6. n   7. a   8. e   9. b   10. k

Table: 1. Cellular level   2. Tissue level   3. Organ level   4. The 10 organ systems: integumentary, skeletal, muscle, nervous, circulatory, digestive, respiratory, urinary, endocrine, reproductive   5. Organ systems functioning together

Multiple Choice: 1. c,e   2. b   3. d   4. a,d,e   5. a   6. c   7. a,b,e   8. d   9. d   10. a   11. c,d   12. b

Visual Foundations: 1. d   2. b,d   3. a   4. c

Chapter 2

Concept Questions: 1. chemical reactions (ordinary chemical means)   2. (a)92, (b)uranium, (c)hydrogen, (d)carbon, hydrogen, oxygen, nitrogen   3. (a)protons, neutrons, electrons, (b)protons and neutrons, (c)electrons   4. (a)protons, (b)neutrons, (c)electrons   5. atomic number   6. atomic mass   7. (a)atomic mass (or numbers of neutrons), (b)atomic number (or numbers of protons)   8. (a)no (zero), (b)protons and electrons   9. electron shell (or energy level) 10. molecule   11. two or more different elements   12. (a)structural, (b)chemical   13. atomic masses of its constituent atoms   14. (a)reactants, (b)products   15. (a)left, (b)right   16. valence   17. potential   18. geometric shape   19. (a)nonpolar, (b)polar   20. (a)cations, (b)anions   21. electronegative   22. polar molecules   23. nonpolar   24. (a)oxidized, (b)reduced, (c)redox   25. hydrogen   26. hydrogen bonding   27. specific heat   28. (a)OH^- and a cation, (b)H^+ and an anion   29. (a)0-14, (b)7, (c)base, (d)acid   30. (a)buffers, (b)acids or bases   31. (a)hydrogen ion, (b)mineral

Building words: 1. neutron   2. radionuclide   3. equilibrium   4. tetrahedron   5. hydration   6. hydrophobic   7. nonelectrolyte   8. hydrophilic

Matching: 1. l   2. c   3. g   4. b   5. f   6. o   7. a   8. j   9. e   10. d

Table: **1.** oxygen **2.** O **3.** covalent bond **4.** carbon **5.** 12.01 **6.** 4 **7.** covalent bond **8.** nitrogen **9.** 7 **10.** 14.00 **11.** phosphorus **12.** 15 **13.** 5 **14.** sodium **15.** 11 **16.** ionic bond **17.** Cl **18.** 7 **19.** ionic bond

Multiple Choice: **1.** b **2.** c,d **3.** b **4.** a-e **5.** d **6.** a **7.** b **8.** a **9.** a,b **10.** e **11.** c,d **12.** c,d,e **13.** a,b,d **14.** a,e **15.** a **16.** b,e **17.** a,b,d,e **18.** b **19.** b,d **20.** e

Visual Foundations: **1a.** carbon 12 **1b.** carbon 14 **2.** b,e

Chapter 3

Concept Questions: **1.** carbohydrates, lipids, proteins, nucleic acids **2.** single, double, triple **3.** tetrahedral **4.** structures **5.** functional **6.** polymers **7.** carbon, hydrogen, oxygen **8.** glucose **9.** glycosidic **10.** simple sugars (monosaccharides) **11.** (a)glycoprotein, (b)glycolipids **12.** carbon, hydrogen, oxygen **13.** (a)glycerol, (b)fatty acid **14.** amphipathic **15.** carbon, hydrogen, oxygen, nitrogen, sulfur **16.** (a)amino and carboxyl, (b)R group (radical) **17.** (a)amino acids, (b)peptide **18.** secondary, tertiary, quaternary **19.** (a)amino acid, (b)three-dimensional (tertiary) **20.** (a)purine or pyrimidine, (b)sugar, (c)phosphate **21.** ATP

Building words: **1.** isomer **2.** macromolecule **3.** monomer **4.** hydrolysis **5.** monosaccharide **6.** hexose **7.** disaccharide **8.** polysaccharide **9.** amyloplast **10.** monoglyceride **11.** diglyceride **12.** triglyceride **13.** amphipathic **14.** dipeptide **15.** polypeptide

Matching: **1.** i **2.** c **3.** l **4.** n **5.** a **6.** o **7.** g **8.** k **9.** m **10.** e

Table: **1.** proteins **2.** R—NH$_2$ **3.** proteins, nucleic acids **4.** R—OH **5.** carbohydrates, proteins, lipids, nucleic acids **6.** polar **7.** phosphate group **8.** nucleic acids, proteins **9.** R—CH$_3$ **10.** nonpolar

Multiple Choice: **1.** c,e **2.** a,b,d **3.** b,c,d,e **4.** b,d **5.** b,c,d **6.** a,b,d **7.** a,c,d **8.** b,c,e **9.** a,c,e **10.** a,d **11.** a,c,d **12.** a,d,e **13.** c,d,e

Visual Foundations: **1.** d **2.** b **3.** a, b **4.** c

Chapter 4

Concept Questions: **1.** the activities of life **2.** (a)living organisms, (b)preexisting cells **3.** plasma membrane **4.** (a)surface-to-volume ratio, (b)rate of diffusion (diffusion of molecules) **5.** (a)electron, (b)cell fractionation **6.** prokaryotic **7.** eukaryotic **8.** genes **9.** chromatin **10.** endoplasmic reticulum **11.** ribosomes **12.** transport vesicles **13.** Golgi complex **14.** lysosomes **15.** mitochondria **16.** chloroplasts **17.** thylakoid **18.** (a)stroma, (b)thylakoids **19.** shape and movement **20.** microfilaments and microtubules **21.** centrioles and basal bodies **22.** cilia and flagella **23.** (a)actin, (b)myosin **24.** cellulose **25.** glycocalyx

Building words: **1.** chlorophyll **2.** chromoplasts **3.** cytoplasm **4.** cytoskeleton **5.** glyoxysome **6.** leucoplast **7.** lysosome **8.** microfilaments **9.** microtubules **10.** microvilli **11.** myosin **12.** peroxisome **13.** proplastid **14.** prokaryotes **15.** cytosol **16.** eukaryote

Matching: **1.** b **2.** h **3.** f **4.** m **5.** n **6.** c **7.** e **8.** l **9.** d **10.** j

Table: **1.** nucleus **2.** hereditary information **3.** eukaryotes **4.** cytoplasm **5.** captures light energy **6.** cytoplasm **7.** ATP synthesis **8.** eukaryotes) **9.** cytoplasm **10.** active in cell division **11.** eukaryotes (animals) **12.** encloses entire cell, including the plasma membrane **13.** protection, support **14.** prokaryotes, eukaryotes (plants) **15.** plasma membrane **16.** boundary of cell **17.** prokaryotes, eukaryotes

Multiple Choice: **1.** a,b,c,e **2.** a,d **3.** a,b,c,e **4.** c,e **5.** d,e **6.** b,c **7.** a-e **8.** a-e **9.** a,b,e **10.** d,e **11.** a,c,e **12.** a,b,d,e **13.** a,c,d

Visual Foundations: **1.** prokaryotic **2.** nuclear area **3.** plant (eukaryotic) **4.** plasma membrane, prominent vacuole, chloroplast, mitochondrion, internal membrane system **5.** cell wall, prominent vacuole, chloroplast **6.** animal (eukaryotic) **7.** plasma membrane, centriole, lysosome, mitochondrion, internal membrane system **8.** centriole, lysosome

Chapter 5

Concept Questions: **1.** plasma membrane **2.** (a)fluid mosaic, (b)lipid bilayer, (c)proteins **3.** lipids **4.** amphipathic **5.** (a)hydrophilic, (b)hydrophobic **6.** (a)head groups, (b)fatty acid chains **7.** hydrocarbon chains **8.** peripheral membrane proteins **9.** integral membrane proteins **10.** proteins **11.** protein **12.** selectively permeable **13.** size, shape, electrical charge **14.** (a)dialysis, (b)osmosis **15.** osmotic pressure **16.** facilitated diffusion and active transport **17.** carrier-mediated active transport **18.** (a)fewer, (b)negatively **19.** concentration gradient **20.** desmosomes, tight junctions, gap junctions **21.** (a)proteins, (b)pores **22.** gap junctions

Building words: **1.** endocytosis **2.** exocytosis **3.** hypertonic **4.** hypotonic **5.** isotonic **6.** phagocytosis **7.** pinocytosis **8.** desmosome **9.** plasmodesma

Matching: **1.** g **2.** i **3.** f **4.** b **5.** a **6.** k **7.** e **8.** m **9.** l **10.** h

Table: **1.** osmosis **2.** dialysis **3.** transfer of solutes by proteins located within the membrane **4.** a form of carrier-mediated transport in which the cell expends energy to move solutes against a concentration gradient **5.** facilitated diffusion **6.** endocytosis **7.** transport of waste products or secretions outside the cell

Multiple Choice: **1.** b,e **2.** d **3.** a **4.** c,e **5.** a,c,e **6.** d,e **7.** c **8.** a,c,e **9.** b,c **10.** b,e **11.** a,b,d,e **12.** b,c,d,e **13.** b,d **14.** b,c,d,e

Visual Foundations: **1.** fluid mosaic model **2.** glycolipids, glycoproteins

Chapter 6

Concept Questions: **1.** producers **2.** work **3.** (a)potential energy, (b)kinetic energy **4.** thermodynamics **5.** (a)created nor destroyed, (b)transferred **6.** randomness or disorder **7.** heat **8.** total bond energies **9.** exergonic reactions **10.** (a)exergonic, (b)endergonic **11.** zero **12.** (a)release, (b)require input of **13.** adenine, ribose, three phosphates **14.** hydrolysis **15.** enzyme-substrate complex **16.** lock-and-key, induced-fit **17.** active sites **18.** coenzyme **19.** temperature, pH, ion concentration **20.** (a)product, (b)substrate **21.** shape and amount **22.** reversible **23.** competitive **24.** noncompetitive

Building words: **1.** kilocalorie **2.** anabolism **3.** catabolism **4.** exergonic **5.** endergonic **6.** allosteric

Matching: **1.** e **2.** b **3.** k **4.** i **5.** l **6.** a **7.** f **8.** d **9.** j **10.** c

Table: **1.** catabolic **2.** exergonic **3.** anabolic **4.** endergonic **5.** catabolic **6.** exergonic **7.** catabolic **8.** exergonic **9.** anabolic **10.** endergonic **11.** anabolic **12.** endergonic

Multiple Choice: **1.** a,b,c **2.** a,b,d **3.** a,b,e **4.** a,d,e **5.** a,d **6.** b,d,e **7.** c **8.** b,c,e **9.** b,c,d **10.** a-e **11.** b,c,e

Chapter 7

Concept Questions: **1.** anabolism **2.** catabolism **3.** adenosine triphosphate (ATP) **4.** aerobic respiration **5.** (a)carbon dioxide and water, (b)energy **6.** dehydrogenations, decarboxylations **7.** cytosol (or cytoplasm) **8.** glyceraldehyde-3-phosphate (PGAL) **9.** pyruvate **10.** two **11.** (a)NAD^+, (b)CO_2 **12.** (a)three, (b)two, (c)eight **13.** two **14.** (a)NAD^+ and FAD, (b)the electron transport system and chemiosmotic phosphorylation **15.** (a)electron transport system, (b)molecular oxygen **16.** (a)intermembrane space, (b)ATP synthetase **17.** fatty acids **18.** deamination **19.** glycerol and fatty acid **20.** beta oxidation **21.** anaerobic respiration **22.** organic compound **23.** ethyl alcohol **24.** lactate **25.** (a)two, (b)36-38

Building words: **1.** aerobe **2.** anaerobe **3.** dehydrogenation **4.** decarboxylation **5.** deamination **6.** glycolysis

Matching: **1.** c **2.** d **3.** h **4.** i **5.** a **6.** e **7.** k **8.** l **9.** f **10.** n

Table: **1.** none **2.** 1 ATP **3.** 1 ATP **4.** none **5.** 1 ATP **6.** 1 ATP **7.** NAD⁺ **8.** 3 ATP
9. 0 ATP **10.** NAD⁺ **11.** 3 ATP **12.** 0 ATP **13.** FAD **14.** 2 ATP **15.** 0 ATP **16.** NAD⁺
17. 3 ATP **18.** 0 ATP

Multiple Choice: **1.** b,e **2.** a,d **3.** c,d,e **4.** a,d,e **5.** c,d **6.** a-e **7.** a,b,e **8.** b,c,d **9.** b,c

Chapter 8

Concept Questions: **1.** (a)carbon dioxide and water, (b)light (solar), (c)chemical **2.** heterotrophs **3.** (a)stroma, (b)mesophyll **4.** thylakoids **5.** chlorophyll **6.** (a)NADPH and ATP, (b)carbohydrate **7.** (a)ground state, (b)excited (energized) **8.** (a)fluorescence, (b)electron acceptor **9.** blue and red **10.** (a)chlorophyll *a*, (b)chlorophyll *b*, (c)carotenoids **11.** (a)absorption spectrum, (b)spectrophotometer **12.** (a)action spectrum, (b)accessory pigments **13.** photophosphorylation **14.** (a)thylakoid membrane, (b)proton gradient, (c)chemiosmosis **15.** ATP synthetase (CF₀-CF₁) **16.** (a)NADH, (b)ATP **17.** P700 **18.** carbon fixation **19.** (a)glucose, (b)ribulose biphosphate (RuBP), (c)phosphoglycerate (PGA), (d)glyceraldehyde-3-phosphate (PGAL) **20.** (a)CO₂, (b)NADPH, (c)ATP **21.** oxaloacetate **22.** (a)RuBP carboxylate, (b)oxygen

Building words: **1.** mesophyll **2.** photosynthesis **3.** photolysis **4.** photophosphorylation **5.** chloroplast **6.** chlorophyll **7.** autotroph **8.** heterotroph

Matching: **1.** m **2.** n **3.** h **4.** e **5.** a **6.** c **7.** k **8.** f **9.** j **10.** d

Table: **1.** anabolic **2.** catabolic **3.** oxidation/reduction (redox) **4.** oxidation/reduction (redox) **5.** thylakoid membrane **6.** cristae of mitochondria **7.** primary hydrogen (electron) acceptor **8.** source of hydrogen (electrons) **9.** NADP⁺ **10.** O₂

Multiple Choice: **1.** a,c,d **2.** a,b,d **3.** a,c,d,e **4.** c **5.** a,c,d **6.** a,c,d **7.** c **8.** b,d,e **9.** a **10.** c,e **11.** a,c,e **12.** a,b,d **13.** a-e **14.** c

Chapter 9

Concept Questions: **1.** chromatin fibers **2.** (a)RNA, (b)polypeptides **3.** cell cycle **4.** (a)first gap, (b)synthesis, (c)second gap **5.** prophase **6.** metaphase **7.** anaphase **8.** telophase **9.** asexual **10.** (a)sexual, (b)diploid, (c)zygote **11.** size and structure **12.** (a)homologous chromosome, (b)diploidy **13.** (a)gametes or spores, (b)somatic (body) cells **14.** (a)prophase, (b)genetic variability **15.** anaphase **16.** anaphase **17.** (a)mitosis, (b)meiosis **18.** (a)somatic, (b)gametes (sex cells) **19.** (a)sporophyte, (b)gametophyte

Building words: **1.** chromosome **2.** interphase **3.** haploid **4.** diploid **5.** centromere **6.** gametogenesis **7.** spermatogenesis **8.** oogenesis **9.** gametophyte **10.** sporophyte

Matching: **1.** n **2.** l **3.** f **4.** c **5.** h **6.** b **7.** g **8.** j **9.** i **10.** d

Table: **1.** chromosomes line up along an equatorial plane of the cell **2.** anaphase **3.** anaphase II **4.** interphase **5.** premeiotic interphase **6.** homologous chromosomes move to opposite poles **7.** telophase **8.** telophase I, telophase II **9.** does not occur **10.** prophase I

Multiple Choice: **1.** j **2.** b **3.** b **4.** a,g,h **5.** g **6.** d **7.** c **8.** h **9.** a,i **10.** e **11.** c,d **12.** b,c,e **13.** a,d **14.** c,d,e **15.** b **16.** h **17.** c **18.** b **19.** b **20.** b **21.** b,f **22.** d **23.** g **24.** d **25.** d

Chapter 10

Concept Questions: **1.** (a)first filial, (b)F₂ or second filial **2.** (a)dominant, (b)recessive **3.** segregate **4.** other pairs of genes **5.** locus **6.** F₁ and homozygous recessive **7.** (a)specific events, (b)zero, (c)one **8.** product **9.** sum of their separate probabilities **10.** two loci **11.** linked **12.** crossing over **13.** recombination **14.** autosomes **15.** (a)XY male, (b)XX female **16.** (a)hyperactive, (b)inactivation **17.** (a)three or more, (b)locus **18.** pleiotropy **19.** epistasis **20.** polygenic inheritance **21.** homozygosity **22.** (a)heterozygosity, (b)hybrid vigor

Building words: **1.** polygene   **2.** homozygous   **3.** heterozygous   **4.** dihybrid   **5.** monohybrid   **6.** heterogametic   **7.** homogametic   **8.** hemizygous

Matching: **1.** d   **2.** a   **3.** k   **4.** l   **5.** b   **6.** m   **7.** i   **8.** g   **9.** c   **10.** f

Table: **1.** tall plant with yellow seeds(TT Yy))   **2.** tall plant with green seeds(TT yy)   **3.** tall plant with yellow seeds (Tt Yy)   **4.** tall plant with green seeds(Tt yy)   **5.** tall plant with yellow seeds (Tt YY)   **6.** tall plant with yellow seeds (Tt Yy)   **7.** short plant with yellow seeds (tt YY)   **8.** short plant with yellow seeds (tt Yy) (primates)   **9.** tall plant with yellow seeds (Tt Yy)   **10.** tall plant with green seeds(Tt yy)   **11.** short plant with yellow seeds (tt Yy)   **12.** short plant with green seeds (tt yy)

Multiple Choice: **1.** both j   **2.** l,m,n,r,s,t   **3.** b,c,d,e,f   **4.** RrTT   **5.** j   **6.** c/n and d/r   **7.** b/m and e/s   **8.** g   **9.** g   **10.** d   **11.** b or d   **12.** c   **13.** b   **14.** e   **15.** b,c,e   **16.** b   **17.** a,b,e   **18.** a,c,d   **19.** a,e   **20.** a,c,d

Visual Foundations: **1.** incomplete dominance   **2.** all pink   **3.** 1:2:1, red:pink:white

Chapter 11

Concept Questions: **1.** Garrod and Beadle and Tatum   **2.** transformation   **3.** viruses (bacteriophages)   **4.** sugar (deoxyribose), phosphate, nitrogenous base   **5.** X-ray diffraction   **6.** antiparallel nucleotide strands   **7.** (a)pyrimidines, (b)cytosine, (c)thymine   **8.** template   **9.** (a)5, (b)3, (c)Okazaki   **10.** DNA polymerases   **11.** origin(s) of replication   **12.** prokaryotic   **13.** (a)nucleosome, (b)histones   **14.** scaffolding

Building Words: **1.** avirulent   **2.** antiparallel

Matching: **1.** l   **2.** a   **3.** d   **4.** k   **5.** i   **6.** f   **7.** g   **8.** h   **9.** b   **10.** e

Table: **1.** Chargaff   **2.** demonstrated that DNA is replicated by a semiconservative mechanism   **3.** developed a model for the structure of DNA   **4.** Hershey-Chase   **5.** Avery, MacLeod and McCarty   **6.** Franklin and Wilkins

Multiple Choice: **1.** a,c,d   **2.** a,c,d   **3.** a-e   **4.** d   **5.** a,b,d   **6.** b,c   **7.** d   **8.** b,c,e   **9.** b   **10.** a,c   **11.** d   **12.** c,e   **13.** c   **14.** e   **15.** a   **16.** d,e   **17.** b   **18.** d

Chapter 12

Concept Questions: **1.** (a)transcription, (b)translation   **2.** codon   **3.** (a)anticodon, (b)amino acid   **4.** ribosomal (rRNA), transfer (tRNA), messenger (mRNA)   **5.** (a)RNA polymerase, (b)nucleoside triphosphate   **6.** promoter or initiator   **7.** (a)upstream, (b)downstream   **8.** (a)leader sequence, (b)termination signals   **9.** aminoacyl-tRNA synthetases   **10.** (a)anticodons, (b)peptide   **11.** AUG   **12.** elongation   **13.** (a)amino, (b)carboxyl   **14.** termination or stop codons   **15.** 7-methyl guanylate   **16.** (a)intervening sequences, (b)expressed sequences   **17.** (a)sixty-one, (b)forty   **18.** nucleotides   **19.** nucleotide sequence   **20.** (a)point or base-substitution, (b)missense, (c)nonsense   **21.** hot spots

Building Words: **1.** anticodon   **2.** ribosome   **3.** polysome   **4.** mutagen   **5.** carcinogen

Matching: **1.** g   **2.** j   **3.** h   **4.** m   **5.** b   **6.** c   **7.** l   **8.** k   **9.** d   **10.** f

Table: **1.** 3'—TTT—5'   **2.** lysine   **3.** 3'—UUU—5'   **4.** 5'—UGG—3'   **5.** 3'—ACC —5'   **6.** 3'— ACC —5'   **7.** 3'—CCT—5'   **8.** glycine   **9.** 3'—CCA—5'   **10.** glycine   **11.** methionine (start codon)   **12.** 5'—UAA—3', 5'—UAG—3', 5'—UGA—3'   **13.** 3'—ATT—5', 3'—ATC—5', 3'—ACT—5'   **14.** 3'—AUU—5', 3'—AUC—5', 3'—ACU—5'

Multiple Choice: **1.** a   **2.** d,e   **3.** a,c,d   **4.** b,c   **5.** c,e   **6.** a,c   **7.** c,d,e   **8.** d   **9.** a,c,e   **10.** a   **11.** b,d   **12.** b,c,e   **13.** a,c,e   **14.** a,b

Chapter 13

Concept Questions: 1. constitutive genes   2. promoter region   3. operator   4. repressor protein   5. (a)inducer molecule, (b)inducible   6. corepressor   7. regulon   8. feedback inhibition   9. TATA box   10. upstream promoter elements (UPEs)   11. (a)heterochromatin, (b)euchromatin   12. differential nuclear RNA processing   13. chemical modification

Building Words: 1. euchromatin   2. heterochromatin   3. corepressor   4. heterodimer

Matching: 1. f   2. a   3. j   4. h   5. l   6. d   7. b   8. g   9. c   10. e

Table: 1. prokaryotic   2. eukaryotic   3. eukaryotic   4. prokaryotic   5. prokaryotic   6. eukaryotic

Multiple Choice: 1. a,d   2. b,d   3. c   4. e   5. e   6. b   7. d,e   8. b,d   9. a,b,c,e   10. a,d,e   11. a,c,d,e   12. b,c,d   13. a,c,d   14. b,c,d   15. a,e

Chapter 14

Concept Questions: 1. palindromic   2. restriction enzymes   3. plasmids or bacteriophages   4. genetic probes   5. genomic library   6. reverse transcriptase   7. (a)DNA polymerase, (b)heat   8. restriction maps   9. restriction fragment length polymorphisms (RFLPs)   10. promoter and regulator   11. (a)fertilized egg, (b)retroviruses

Building Words: 1. transgenic   2. retrovirus

Matching: 1. k   2. f   3. c   4. a   5. i   6. g   7. l   8. h   9. d   10. e

Table: 1. RFLPs   2. used to devise new combinations of genes that produce improved pharmaceutical and agricultural products   3. cloning   4. genetic probe   5. gel electrophoresis   6. establishes landmarks in a cloned DNA fragment

Multiple Choice: 1. a,c,d   2. a   3. b,d,e   4. b,d,e   5. b   6. c,d   7. c,e   8. c,e   9. b,e   10. d   11. d   12. b,c   13. c,d,e   14. a,c,e   15. c,d,e   16. a,c

Chapter 15

Concept Questions: 1. isogenic strains   2. congenital   3. cytogenetics   4. photomicrograph   5. polyploidy   6. (a)aneuploidy, (b)disomic, (c)trisomic, (d)monosomic   7. (a)nondisjunction, (b)Down syndrome   8. (a)nuclear sexing, (b)Barr body   9. Klinefelter syndrome   10. Turner syndrome   11. inborn errors of metabolism   12. sickle cell anemia   13. cystic fibrosis   14. amniocentesis   15. chorionic villus sampling (CVS)   16. severe combined immune deficiency (SCID)   17. (a)I^AI^A and I^Ai^o, (b)I^BI^B and I^Bi^o, (c)I^AI^B, (d)i^oi^o 18. erythroblastosis fetalis   19. consanguineous matings

Building Words: 1. isogenic   2. cytogenetics   3. polyploidy   4. trisomy   5. translocation

Matching: 1. a   2. j   3. f   4. h   5. l   6. e   7. d   8. i   9. c   10. b

Table: 1. cystic fibrosis   2. autosomal recessive trait   3. . autosomal recessive trait   4. lack enzyme that breaks down a brain membrane lipid   5. autosomal dominant trait   6. late onset of severe mental and physical deterioration   7. Down syndrome   8. 47 chromosomes because have 3 copies of chromosome 21   9. Hemophilia A   10. X-linked recessive trait   11. lack enzyme that converts phenylalanine to tyrosine

Multiple Choice: 1. c,e   2. c   3. b   4. a,b,c   5. a,d   6. b,c,d   7. a,d,e   8. b,c,d   9. a,b,c   10. b,e   11. b   12. e

Visual Foundations: 1. c   2. a   3. d

Chapter 16

Concept Questions: **1.** (a)determination, (b)differentiation   **2.** (a)morphogenesis, (b)pattern formation   **3.** nuclear equivalence   **4.** totipotent   **5.** polytene   **6.** puffs   **7.** maternal effect   **8.** (a)zygotic genes, (b)gap genes (c)homeotic genes   **9.** homeobox   **10.** stem or founder cells   **11.** mosaic   **12.** chimera   **13.** regulative   **14.** genomic rearrangements

Building Words: **1.** morphogenesis   **2.** polytene   **3.** chronogene

Matching: **1.** j   **2.** k   **3.** b   **4.** i   **5.** e   **6.** d   **7.** c   **8.** a   **9.** f   **10.** g

Table: **1.** ectoderm   **2.** nervous system   **3.** germ line   **4.** gamete-producing cells   **5.** mesoderm   **6.** mesoderm   **7.** skeletal muscles   **8.** exocrine cell   **9.** endoderm   **10.** epithelial cell   **11.** blood

Multiple Choice: **1.** a,d   **2.** a,c,d   **3.** c,e   **4.** c,e   **5.** b   **6.** c   **7.** a,c   **8.** d   **9.** c,e   **10.** a   **11.** c   **12.** b   **13.** d   **14.** c,e   **15.** b,d

Chapter 17

Concept Questions: **1.** descent with modification   **2.** populations   **3.** gene pool   **4.** Jean Baptiste de Lamarck (Lamarck)   **5.** (a)H.M.S. Beagle, (b)Galapagos Islands   **6.** artificial selection   **7.** Thomas Malthus   **8.** Alfred Russell Wallace   **9.** *Origin of Species by Means of Natural Selection*   **10.** overproduction, variation, limits on population growth (a struggle for existence), and differential reproductive success ("survival of the fittest")   **11.** mutation   **12.** index fossils   **13.** homologous   **14.** analogous   **15.** vestigial organs   **16.** biogeography   **17.** center of origin

Building Words: **1.** homologous   **2.** biogeography

Matching: **1.** a   **2.** d   **3.** f   **4.** h   **5.** j   **6.** l   **7.** k   **8.** i   **9.** e   **10.** b

Table: **1.** homologous structures   **2.** analogous structures   **3.** the presence of useless structures is to be expected as a species adapts to a changing mode of life   **4.** distribution of plants and animals   **5.** all organisms use a genetic code that is virtually identical

Multiple Choice: **1.** c   **2.** b   **3.** a   **4.** a,b   **5.** a   **6.** c,d   **7.** c   **8.** e   **9.** b,e   **10.** c   **11.** a   **12.** c,e   **13.** d

Chapter 18

Concept Questions: **1.** microevolution   **2.** breeding population   **3.** (a)$p^2 + 2pq + q^2$, (b)Hardy-Weinberg equilibrium   **4.** large population, isolation, no mutations, no selection, random mating   **5.** mutation   **6.** gene or allele frequencies   **7.** genetic bottleneck   **8.** founder effect   **9.** gene flow   **10.** natural selection   **11.** polymorphism   **12** (a)more common, (b)less common

Building Words: **1.** microevolution

Matching: **1.** l   **2.** i   **3.** g   **4.** e   **5.** b   **6.** a   **7.** d   **8.** f   **9.** h   **10.** k

Table: **1.** alleles in the genotypes of a breeding population   **2.** offspring of a single mating   **3.** many generations   **4.** genotype   **5.** gene pool   **6.** genetic variation studied   **7.** none   **8.** possible through a change in allele frequency

Multiple Choice: **1.** c,d   **2.** c   **3.** a   **4.** c   **5.** c,e   **6.** c   **7.** a,e   **8.** d   **9.** b,d,e   **10.** g   **11.** k   **12.** h   **13.** c   **14.** j   **15.** y

Chapter 19

Concept Questions: **1.** Linnaeus   **2.** (a)reproductively, (b)gene pool   **3.** fertilization   **4.** (a)temporal isolation, (b) behavioral isolation, (c)mechanical isolation, (d)gametic isolation   **5.** hybrid inviability   **6.** hybrid sterility   **7.** speciation   **8.** geographically isolated   **9.** hybridization and polyploidy   **10.** (a)polyploidy, (b)allopolyploidy   **11.** punctuated equilibrium   **12.** gradualism   **13.** (a)allometric growth, (b)paedomorphosis   **14.** (a)adaptive zones, (b)adaptive radiation   **15.** (a)background, (b)mass

Building Words: **1.** polyploidy   **2.** allopatric   **3.** allopolyploidy   **4.** macroevolution   **5.** allometric

Matching: **1.** g   **2.** k   **3.** n   **4.** a   **5.** d   **6.** e   **7.** i   **8.** m   **9.** b   **10.** c

Table: **1.** temporal isolation   **2.** black sage and white sage   **3.** hybrid inviability   **4.** nearly all hybrids die in the embryonic stage   **5.** gametic isolation   **6.** basket sponge (aquatic animals that release gametes in the water)   **7.** behavioral isolation   **8.** highly specialized courtship behaviors   **9.** temporal isolation   **10.** *Drosophila pseudoobscura and Drosophila persimilis*   **11.** hybrid sterility   **12.** geographical isolation   **13.** pupfish

Multiple Choice: **1.** d   **2.** e   **3.** c   **4.** b,d   **5.** b,d   **6.** a,c,e   **7.** a   **8.** c,d   **9.** e   **10.** a,c,e   **11.** a,b,e

Chapter 20

Concept Questions: **1.** absence of free oxygen, energy, chemical building blocks, time   **2.** (a)Oparin, (b)Haldane   **3.** (a)Miller and Urey, (b)hydrogen (H_2), methane (CH_4), ammonia (NH_3)   **4.** protobionts   **5.** (a)RNA, (b)ribosomes   **6.** (a)anaerobic, (b)prokaryotic   **7.** (a)heterotrophs, (b)autotrophs   **8.** oxygen   **9.** endosymbiont   **10.** (a)570 million, (b)Ordovician, (c)ostracoderm   **11.** (a)248 million, (b)Triassic, Jurassic, Cretaceous, (c)Triassic, (d)Jurassic   **12.** 65 million

Building Words: **1.** autotroph   **2.** protobiont   **3.** heterotroph   **4.** precambrian   **5.** aerobe   **6.** anaerobe

Matching: **1.** j   **2.** e   **3.** a   **4.** l   **5.** h   **6.** d   **7.** b   **8.** c   **9.** g   **10.** k

Table: **1.** 120 million years ago   **2.** cretaceous   **3.** rise of flowering plants, rise and fall of dinosaurs   **4.** formation of Pangaea   **5.** land mostly covered by sea   **6.** ordovician   **7.** marine algae, invertebrates, first fishes   **8.** Triassic   **9.** gymnosperms, ferns, first dinosaurs, egg-laying mammals   **10.** 50 million years from present   **11.** unknown   **12.** unknown

Multiple Choice: **1.** b,c,d   **2.** c,d,e   **3.** a,d   **4.** e   **5.** d   **6.** a,d   **7.** a,e   **8.** c,e   **9.** d   **10.** a,b,c,d   **11.** b,d   **12.** d,e   **13.** d   **14.** c

Chapter 21

Concept Questions: **1.** (a)prosimians, (b)anthropoids   **2.** (a)prosimian, (b)Oligocene   **3.** hominids   **4.** gibbons (*Hylobates*), orangutans (*Pongo*), gorillas (*Gorilla*), chimpanzees (*Pan*)   **5.** 4.4 million   **6.** *Australopithecus afarensis*   **7.** 2.5 million   **8.** 2.0 million   **9.** Africa   **10.** knowledge   **11.** development of hunter/gatherer societies, development of agriculture and the Industrial Revolution

Building Words: **1.** anthropoid   **2.** quadrupedal   **3.** bipedal   **4.** prosimian   **5.** hominoid   **6.** supraorbital

Matching: **1.** e   **2.** i   **3.** a   **4.** b   **5.** k   **6.** l   **7.** j   **8.** g   **9.** c   **10.** d

Table: **1.** *Australopithecus afarensis*   **2.** 3-4 million years ago   **3.** 2.5 million years ago   **4.** more modern hominid features than *Australopithecus*, first primate to design tools   **5.** 2 million years ago   **6.** larger brain than *H. habilis*, made more sophisticated tools, used fire   **7.** Cro-magnon   **8.** 40,000 years ago   **9.** Neanderthal

Multiple Choice: **1.** b,e   **2.** c,e   **3.** d   **4.** b,c,e   **5.** e   **6.** d,e   **7.** a,d,e   **8.** b   **9.** c,d,e   **10.** a

Chapter 22

Concept Questions: **1.** taxonomy   **2.** (a)binomial system, (b)genus, (c)species (specific epithet)   **3.** (a)family, (b)class, (c)phylum(a)   **4.** taxon   **5.** Prokaryotae, Protista, Fungi, Animalia, Plantae   **6.** evolutionary relationships   **7.** (a)monophyletic, (b)clade   **8.** polyphyletic   **9.** (a)ancient, (b)recent   **10.** DNA and protein   **11.** phenetics, cladistics, classical evolutionary taxonomy   **12.** phenotypic characteristics   **13.** monophyletic

Building Words: **1.** subspecies   **2.** monophyletic   **3.** polyphyletic   **4.** prokaryote   **5.** eukaryote

Matching: **1.** i   **2.** g   **3.** l   **4.** a   **5.** b   **6.** k   **7.** j   **8.** d   **9.** e   **10.** h

Table: **1.** phenetic system   **2.** cladistic approach   **3.** classical evolutionary taxonomy

Multiple Choice: **1.** c,e   **2.** b   **3.** d   **4.** d   **5.** d   **6.** c   **7.** b,d   **8.** a   **9.** a,c,d

Chapter 23

Concept Questions: **1.** (a)nucleic acid (DNA or RNA), (b)capsid   **2.** "bits" of nucleic acids that "escaped" from cells   **3.** (a)virulent (or lytic), (b)temperate (or lysogenic)   **4.** attachment, penetration, replication, assembly, release   **5.** (a)prophage, (b)lysogenic   **6.** lysogenic conversion   **7.** transduction   **8.** retroviruses   **9.** viroids   **10.** (a)cocci, (b)diplococci, (c)streptococci, (d)staphylococci, (e)bacilli, (f)spirilla   **11.** peptidoglycan   **12.** (a)peptidoglycan, (b)crystal violet   **13.** plasmids   **14.** (a)transformation, (b)transduction, (c)conjugation   **15.** (a)saprobes, (b)chemosynthetic autotrophs (chemoautotrophs)   **16.** carbon dioxide and hydrogen   **17.** axial filaments   **18.** (a)cyanobacteria, (b)photosynthetic lamellae   **19.** arthropods   **20.** streptococci   **21.** staphylococci   **22.** clostridia   **23.** (a)actinomycetes, (b)actinospores

Building Words: **1.** bacteriophage   **2.** exotoxin   **3.** endotoxin   **4.** archaebacteria   **5.** halobacteria   **6.** methanogen   **7.** endospore   **8.** eubacteria   **9.** pathogen   **10.** anaerobe

Matching: **1.** l   **2.** a   **3.** g   **4.** e   **5.** b   **6.** c   **7.** h   **8.** d   **9.** k   **10.** i

Table: **1.** single-stranded RNA   **2.** capsid   **3.** rabies   **4.** 2 identical single-strands of RNA   **5.** capsid   **6.** parvovirus   **7.** DNA   **8.** gram-negative cell wall   **9.** none   **10.** mycobacteria   **11.** Clostridium botulinum   **12.** gram-positive cell wall

Multiple Choice: **1.** c   **2.** a,b   **3.** e   **4.** d   **5.** c   **6.** a,b   **7.** a,b,d   **8.** a   **9.** a,b,e   **10.** b   **11.** c   **12.** a,c   **13.** d   **14.** d   **15.** b,d,e   **16.** b,c,d,e   **17.** a

Chapter 24

Concept Questions: **1.** coenocytic   **2.** plankton   **3.** pseudopodia   **4.** *Entamoeba histolytica*   **5.** axopods   **6.** *Trypanosoma*   **7.** choanoflagellates   **8.** trichocysts   **9.** (a)micronuclei, (b)macronucleus   **10.** conjugation   **11.** *Plasmodium*   **12.** Zooxanthellae   **13.** diatomaceous earth   **14.** (a)chlorophyll *a*, chlorophyll *b*, carotenoids, (b)paramylon   **15.** lichen   **16.** (a)isogamous, (b)anisogamous, (c)oogamous   **17.** (a)holdfast, (b)phycoerythrin, (c)phycocyanin   **18.** (a)blades, (b)stipes   **19.** sporangia   **20.** (a)swarm cell, (b)myxamoeba   **21.** pseudoplasmodium (slug)   **22.** (a)mycelium, (b)zoospores, (c)oospores

Building Words: **1.** syngamy   **2.** protozoa   **3.** cytopharynx   **4.** pseudopod   **5.** trichocyst   **6.** micronucleus   **7.** macronucleus   **8.** pseudoplasmodium   **9.** isogamy

Matching: **1.** b   **2.** k   **3.** g   **4.** i   **5.** d   **6.** c   **7.** f   **8.** e   **9.** j   **10.** l

Table: **1.** flagellates   **2.** heterotrophic (ingest food)   **3.** parasitic   **4.** nonmotile   **5.** free living, some endosymbiotic   **6.** autotrophic (photosynthetic)   **7.** free living, some symbiotic   **8.** autotrophis (photosynthetic)   **9.** most flagellated, some nonmotile   **10.** free living   **11.** heterotrophic (ingest food)   **12.** heterotrophic (absorb nutrients)

Multiple Choice: **1.** b   **2.** b,d,e   **3.** c   **4.** b   **5.** c,d   **6.** a,c,e   **7.** b   **8.** c,d,e   **9.** a,d,e   **10.** b   **11.** a

Chapter 25

Concept Questions: 1. chitin 2. (a)yeasts, (b)molds 3. (a)dikaryotic, (b)monokaryotic 4. (a)sexual spores and fruiting bodies, (b)Zygomycota, Ascomycota, Basidiomycota 5. (a)*Rhizopus stolonifer*, (b)heterothallic 6. (a)asci, (b)conidia, (c)conidiophores 7. ascocarps 8. (a)basidium, (b)basidiospores 9. (a)button, (b)basidiocarp 10. sexual stage 11. (a)crustose, (b)foliose, (c)fruticose 12. (a)green alga or cyanobacterium, (b)ascomycete 13. soredia 14. (a)saprophytes, (b)carbon dioxide, (c)minerals (plant nutrients) 15. mycorrhizae (fungus-roots) 16. (a)yeasts, (b)fruit sugars, (c)grains, (d)carbon dioxide 17. (a)*Penicillium*, (b)*Aspergillus tamarii* 18. *Amanita* 19. psilocybin 20. (a)*Penicillium notatum*, (b)penicillin 21. (a)ergot, (b)ergotism 22. (a)cutinase, (b)haustoria

Building Words: 1. saprotroph 2. coenocytic 3. monokaryotic 4. heterothallic 5. conidiophore 6. homothallic

Matching: 1. j 2. k 3. b 4. c 5. m 6. l 7. e 8. d 9. h 10. f

Table: 1. ascospores 2. yeasts, cup fungi, morels, truffles, and green and pink molds 3. Basidiomycota 4. Deuteromycota 5. sexual stage not observed

Multiple Choice: 1. d 2. b,d 3. c 4. b,d 5. b 6. a,c,e 7. a-e 8. e 9. a,c,e 10. a,b,d 11. a-e 12. c,d,e 13. a,d 14. a-e

Chapter 26

Concept Questions: 1. (a)green algae; (b)chlorophylls *a* and *b*, carotenes, xanthophylls; (c)starch, (d)cellulose, (e)cell plate 2. (a)mosses and other bryophytes, (b)ferns and their allies, (c)gymnosperms, (d)angiosperms (flowering plants) 3. (a)waxy cuticle, (b)stomata 4. (a)gametangia, (b)antheridium, (c)archegonium 5. lignin 6. (a)zygote (fused gametes), (b)spores 7. mosses, liverworts, hornworts 8. vascular tissues 9. (a)*Sphagnum*, (b)peat mosses 10. gemmae 11. (a)microphyll, (b)megaphyll 12. (a)rhizome, (b)fronds 13. (a)sporangia, (b)sori, (c)prothallus 14. (a)upright stem, (b)dichotomous, (c)fungus 15. *Equisitum* 16. (a)homospory, (b)heterospory 17. (a)microsporangia, (b)microsporocytes, (c)microspores, (d)megaspores, (e)megasporocytes

Building Words: 1. xanthophyll 2. gametangium 3. archegonium 4. gametophyte 5. sporophyte 6. microphyll 7. megaphyll 8. sporangium 9. homospory 10. heterospory 11. megaspore 12. microspore 13. bryophyte

Matching: 1. d 2. c 3. g 4. i 5. b 6. e 7. a 8. k 9. m 10. h

Table: 1. nonvascular 2. seedless, reproduce by spores 3. gametophyte 4. seedless, reproduce by spores 5. vascular 6. sporophyte 7. nonvascular 8. seedless, reproduce by spores 9. seedless, reproduce by spores 10. sporophyte 11. vascular 12. sporophyte 13. vascular 14. naked seeds

Multiple Choice: 1. a,c,d 2. a,b,c 3. a,e 4. a,c,d 5. b 6. a-e 7. b,d 8. c,d 9. a,e 10. a,b,c 11. b,d 12. b,d 13. b,c,e 14. b,e 15. a,c,e

Chapter 27

Concept Questions: 1. (a)gymnosperms, (b)angiosperms 2. (a)xylem, (b)phloem 3. Coniferophyta (conifers), Ginkgophyta, Cycadophyta, Gnetophyta 4. monoecious 5. pines, spruces, firs, larch, cypress, hemlock, etc. 6. (a)sporophylls, (b)microsporocytes, (c)male gametophytes (pollen grains) 7. (a)megasporangia, (b)haploid megaspores, (c)female gametophyte 8. dioecious 9. vessels 10. (a)three, (b)one, (c)endosperm 11. (a)four or five, (b)two, (c)cotyledons 12. (a)palms, grasses, orchids, lilies, etc. (b)oaks, roses, cacti, blueberries, sunflowers, etc. 13. (a)sepals, petals, stamens, carpels; (b)stamens; (c)carpels; (d)perfect 14. (a)anther, (b)ovary 15. (a)megaspores, (b)embryo sac 16. (a)microspores, (b)male gametophyte (pollen grain), (c)sperm nuclei 17. zygote, endosperm tissue 18. (a)progymnosperms, (b)seed ferns, (c)gymnosperms

Building Words: **1.** angiosperm **2.** gymnosperm **3.** monoecious **4.** dioecious

Matching: **1.** a **2.** e **3.** f **4.** i **5.** k **6.** m **7.** h **8.** l **9.** b **10.** c

Table: **1.** endosperm **2.** cotyledons **3.** 1 cotyledon **4.** vascular bundles arranged in a circle **5.** parallel venation **6.** netted venation **7.** long tapering blades **8.** herbaceous or woody

Multiple Choice: **1.** a-e **2.** b,e **3.** b **4.** b,c **5.** a,b,d,e **6.** a,c,d,e **7.** a-e **8.** a,e **9.** b **10.** c,d,e

Chapter 28

Concept Questions: **1.** (a)sea (salt) water, (b)fresh water, (c)land (terrestrial) **2.** (a)anterior, (b)posterior, (c)dorsal, (d)ventral, (e)medial, (f)lateral, (g)cephalic, (h)caudal **3.** (a)acoelomates, (b)pseudocoelomates, (c)coelomates **4.** (a)mouth, (b)anus, (c)radial cleavage, (d)spiral cleavage **5.** choanoflagellates **6.** (a)Calcarea, (b)Hexactinellida, (c)Demospongia, (d)spongin **7.** (a)spongocoel, (b)osculum **8.** (a)cnidocytes, (b)epidermis, (c)gastrodermis, (d)mesoglea **9.** (a)Hydrozoa, (b)Scyphozoa, (c)Anthozoa **10.** (a)cnidocytes, (b)nematocysts **11.** (a)medusa, (b)polyp **12.** (a)biradially, (b)eight, (c)glue cells **13.** (a)bilateral, (b)three, (c)protonephridia **14.** (a)Turbellaria, (b)Trematoda and Monogenea, (c)Cestoda **15.** pharynx **16.** suckers **17.** (a)scolex (head), (b)proglottid **18.** proboscis **19.** (a)three, (b)pseudocoelom, (c)bilateral **20.** hookworm **21.** (a)trichina worms, (b)cysts **22.** pinworms **23.** (a)cell constancy, (b)cilia

Building Words: **1.** pseudocoelom **2.** ectoderm **3.** mesoderm **4.** protostome **5.** deuterostome **6.** schizocoely **7.** enterocoely **8.** spongocoel **9.** amebocyte **10.** gastrodermis

Matching: **1.** f **2.** k **3.** h **4.** a **5.** d **6.** e **7.** c **8.** m **9.** j **10.** l

Table: **1.** multicellular, cells loosely arranged **2.** organs **3.** organ systems **4.** not present **5.** acoelomate **6.** irritability of cytoplasm **7.** cephalization, ladder-type system, sense organs **8.** diffusion **9.** diffusion **10.** intracellular **11.** complete digestive tract **12.** asexual: budding; sexual: most are hermaphroditic **13.** sexual: separate sexes **14.** carnivores, parasites **15.** carnivores, parasites

Multiple Choice: **1.** a,d,e **2.** d **3.** a,c,e **4.** c,e **5.** c,e **6.** c **7.** d,e **8.** a,b,c **9.** d **10.** c **11.** a,d **12.** e **13.** a **14.** b

Chapter 29

Concept Questions: **1.** (a)mouth, (b)mesoderm **2.** independent movement of digestive tract (food) and body, coelomic fluid bathes coelomic cells and organs (transports food, oxygen, wastes), acts as hydrostatic skeleton, space for development and function of organs **3.** Arthropoda **4.** (a)desiccation (fluid loss), (b)gravity **5.** (a)mantle, (b)Arthropoda **6.** hemocoel **7.** (a)trochophore, (b)veliger **8.** eight **9.** insects **10.** torsion **11.** (a)mantle, (b)calcium carbonate **12.** adductor muscle **13.** (a)tentacles, (b)ten, (c)eight **14.** septa **15.** setae **16.** metanephridia **17.** parapodia **18.** (a)Annelida, (b)Oligochaeta, (c)*Lumbricus terrestris* **19.** (a)crop, (b)gizzard **20.** (a)hemoglobin, (b)metanephridia, (c)skin **21.** (a)jointed appendages, (b)exoskeleton, (c)chitin **22.** head, thorax, abdomen **23.** tracheae **24.** (a)Chelicerata, (b)Crustacea, (c)Uniramia **25.** (a)trilobites, (b)chelicerates **26.** (a)Merostomata, (b)Arachnida **27.** (a)cephalothorax, (b)six, (c)four **28.** book lungs **29.** spinnerets **30.** (a)mandibles, (b)biramous, (c)two, (d)five **31.** barnacles **32.** Decapoda **33.** (a)maxillae, (b)maxillipeds, (c)chelipeds, (d)walking legs **34.** (a)reproductive, (b)swimmerets **35.** (a)jointed, (b)possess tracheal tubes, (c)have six feet **36.** (a)three, (b)one or two, (c)one, (d)Malpighian tubules **37.** (a)Chilopoda, (b)one, (c)Diplopoda, (d)two

Building Words: **1.** endoskeleton **2.** exoskeleton **3.** bivalve **4.** biramous **5.** uniramous **6.** trilobite **7.** cephalothorax **8.** hexapod **9.** arthropod

Matching: **1.** c **2.** i **3.** a **4.** h **5.** g **6.** b **7.** j **8.** k **9.** f **10.** e

Table: **1.** organ system **2.** organ system **3.** coelomate protostomes **4.** coelomate protostomes **5.** 3 pairs of ganglion, sense organs **6.** brain, ventral nerve cords, well-developed sense organs **7.** closed system **8.** open system **9.** complete digestive tract **10.** complete digestive tract **11.** sexual: hermaphroditic **12.** herbivores, carnivores, scavengers, suspension feeders

Multiple Choice: **1.** b,d,e **2.** b,c,e **3.** b **4.** d **5.** b,d,e **6.** a,b,d **7.** b,d **8.** a,c,e **9.** a,c,d **10.** a,d

Chapter 30

Concept Questions: **1.** (a)anus, (b)mouth **2.** (a)spiny, (b)bilateral, (c)radial (pentaradial), (d)coelom **3.** dorsal (upper) **4.** (a)central disk, (b)tube feet **5.** test **6.** tube feet **7.** bilateral **8.** (a)notochord, (b)nerve cord, (c)pharyngeal gill slits **9.** cellulose **10.** *Amphioxus* **11.** pharynx **12.** (a)vertebral column, (b)cranium, (c)cephalization **13.** (a)Agnatha, (b)Chondrichthyes, (c)Osteichthyes, (d)Amphibia, (e)Reptilia, (f)Aves, (g)Mammalia **14.** ostracoderms **15.** placoid scales **16.** (a)lateral line organs, (b)electroreceptors, **17.** (a)oviparous, (b)ovoviviparous, (c)viviparous **18.** (a)sacropterygians, (b)coelacanths **19.** (a)ray-finned, (b)sarcopterygians, (c)teleosts, (d)swim bladders **20.** (a)Urodela, (b)Anura, (c)Apoda **21.** neoteny **22.** (a)moist skin, (b)three, (c)mucous glands **23.** (a)three, (b)uric acid, (c)ectothermic **24.** (a)Chelonia, (b)Squamata, (c)Crocodilia **25.** Mesozoic **26.** (a)four, (b)endothermic, (c)uric acid, (d)feathers **27.** (a)crop, (b)proventriculus, (c)gizzard **28.** *Archaeopteryx* **29.** hair, mammary glands, differentiated teeth **30.** (a)therapsids, (b)Triassic period **31.** monotremes **32.** (a)marsupials, (b)marsupium **33.** oxygen and nutrients

Building Words: **1.** agnathan **2.** anuran **3.** apodal **4.** Chondrichthyes **5.** Osteichthyes **6.** tetrapod **7.** Urodela **8.** echinoderm

Matching: **1.** g **2.** h **3.** f **4.** i **5.** a **6.** l **7.** m **8.** k **9.** n **10.** o

Table: **1.** sharks, rays, skates **2.** cartilage **3.** jawed, gills, placoid scales, well-developed sense organs **4.** Osteichthyes **5.** two-chambered heart **6.** bone **7.** salamanders, frogs and toads, caccilians **8.** three chambered heart **9.** Reptilia **10.** turtles, lizards, snakes, alligators **11.** bone **12.** Aves **13.** four chambered heart **14.** light hollow bone with air spaces **15.** Mammalia **16.** bone **17.** tetrapods with hair, mammary glands, constant body temperature and a highly developed nervous system

Multiple Choice: **1.** c **2.** d,e **3.** a,c,d **4.** d,e **5.** a,b **6.** c,e **7.** c,e **8.** b,c **9.** d **10.** b,e **11.** c,d,e **12.** c,d,e **13.** b,c,d **14.** a-e **15.** d **16.** d **17.** a,b **18.** a,d **19.** c,e **20.** b,c,d **21.** d

Chapter 31

Concept Questions: **1.** (a)perennials, (b)annuals, (c)biennials **2.** (a)tissues, (b)organs **3.** photosynthesis, storage, secretion **4.** support **5.** (a)water and dissolved minerals, (b)parenchyma cells, (c)fibers **6.** food **7.** (a)sieve tube members, (b)sieve plates **8.** (a)protection, (b)cuticle, (c)stomata **9.** protection **10.** (a)meristems, (b)division, elongation, differentiation (specialization) **11.** (a)primary, (b)secondary **12.** (a)area of cell division, (b)area of cell elongation, (c)area of cell maturation **13.** (a)leaf primordia, (b)bud primordia **14.** (a)vascular cambium, (b)cork cambium

Building Words: **1.** trichome **2.** biennial **3.** epidermis **4.** stoma

Matching: **1.** o **2.** m **3.** a **4.** n **5.** c **6.** k **7.** b **8.** h **9.** f **10.** g

Table: **1.** ground tissue **2.** stems and leaves **3.** sclerenchyma **4.** structural support **5.** xylem **6.** vascular tissue **7.** conducts sugar in solution **8.** extents throughout plant body **9.** dermal tissue **10.** protection of plant body, controls gas exchange and water loss on stems, leaves **11.** covers body of herbaceous plants **12.** periderm **13.** protection of plant body

Multiple Choice: **1.** c **2.** c,d,e **3.** a,b,c,e **4.** e **5.** d **6.** c **7.** a,d **8.** a,b,e

Chapter 32

Concept Questions: **1.** (a)mesophyll, (b)palisade, (c)spongy **2.** (a)xylem, (b)phloem, (c)bundle sheath **3.** (a)netted or palmate, (b)parallel, (c)subsidiary **4.** amount of light, [CO_2], a circadian rhythm, humidity **5.** (a)vacuoles, (b)guard, (c)open, (d)close **6.** (a)guttation, (b)transpiration, (c)temperature, light, wind, relative humidity **7.** hormones **8.** (a)abscission zone, (b)fibers, **9.** (a)spines, (b)tendrils

Building Words: **1.** abscission **2.** circadian - **3.** mesophyll **4.** transpiration

Matching: **1.** b **2.** j **3.** g **4.** k **5.** c **6.** h **7.** f **8.** d **9.** m **10.** e

Table: **1.** efficient capturing of sunlight **2.** helps reduce or control water loss **3.** stomata **4.** allows light to penetrate to photosynthetic tissue **5.** mesophyll **6.** air space in mesophyll tissue **7.** bundle sheaths **8.** provide water and minerals from roots **9.** phloem in veins

Multiple Choice: **1.** a,c,d **2.** a-e **3.** d **4.** a,b,e **5.** a,b,c **6.** a,d,e **7.** b,c,e **8.** c **9.** b,c,e

Chapter 33

Concept Questions: **1.** support, conduct, produce new stem tissue **2.** (a)vascular bundles, (b)xylem, phloem, (c)vascular cambium, (d)pith, (e)cortex **3.** ground tissue (parenchyma) **4.** (a)vascular cambium, (b)cork cambium **5.** (a)rays, (b)parenchyma **6.** (a)cork cells, (b)cork parenchyma **7.** (a)heartwood, (b)sapwood **8.** (a)springwood, (b)late summerwood **9.** (a)water potential, (b)less, (c)more, (d)less **10.** soil and root cells **11.** (a)tension-cohesion mechanism, (b)transpiration **12.** (a)pressure flow, (b)source, (c)sink

Building Words: **1.** translocation

Matching: **1.** a **2.** g **3.** h **4.** b **5.** c **6.** l **7.** m **8.** j **9.** i **10.** f

Table: **1.** secondary phloem **2.** conducts dissolved sugar **3.** produced by cork cambium **4.** storage **5.** cork cells **6.** periderm **7.** vascular cambium **8.** produces secondary xylem and secondary phloem **9.** a lateral meristem, usually arises from parenchyma **10.** produces periderm (secondary growth)

Multiple Choice: **1.** a,b,d **2.** a,b,c,d **3.** a,c,e **4.** a,b,d **5.** a,d **6.** b **7.** b,d,e **8.** c **9.** d **10.** a-e

Chapter 34

Concept Questions: **1.** anchorage, absorption, conduction, storage **2.** (a)endodermis, (b)Casparian strip **3.** (a)pericycle, (b)vascular **4.** (a)cell walls, (b)cellulose, (c)endodermis, (d)epidermis, (e)endodermis, (f)xylem **5.** (a)storage, (b)pericycle **6.** (a)adventitious, (b)prop, (c)contractile **7.** minerals, organic matter, organisms, soil air, soil water **8.** (a)sand and silt, (b)clay, (c)humus, (d)soil air and water **9.** hydroponics **10.** (a)essential, (b)carbon, oxygen, hydrogen, nitrogen, potassium, phosphorus, sulfur, magnesium, calcium, (c)iron, boron, manganese, copper, zinc, molybdenum, chlorine **11.** nitrogen, phosphorus, potassium

Definitions: **1.** hydroponics **2.** macronutrient **3.** micronutrient

Building Words: **1.** hydroponics **2.** macronutrient **3.** micronutrient

Matching: **1.** i **2.** g **3.** d **4.** a **5.** c **6.** k **7.** j **8.** e **9.** m **10.** l

Table: **1.** area of dividing cells that causes an increase in length of the root **2.** root hairs **3.** protects root **4.** storage **5.** endodermis **6.** pericycle **7.** xylem **8.** conducts dissolved sugars

Multiple Choice: **1.** c **2.** a,b,c **3.** b,d **4.** a,b,d,e **5.** a,b,d **6.** a,c,e **7.** a,d **8.** c **9.** c **10.** b **11.** c **12.** c **13.** b,e

Chapter 35

Concept Questions: **1.** simple, aggregate, multiple, accessory   **2.** (a)simple, (b)berries, (c)drupes, (d)follicle, (e)legumes, (f)capsules   **3.** (a)aggregate, (b)multiple   **4.** (a)accessory, (b)receptacle, (c)floral tube   **5.** wind, animals, water, explosive dehiscence   **6.** (a)rhizomes, (b)tubers, (c)bulbs, (d)corms, (e)stolons   **7.** suckers   **8.** apomixis

Building Words: **1.** endosperm   **2.** hypocotyl

Matching: **1.** d   **2.** g   **3.** l   **4.** b   **5.** a   **6.** k   **7.** j   **8.** m   **9.** c   **10.** f

Table: **1.** accessory   **2.** dry: does not open   **3.** nut: hard, thick fruit wall   **4.** aggregate   **5.** develops from a flower with many separate ovaries   **6.** simple   **7.** fleshy   **8.** drupe: hard, stony pit, one seed   **9.** legume: splits along 2 sides   **10.** grain: single seed fused to fruit wall   **11.** multiple   **12.** formed from the fused ovaries of many flowers

Multiple Choice: **1.** d   **2.** d   **3.** b,d,e   **4.** b,d,e   **5.** b,d,e   **6.** e   **7.** e   **8.** d

Chapter 36

Concept Questions: **1.** circadian rhythms   **2.** (a)pulvinus, (b)potassium, (c)turgor movements   **3.** solar tracking   **4.** (a)phototropism, (b)gravitropism, (c)thigmotropism   **5.** auxins, gibberellins, cytokinins, ethylene, abscisic acid   **6.** (a)curvature, (b)indoleacetic acid (IAA), (c)shoot tip, (d)polar transport   **7.** flowering and germination   **8.** (a)cell division and differentiation, (b)senescence   **9.** fruit ripening and leaf abscission   **10.** water

Building Words: **1.** phototropism   **2.** gravitropism   **3.** phytochrome   **4.** photoperiodism

Matching: **1.** b   **2.** i   **3.** a   **4.** m   **5.** j   **6.** d   **7.** c   **8.** f   **9.** l   **10.** e

Table: **1.** auxins   **2.** shoot apical meristem, young leaves, seeds   **3.** cytokinin   **4.** gibberellin   **5.** shoot and root apical meristems, young leaves, embryo in seeds   **6.** abscisic acid   **7.** older leaves, root cap, stem   **8.** ethylene   **9.** roots   **10.** auxins

Multiple Choice: **1.** a-e   **2.** e   **3.** c   **4.** a,d   **5.** e   **6.** a   **7.** b   **8.** a   **9.** d

Chapter 37

Concept Questions: **1.** number   **2.** (a)tissue, (b)organs, (c)organ systems   **3.** (a)sheets (continuous layers), (b)basement membrane   **4.** protection, absorption, secretion, sensation   **5.** (a)simple, (b)stratified, (c)squamous, (d)cuboidal, (e)columnar   **6.** loose, dense, elastic, reticular, adipose, cartilage, bone, blood, lymph, blood cell-producing cells   **7.** (a)intercellular substance, (b)fibers, (c)matrix   **8.** intercellular substance   **9.** (a)collagen, (b)elastin, (c)collagen and glycoprotein   **10.** (a)fibers and matrix, (b)macrophages   **11.** (a)mast cells, (b)plasma cells   **12.** (a)areolar, (b)subcutaneous layer   **13.** (a)collagen, (b)regular dense, (c)tendons, (d)irregular dense   **14.** arteries   **15.** liver, spleen, lymph nodes   **16.** energy storage, cushioning internal organs   **17.** skeleton (endoskeleton)   **18.** (a)cartilage, (b)bone, (c)Chondrichthyes (sharks, rays, etc.)   **19.** (a)chondrocytes, (b)matrix, (c)lacunae   **20.** (a)matrix, (b)osteocytes, (c)vascularized   **21.** (a)hydroxyapatite crystals, (b)canaliculi   **22.** (a)osteon, (b)Haversian canal, (c)lamellae   **23.** plasma   **24.** (a)erythrocytes, (b)hemoglobin, (c)biconcave disc   **25.** leukocytes   **26.** bone marrow   **27.** fibers   **28.** (a)actin, (b)myosin, (c)myofibrils   **29.** (a)skeletal, (b)smooth, (c)cardiac, (d)smooth, cardiac, (e)skeletal   **30.** (a)neurons, (b)glial   **31.** synapses   **32.** nerve   **33.** (a)cell body, (b)dendrite(s), (c)axon(s)   **34.** integumentary, skeletal, muscle, nervous, circulatory, digestive, respiratory, urinary, endocrine, reproductive

Building Words: **1.** multicellular   **2.** pseudostratified   **3.** fibroblast   **4.** intercellular   **5.** macrophage   **6.** chondrocyte   **7.** osteocyte   **8.** myofibril   **9.** homeostasis

Matching: **1.** l   **2.** a   **3.** k   **4.** c   **5.** i   **6.** e   **7.** b   **8.** g   **9.** f   **10.** d

Table: **1.** simple squamous epithelium **2.** air sacs of lungs, lining of blood vessels **3.** epithelial tissue **4.** some respiratory passages, ducts of many glands **5.** secretion, movement of mucus, protection **6.** connective tissue **7.** energy storage, protection of some organs, insulation **8.** bone **9.** support and protection of internal organs, calcium reserve, site of skeletal muscle attachments **10.** connective tissue **11.** within organs of the cardiovascular system **12.** transport of oxygen, nutrients, waste product and other materials **13.** walls of the heart **14.** skeletal muscle **15.** nervous tissue

Multiple Choice: **1.** a,c,d **2.** a,b,c,e **3.** d **4.** a,c,e **5.** b **6.** d **7.** a,c,d **8.** b,e **9.** b **10.** c **11.** a,c,d **12.** c,e **13.** a,c **14.** c **15.** a **16.** b,c,d **17.** d **18.** a,b,e **19.** a,c,e **20.** c

Chapter 38

Concept Questions: **1.** skeletal **2.** epithelial **3.** secretory **4.** skin **5.** skin **6.** (a)epidermis, (b)strata **7.** stratum basale **8.** (a)keratin, (b)strength, flexibility, waterproofing **9.** (a)connective tissue, (b)subcutaneous **10.** muscle **11.** exoskeleton **12.** (a)hydrostatic skeleton, (b)longitudinally, (c)circularly, (d)outer, (e)inner **13.** septa **14.** protection **15.** molting **16.** (a)endoskeletons, (b)chordates **17.** echinoderms **18.** (a)axial, (b)appendicular **19.** (a)skull, (b)vertebral column, (c)rib cage, (d)centrum, (e)neural arch **20.** (a)pectoral girdle, (b)pelvic girdle, (c)limbs, (d)opposable digit (thumb) **21.** (a)periostium, (b)epiphyses, (c)diaphysis, (d)metaphysis, (e)epiphyseal line, (f)compact bone, (g)cancellous **22.** (a)endochondral, (b)intramembranous **23.** (a)joints, (b)immovable joints, (c)slightly movable joints, (d)freely movable joints **24.** (a)actin, (b)myosin **25.** (a)smooth, (b)striated **26.** fascicles **27.** (a)actin and myosin, (b)sarcomere **28.** inward **29.** (a)acetylcholine, (b)action potential **30.** (a)calcium ions, (b)actin, (c)myosin **31.** shortening **32.** (a)ATP, (b)creatine phosphate, (c)glycogen **33.** (a)tendons, (b)antagonistically, (c)antagonist **34.** (a)smooth, (b)cardiac, (c)skeletal, (d)slow-twitch, (e)fast-twitch

Building Words: **1.** epidermis **2.** periosteum **3.** endochondral **4.** osteoblast **5.** osteoclast **6.** myofilament

Matching: **1.** h **2.** f **3.** j **4.** l **5.** g **6.** b **7.** d **8.** m **9.** i **10.** a

Table: **1.** chitin **2.** exoskeleton **3.** external covering jointed for movement, does not grow so animal must periodically molt **4.** echinoderms **5.** calcium salts **6.** cartilage or bone **7.** endoskeleton **8.** internal, composed of living tissue and grows with the animal

Multiple Choice: **1.** c **2.** a,b **3.** e **4.** c,d **5.** a **6.** d **7.** d **8.** c,e **9.** a **10.** a,b,d **11.** b,c **12.** a,c **13.** b,e **14.** b **15.** b,e

Chapter 39

Concept Questions: **1.** (a)endocrine, (b)nervous **2.** stimuli **3.** (a)reception, (b)afferent, (c)transmit, (d)interneurons, (e)integration, (f)efferent, (g)effectors **4.** neuron **5.** neuroglia **6.** (a)dendrites, (b)cell body, (c)axon **7.** (a)axon terminals, (b)synaptic knobs, (c)neurotransmitter **8.** (a)Schwann cells, (b)cellular sheath, (c)myelin sheath, (d)myelin **9.** (a)nerve, (b)tract (pathway) **10.** (a)ganglia, (b)nuclei **11.** (a)resting potential, (b)-70 mV **12.** inner **13.** (a)inside, (b)outside **14.** (a)sodium-potassium pump, (b)ion-specific channels, (c)protein anions **15.** (a)action potential, (b)sodium ions, (c)+35 mV **16.** (a)voltage-activated ion pumps, (b)threshold level, (c)-55 mV **17.** (a)spike, (b)voltage-activated sodium channels **18.** wave of depolarization **19.** repolarization **20.** node of Ranvier **21.** all-or-none **22.** (a)synapse, (b)neuroglandular, (c)neuromuscular (motor end plates) **23.** (a)presynaptic, (b)postsynaptic **24.** (a)electrical synapses, (b)chemical synapses, (c)neurotransmitter molecules, (d)synaptic cleft **25.** (a)receptors, (b)chemically activated ion channels **26.** (a)excitatory postsynaptic potential (EPSP), (b)inhibitory postsynaptic potential (IPSP) **27.** gradual potentials **28.** (a)summation, (b)temporal summation, (c)spatial summation **29.** (a)acetylcholine, (b)neuromuscular **30.** (a)adrenergic neurons, (b)catecholamine (biogenic amine) **31.** further apart **32.** dendrites and cell body **33.** convergence, divergence, facilitation

Building Words: **1.** interneuron **2.** neuroglia **3.** neurilemma **4.** neurotransmitter **5.** multipolar **6.** postsynaptic **7.** presynaptic

Matching: **1.** d **2.** j **3.** f **4.** b **5.** i **6.** k **7.** a **8.** h **9.** e **10.** n

Table: **1.** -70mV **2.** stable **3.** sodium-potassium pump active, sodium and potassium channels open **4.** threshold potential **5.** voltage-activated sodium ion channels and potassium ion channels open **6.** +35mV **7.** rise: depolarization; fall: repolarization **8.** depolarization **9.** chemically-activated sodium channels open **10.** hyperpolarization

Multiple Choice: **1.** b,c,e **2.** b **3.** b,c **4.** c,d,e **5.** c,d **6.** b **7.** a,d **8.** a,c **9.** c **10.** b,c **11.** d **12.** c,e **13.** a

Chapter 40

Concept Questions: **1.** (a)nerve net, (b)cnidarians **2.** nerve ring **3.** (a)ladder-type, (b)cerebral ganglia **4.** (a)ganglia, (b)ventral nerve cords **5.** (a)central nervous (CNS), (b)peripheral nervous (PNS), (c)sympathetic and parasympathetic **6.** (a)neural tube, (b)forebrain, midbrain, hindbrain **7.** rhombencephalon **8.** (a)cerebellum, (b)pons, (c)metencephalon, (d)medulla, (e)myelencephalon **9.** brainstem **10.** association **11.** (a)superior colliculi, (b)inferior colliculi, (c)red nucleus **12.** (a)prosencephalon, (b)thalamus and hypothalamus, (c)cerebrum **13.** (a)lateral ventricles, (b)third ventricle **14.** (a)thalamus, (b)hypothalamus, (c)pituitary **15.** corpus striatum **16.** (a)hemispheres, (b)white matter, (c)cerebral cortex, (d)gyri, (e)sulci **17.** (a)dura matter, arachnoid, pia mater, (b)choroid plexus **18.** second lumbar **19.** (a)white matter, (b)tracts **20.** (a)ascending tracts, (b)descending tracts **21.** (a)reflex action, (b)sensory, (c)association, (d)motor **22.** (a)sensory, (b)motor, (c)association **23.** (a)frontal lobes, (b)parietal, (c)central sulcus, (d)temporal, (e)occipital **24.** brain stem and thalamus **25.** cerebrum and diencephalon **26.** (a)electroencephalogram (EEG), (b)delta, (c)alpha, (d)beta **27.** (a)rapid eye movement (REM), (b)non-REM **28.** (a)sensory, (b)short term, (c)long term **29.** cortex **30.** (a)sense receptors, (b)sensory afferent neurons, (c)motor efferent neurons **31.** (a)cranial, (b)spinal **32.** (a)dorsal, (b)ventral **33.** (a)sympathetic, (b)parasympathetic

Building Words: **1.** hypothalamus **2.** postganglionic **3.** preganglionic **4.** paravertebral

Matching: **1.** g **2.** d **3.** j **4.** c **5.** h **6.** a **7.** e **8.** k **9.** b **10.** i

Table: **1.** pons **2.** midbrain **3.** center for visual and auditory reflexes **4.** diencephalon **5.** relay center for motor and sensory information **6.** hypothalamus **7.** diencephalon **8.** metencephalon **9.** cerebrum **10.** reticular activating system **11.** limbic system

Multiple Choice: **1.** a,c **2.** d **3.** a,c **4.** a,b,d **5.** a,c,d,e **6.** a-e **7.** b **8.** b **9.** b **10.** e **11.** b,d **12.** d **13.** b **14.** a,d,e

Chapter 41

Concept Questions: **1.** neuron **2.** (a)touch, smell, taste, sight, hearing, (b)balance, pressure, pain, temperature, proprioception **3.** (a)exteroceptors, (b)proprioceptors, (c)interoceptors **4.** (a)mechanoreceptors, (b)electroreceptors, (c)thermoreceptors, (d)chemoreceptors **5.** (a)electrical, (b)receptor **6.** (a)graded, (b)depolarize, (c)action potential, (d)sensory **7.** more **8.** brain **9.** (a)number and identity, (b)frequency and total number **10.** sensory adaptation **11.** deformed **12.** mechanoreceptors **13.** (a)Pacinian corpuscles, (b)Meissner's corpuscles, Ruffini's end organs, Merkel's disks (c)pain **14.** (a)statoliths, (b)hair cells **15.** (a)canal, (b)receptor cells, (c)cupula **16.** muscle spindles, Golgi tendon organs, joint receptors **17.** (a)otoliths, (b)saccule and utricle **18.** (a)angular acceleration, (b)endolymph, (c)crista **19.** (a)cochlea, (b)mechanoreceptors **20.** (a)tympanic membrane, (b)malleus, incus, stapes, (c)oval window **21.** (a)basilar membrane, (b)organ of Corti, (c)cochlear nerve **22.** (a)gustation, (b)olfaction, (c)chemoreceptive **23.** sweet, sour, salty, bitter **24.** (a)olfactory epithelium, (b)50 **25.** pit vipers and boas **26.** hypothalamus **27.** rhodopsins **28.** eyespots (ocelli) **29.** (a)light, (b)vision, (c)lens **30.** ommatidia **31.** (a)sclera, (b)shape (rigidity), (c)cornea **32.** (a)lens, (b)retina, (c)rods, (d)rhodopsin **33.** (a)cones, (b)fovea

Building Words: **1.** proprioceptor **2.** otolith **3.** endolymph

Matching: **1.** l **2.** b **3.** j **4.** i **5.** n **6.** k **7.** o **8.** d **9.** a **10.** c

Table: **1.** electroreceptor **2.** exteroceptor **3.** pressure waves **4.** mechanoreceptor **5.** interoceptor **6.** receptor in the human hypothalamus **7.** muscle contraction **8.** proprioceptor **9.** thermoreceptor **10.** pit organ of a viper **11.** light energy **12.** photoreceptor **13.** interoceptor **14.** chemoreceptor **15.** mammalian taste buds **16.** chemoreceptor **17.** exteroceptor

Multiple Choice: **1.** c,d **2.** b,d,e **3.** b **4.** b,d **5.** a **6.** c **7.** c **8.** d **9.** c

Chapter 42

Concept Questions: **1.** sponges, cnidarians, ctenophores, flatworms, nematodes **2.** (a)circulatory, (b)diffusion **3.** (a)blood, (b)heart, (c)vessels **4.** (a)open circulatory, (b)hemocoel, (c)closed circulatory **5.** gastrovascular cavity **6.** (a)hemolymph, (b)interstitial **7.** (a)arthropods, most mollusks (b)sinuses **8.** (a)hemocyanin, (b)copper **9.** (a)dorsal, (b)ventral, (c)five **10.** plasma **11.**(a)closed, (b)heart, (c)blood **12.** (a)nutrients, oxygen, metabolic wastes, hormones, (b)maintenance of fluid balance, defense, (c)endotherms **13.**(a)red blood cells, white blood cells, platelets, (b)plasma **14.** (a)interstitial, (b)intracellular **15.** (a)fibrinogen, (b)gamma globulins, (c)plasma proteins, (d)lipoproteins **16.** (a)oxygen, (b)red bone marrow, (c)hemoglobin, (d)120 **17.** (a)neutrophils, (b)basophils and eosinophiles, (c)granular leukocytes **18.** (a)lymphocytes, (b)monocytes, (c)macrophages **19.** (a)thrombocytes, (b)hemostasis **20.** (a)arteries, (b)veins **21.** (a)tunica intima, (b)tunica media, (c)tunica adventitia **22.** capillaries **23.** (a)ventricles, (b)atria **24.** (a)one, (b)one, (c)sinus venosus, (d)conus arteriosus **25.** (a)three, (b)sinus venosus, (c)conus arteriosus **26.** (a)more, (b)higher **27.** (a)pericardium, (b)pericardial cavity **28.** (a)interventricular, (b)interatrial septum **29.** (a)four, (b)atrioventricular valve, (c)tricuspid valve, (d)mitral valve, (e)semilunar valves **30.** (a)sinoatrial (SA) node, (b)atrioventricular (AV) node, (c)atrioventricular (AV) bundle **31.** (a)cardiac, (b)systole, (c)diastole **32.** (a)lub, (b)AV valves, (c)dub, (d)semilunar valves, (e)diastole **33.** electrocardiograph (ECG or EKG) **34.** (a)cardiac output (CO), (b)heart rate, (c)stroke volume, (d)five liters/minute **35.** (a)venous, (b)neural and hormonal **36.** cardiac centers **37.** (a)hypertension, (b)increase, (c)salt **38.** diameter of arterioles **39.** valves **40.** (a)baroreceptors, (b)pressure changes, (c)vasomotor centers **41.** (a)pulmonary circuit, (b)systemic circuit **42.** (a)rich, (b)left **43.** (a)aorta, (b)brain, (c)shoulder area, (d)legs **44.** (a)superior vena cava, (b)inferior vena cava **45.** (a)interstitial fluid, (b)fat (lipid) **46.** (a)lymph, (b)interstitial fluids, (c)lymph nodes, (d)subclavian, (e)thoracic, (f)right lymphatic **47.** edema

Building Words: **1.** hemocoel **2.** hemocyanin **3.** erythrocyte **4.** leukocyte **5.** neutrophil **6.** eosinophil **7.** basophil **8.** leukemia **9.** vasoconstriction **10.** vasodilation **11.** pericardium **12.** semilunar **13.** baroreceptor

Matching: **1.** h **2.** i **3.** g **4.** j **5.** d **6.** k **7.** b **8.** l **9.** a **10.** m

Table: **1.** lipoproteins **2.** plasma **3.** cellular **4.** transport of oxygen **5.** blood clotting **6.** monocytes **7.** cellular **8.** albumins and globulins **9.** plasma **10.** defense, principle phagocytic cell in blood

Multiple Choice: **1.** a,b,d **2.** c,d,e **3.** a,c,e **4.** a,c **5.** b,d,e **6.** b,c,d **7.** c,e **8.** a **9.** b **10.** b,c **11.** a,d,e **12.** a,b,c,e **13.** c **14.** c,e **15.** d **16.** c,e **17.** d **18.** d **19.** a **20.** d

Chapter 43

Concept Questions: **1.** antigen **2.** (a)nonspecific defense mechanisms, (b)specific defense mechanisms (immune responses) **3.** immunology **4.** antibodies **5.** (a)phagocytosis, (b)inflammatory response **6.** (a)skin (outer covering), (b)acid secretions and enzymes, (c)mucous lining **7.** viral replication **8.** histamine **9.** redness, heat, edema, pain **10.** (a)hypothalamus, (b)interleuken-1 (IL-1) **11.** neutrophils and macrophages **12.** (a)stem, (b)bone marrow **13.** (a)cytotoxic, (b)helper, (c)suppressor **14.** antigen-presenting cells **15.** thymosin **16.** clonal selection **17.** (a)immunoglobulins, (b)Ig, (c)antigenic determinants (epitopes), (d)binding sites **18.** (a)IgG, IgM, IgA, IgD, IgE, (b)IgGs, (c)IgAs, (d)IgD, (e)IgE **19.** granules **20.** lymphotoxins **21.** (a)macrophage, (b)antigen-MHC, (c)competent T cell, (d)competent T cells, (e)cytotoxic T cells **22.** (a)IgM, (b)IgG **23.** immunization **24.** passive immunity **25.** blocking antibodies

26. T cells   27. (a)allergen, (b)IgE, (c)histamine   28. (a)acquired immune deficiency syndrome, (b)human immunodeficiency virus (HIV), (c)helper T

Building Words: 1. antibody   2. antihistamine   3. lymphocyte   4. monoclonal   5. autoimmune 6. lysozyme

Matching: 1. b   2. o   3. d   4. n   5. e   6. g   7. m   8. l   9. i   10. k

Table: 1. B lymphocytes   2. plasma cells   3. memory B cells   4. T lymphocytes   5. helper T cells   6. cytotoxic T cells   7. memory T cells   8. natural killer cells   9. macrophages   10. neutrophils

Multiple Choice: 1. a,c,d,e   2. b,c,d   3. b,c,e   4. a,b,c,e   5. c   6. a,d   7. c   8. a,e   9. a,c,d   10. b   11. a-d   12. b   13. b,c,e   14. a,d   15. a,c,d,e   16. a-e

Chapter 44

Concept Questions: 1. (a)respiration, (b)ventilation   2. (a)density and viscosity, (b)more, (c)faster   3. (a)thin walls and large surface area, (b)moist, (c)blood vessels (or hemolymph)   4. body surface, tracheal tubes, gills, lungs   5. (a)surface-to-volume, (b)metabolic rate   6. (a)tracheal tubes, (b)spiracles, (c)tracheoles   7. (a)filaments, (b)countercurrent exchange system   8. (a)body surface, (b)body cavity, (c)book lungs   9. (a)heme (iron-porphyrin), (b)globin   10. (a)epiglottis, (b)larynx   11. (a)three, (b)two   12. (a)pharynx, (b)larynx, (c)trachea, (d)bronchi, (e)bronchioles, (f)alveoli   13.(a)inspiration, (b)expiration   14. (a)tidal volume, (b)500 ml, (c)vital capacity   15. (a)pressure (tension), (b)Dalton's law of partial pressure   16. (a)iron, (b)heme, (c)oxyhemoglobin (HbO_2)   17. (a)lower, (b)Bohr effect   18. (a)bicarbonate (HCO_3^-), (b)carbonic anhydrase   19. (a)medulla, (b)chemoreceptors   20. hypoxia   21. decompression sickness   22. chronic obstructive pulmonary diseases

Building Words: 1. hyperventilation   2. hypoxia

Matching: 1. n   2. a   3. f   4. l   5. c   6. j   7. b   8. d   9. i   10. g

Table: 1. body surface   2. body surface   3. tracheal tubes   4. gills   5. gills   6. gills   7. lungs   8. lungs   9. lungs   10. lungs

Multiple Choice: 1. a,b,e   2. a   3. d   4. d   5. c   6. a,c   7. e   8. a,b,d   9. a,d   10. b,c,e   11. e   12. d   13. c,d   14. a,c,e   15. b,d,e   16. a-e   17. a,b,d,e   18. b,c,e   19. a,e   20. b,c,e

Chapter 45

Concept Questions: 1. heterotrophs (consumers)   2. (a)egestion, (b)elimination   3. herbivores   4. (a)carnivores, (b)shorter   5. omnivores   6. (a)mouth, (b)anus, (c)pharynx (throat), (d)esophagus, (e)small intestine   7. (a)submucosa, (b)mucosa, (c)adventitia, (d)muscularis (muscle layer)   8. (a)incisors, (b)canines, (c)molars and premolars, (d)enamel, (e)dentin, (f)pulp cavity   9. amylase   10. (a)peristalsis, (b)bolus   11. (a)simple columnar epithelial, (b)parietal, (c)intrinsic factor, (d)chief, (e)pepsin   12. (a)duodenum, jejunum, ileum, (b)duodenum   13. (a)villi, (b)microvilli   14. (a)glycogen, (b)fatty acids and urea   15. (a)gallbladder, (b)emulsification   16. (a)proteases, (b)lipases, (c)amylases   17. (a)monosaccharides, (b)amino acids, (c)monoacylglycerols (fatty acids and glycerol)   18. (a)villi, (b)blood, (c)lymph   19. (a)cecum, (b)ascending colon, (c)transverse colon, (d)descending colon, (e)sigmoid colon, (f)rectum, (g)anus   20. (a)excretion, (b)elimination   21. (a)Calories, (b)kilocalorie   22. fiber   23. (a)lipoproteins, (b)high-density lipoproteins (HDLs), (c)low-density lipoproteins (LDLs)   24. (a)liver, (b)deamination   25. (a)fat-soluble, (b)water-soluble, (c)B and C   26. sodium, chlorine, potassium, magnesium, calcium, sulfur, phosphorus   27. (a)increases, (b)decreases   28. (a)basal metabolic rate (BMR), (b)total metabolic rate

Building Words: 1. herbivore   2. carnivore   3. omnivore   4. submucosa   5. peritonitis   6. epiglottis 7. microvilli 8. chylomicron

Matching: 1. h   2. e   3. j   4. i   5. o   6. l   7. a   8. c   9. k   10. n

Table: **1.** splits maltose into glucose **2.** small intestine **3.** pepsin **4.** stomach **5.** trypsin **6.** pancreas **7.** RNA to free nucleotides **8.** pancreas **9.** pancreatic lipase **10.** dipeptidase **11.** small intestine **12.** lactose to glucose and galactose

Multiple Choice: **1.** c,d **2.** a,b,c **3.** a **4.** b,e **5.** d **6.** b,c **7.** a,e **8.** d,e **9.** b **10.** b,e **11.** a,b **12.** a,b,c **13.** c **14.** b,c,e **15.** a **16.** a,b,e **17.** a **18.** a,c,d **19.** d **20.** e **21.** a,b,d **22.** c **23.** d **24.** b,c,d **25.** a,e **26.** c,d,e **27.** a-e **28.** c **29.** a,d

Chapter 46

Concept Questions: **1.** osmoregulation and excretion **2.** ammonia, urea, uric acid **3.** (a)ammonia, (b)carbon dioxide **4.** uric acid **5.** (a)osmotic conformers, (b)osmotic regulators **6.** (a)nephridial pores, (b)protonephridia, (c)metanephridia **7.** (a)hemocoel, (b)diffusion and active transport **8.** skin, lungs or gills, digestive system **9.** (a)ureters, (b)urethra **10.** (a)Bowman's capsule, (b)renal tubule, (c)Bowman's capsule, (d)loop of Henle, (e)collecting duct **11.** (a)filtration, (b)reabsorption **12.** Bowman's capsule **13.** renal tubules **14.** urine **15.** (a)cortical, (b)juxtamedullary **16.** (a)descending, (b)ascending, (c)countercurrent mechanism **17.** vasa recta **18.** (a)antidiuretic hormone (ADH), (b)posterior pituitary gland **19.** hypothalamus **20.** (a)adrenal cortex, (b)sodium

Building Words: **1.** protonephridium **2.** podocyte **3.** juxtamedullary

Matching: **1.** e **2.** a **3.** f **4.** n **5.** o **6.** j **7.** m **8.** l **9.** i **10.** b

Table: **1.** diffusion **2.** protonephridia **3.** metanephridia **4.** Malpighian tubules **5.** kidney

Multiple Choice: **1.** a,c,e **2.** b **3.** a,b **4.** a,d,e **5.** c **6.** c,d **7.** b,d **8.** d **9.** c **10.** e **11.** c,e **12.** c **13.** a,c,d **14.** b,d **15.** a,d,e **16.** e **17.** b,c,d **18.** a,c **19.** c,d,e

Chapter 47

Concept Questions: **1.** (a)glands and tissues, (b)nervous **2.** target tissues (organs) **3.** (a)neurohormones, (b)neurosecretory (neuroendocrine) **4.** steroids, proteins, derivatives of fatty acids or amino acids **5.** (a)cholesterol, (b)lipid **6.** (a)DNA protein, (b)mRNA **7.** (a)cyclic AMP, (b)calcium ion **8.** proteins (enzymes) **9.** local hormones **10.** neurons **11.** (a)BH or ecdysiotropin, (b)prothoracic, (c)ecdysone **12.** releasing and release-inhibiting **13.** (a)antidiuretic hormone (ADH), (b)oxytocin, (c)posterior pituitary gland **14.** tropic **15.** prolactin **16.** (a)somatotropin, (b)anterior pituitary, (c)protein synthesis **17.** (a)hypothalamus, (b)growth hormone-inhibiting hormone (GHIH or somatostatin), (c)growth hormone-releasing hormone (GHRH) **18.** (a)gigantism, (b)acromegaly **19.** (a)thyroxine, (b)iodine **20.** (a)anterior pituitary, (b)thyroid-stimulating hormone (TSH) **21.** cretinism **22.** (a)bones, (b)kidney tubules, (c)thyroid, (d)calcitonin **23.** (a)islets of Langerhans, (b)beta, (c)insulin, (d)alpha, (e)glucagon **24.** (a)skeletal muscle and fat cells, (b)glycogenesis **25.** diabetes mellitus **26.** epinephrine, norepinephrine **27.** (a)androgens, mineralocorticoids, glucocorticoids, (b)androgen, (c)dehydroepiandrosterone (DHEA) **28.** aldosterone **29.** cortisol (hydrocortisone) **30.** (a)corticotropin-releasing factor, (b) adrenocorticotropic hormone **31.** thymosin **32.** renin **33.** atrial natriuretic factor **34.** (a)pineal, (b)melatonin

Building Words: **1.** neurohormone **2.** neurosecretory **3.** hypersecretion **4.** hyposecretion **5.** hyperthyroidism **6.** hypoglycemia **7.** hyperglycemia **8.** acromegaly

Matching: **1.** i **2.** d **3.** l **4.** a **5.** g **6.** c **7.** e **8.** n **9.** m **10.** k

Table: **1.** steroid **2.** maintain sodium and potassium balance; increase sodium reabsorption and potassium excretion **3.** amino acid derivative **4.** stimulates metabolic rate; essential to normal growth and development **5.** ADH **6.** glucocorticoids **7.** steroid **8.** prostaglandins **9.** fatty acid derivative **10.** stimulates secretion of adrenal cortical hormones **11.** amino acid derivative **12.** help body cope with stress; increase heart rate, blood pressure, metabolic rate; raise blood sugar level **13.** oxytocin **14.** peptide or protein

Multiple Choice: **1.** a,d **2.** b,c **3.** c,e **4.** a,d **5.** a,c,d **6** b **7** b **8** b,c,d **9** a,d,e

Chapter 48

Concept Questions: **1.** gamete **2.** budding **3.** fragmentation **4.** unfertilized egg **5.** parasitic tapeworms **6.** (a)scrotum, (b)seminiferous tubules **7.** (a)spermatogonium, (b)primary spermatocyte **8.** acrosome **9.** inguinal canal **10.** (a)epididymis, (b)vas deferens **11.** (a)ejaculatory duct, (b)urethra **12.** (a)400 million, (b)seminal vesicles and prostate gland **13.** (a)shaft, (b)glans, (c)prepuce (foreskin), (d)circumcision **14.** (a)cavernous bodies (corpora cavernosa), (b)blood **15.** (a)FSH and LH, (b)testosterone **16.** (a)testosterone, (b)interstitial **17.** (a)oogenesis, (b)oogonia, (c)primary oocytes **18.** (a)follicle, (b)FSH, (c)secondary oocyte **19.** corpus luteum **20.** uterine tube (Fallopian tube, oviduct) **21.** (a)endometrium, (b)menstruation **22.** cervix **23.** labia minora, labia majora, clitoris, vaginal opening (introitus), mons pubis, (some also include hymen, vestibule) **24.** (a)lactation, (b)prolactin, (c)anterior pituitary **25.** ovulation **26.** (a)estrogen (estradiol), (b)FSH and LH, (c)LH **27.** estrogens and progesterone **28.** vasocongestion, increased muscle tension (myotonia) **29.** excitement phase, plateau phase, orgasm, resolution **30.** sterility **31.** progestin (synthetic progesterone) and estrogen **32.** sexually transmitted diseases (STDs) **33.** (a)vasectomy, (b)tubal ligation

Building Words: **1.** parthenogenesis **2.** spermatogenesis **3.** acrosome **4.** circumcision **5.** oogenesis **6.** endometrium **7.** postovulatory **8.** premenstrual **9.** vasectomy **10.** vasocongestion **11.** intrauterine **12.** contraception

Matching: **1.** m **2.** j **3.** k **4.** b **5.** g **6.** c **7.** e **8.** d **9.** i **10.** l

Table: **1.** anterior pituitary **2.** gonad **3.** stimulates development of seminiferous tubules, may stimulate spermatogenesis **4.** stimulates follicle development, secretion of estrogen **5.** stimulates testosterone secretion **6.** stimulates ovulation, corpus luteum growth **7.** gonad **8.** testosterone **9.** gonad **10.** estrogens

Multiple Choice: **1.** a,e **2.** c **3.** c **4.** a,b,c,e **5.** a,c **6.** a,d, **7.** a,c **8.** b,d **9.** a-e **10.** a,c,e **11.** a,e **12.** c **13.** c,d,e **14.** a **15.** d **16.** a,c,e **17.** e **18.** e **19.** a,b,c **20.** c **21.** a,b **22.** b,d **23.** c,e

Chapter 49

Concept Questions: **1.** (a)increases, (b)morphogenesis, (c)cellular differentiation **2.** (a)diploid, (b)sex **3.** (a)zona pellucida, (b)acrosome, (c)vitelline **4.** bindin **5.** (a)fertilization cone, (b)sperm **6.** microtubules **7.** (a)calcium, (b)cortical **8.** cleavage **9.** (a)invertebrates and simple chordates, (b)holoblastic, (c)radial, (d)spiral **10.** (a)meroblastic, (b)blastodisc **11.** (a)gastrulation, (b)germ layers, (c)endoderm, ectoderm, mesoderm **12.** (a)endoderm, (b)ectoderm, (c)mesoderm **13.** archenteron **14.** blastopore **15.** primitive streak **16.** (a)notochord, (b)neural groove, (c)neural tube, (d)central nervous system **17.** (a)pharyngeal pouches, (b)branchial grooves **18.** chorion, amnion, allantois, yolk sac **19.** amnion **20.** allantois **21.** (a)zona pellucida, (b)morula, (c)blastocyst **22.** (a)implantation, (b)seventh **23.** embryonic chorion and maternal **24.** (a)umbilical cord, (b)yolk sac, allantois, umbilical arteries, umbilical vein **25.** second and third **26.** fetus **27.** end of first trimester **28.** (a)gestation period, (b)280, (c)parturition **29.** (a)cervix, (b)baby is delivered, (c)placenta is delivered **30.** (a)ultrasound, (b)amniocentesis **31.** embryo, fetus, neonate, infant, child, adolescent, young adult, middle age, old age

Building Words: **1.** telolecithal **2.** isolecithal **3.** blastomere **4.** blastocyst **5.** blastopore **6.** holoblastic **7.** meroblastic **8.** archenteron **9.** trophoblast **10.** trimester **11.** neonate

Matching: **1.** a **2.** e **3.** j **4.** n **5.** c **6.** f **7.** k **8.** l **9.** i **10.** h

Table: **1.** gastrulation **2.** 24 hours **3.** fertilization **4.** 0 hour **5.** 7 days **6.** blastocyst begins to implant **7.** fertilization **8.** organogenesis **9.** cleavage **10.** 3 days **11.** 266 days **12.** birth

Multiple Choice: **1.** c **2.** b,c,e **3.** d **4.** a-e **5.** c **6.** c,e **7.** a **8.** d **9.** b **10.** c,e **11.** d **12.** c **13.** b,c **14.** b,d

Chapter 50

Concept Questions: 1. signals (stimuli) 2. ethology 3. (a)diurnal, (b)nocturnal, (c)crepuscular 4. biological rhythms 5. nervous, endocrine 6. instinct 7. sign stimulus 8. experience 9. (a)specific conditioned stimulus, (b)unconditioned 10. (a)positive reinforcement, (b)negative reinforcement 11. behavioral bond 12. most important 13. (a)behavioral trigger, (b)orient and navigate 14. optimal foraging 15. foraging strategy 16. travel time, handling time, eating time 17. society 18. pheromones 19. testosterone 20. home range 21. courtship 22. cooperation, suppression of aggression, communication 23. mating, rearing young 24. Hymenoptera, termites 25. altruistic behavior

Building Words: 1. sociobiology

Matching: 1. b 2. e 3. i 4. m 5. l 6. g 7. h 8. a 9. c 10. f

Table: 1. learned 2. classical conditioning 3. operant conditioning 4. habituation 5. imprinting 6. insight learning

Multiple Choice: 1. d 2. e 3. b 4. b 5. a,b 6. d 7. b 8. a 9. e 10. c

Chapter 51

Concept Questions: 1. (a)population, (b)community, (c)ecosystem 2. climate 3. permafrost 4. rain forests 5. 30-49 inches (75-125 cm) 6. Mediterranean 7. allelopathy 8. (a)African, (b)South America and Australia 9. (a)80-180 inches (200-450 cm), (b)vegetation 10. light, minerals, oxygen 11. (a)plankton, (b)nekton, (c)benthos 12. (a)phytoplankton, (b)zooplankton 13. strength of the current 14. (a)littoral, (b)limnetic, (c)profundal, (d)littoral 15. (a)thermal stratification, (b)thermocline, (c)fall turnover, (d)spring turnover 16. (a)marshes, (b)swamps, (c)prairie potholes, (d)peat moss bogs 17. salt marshes 18. intertidal zone 19. (a)neritic province, (b)euphotic 20. oceanic province

Definitions: 1. phytoplankton 2. zooplankton 3. intertidal 4. thermocline 5. biosphere 6. hydrosphere 7. lithosphere

Matching: 1. e 2. n 3. g 4. d 5. i 6. j 7. a 8. f 9. c 10. l

Table: 1. long, harsh winters, short summers, little rain 2. permafrost 3. low-growing plants adapted to extreme cold and short growing season, few species 4. taiga 5. high rainfall 6. rich topsoil, clay in lower layer 7. broad-leaf trees that lose leaves seasonally, variety of large mammals 8. temperate grassland 9. deep, mineral-rich soil 10. wet, mild winters; dry summers 11. thickets of small-leafed shrubs and trees, fire adaptation; mule deer, chipmunks 12. desert 13. low in organic material, often high mineral content 14. savanna 15. low in essential mineral nutrients 16. high rainfall evenly distributed throughout year 17. mineral-poor soil 18. high species diversity

Multiple Choice: 1. a,d 2. e 3. d 4. b,e 5. b,d 6. b 7. e 8. b 9. b 10. c 11. d 12. a,d 13. b,d 14. a,c,e

Chapter 52

Concept Questions: 1. birth rate, death rate, sex ratios, age ranges 2. population density 3. (a)random dispersion, (b)uniform dispersion, (c)clumped (aggregated) dispersion 4. arrival and departure rates (natality and mortality) 5. (a)change in number of individuals, (b)change in time, (c)natality, (d)mortality 6. (a)S-shaped, (b)J-shaped 7. (a)biotic potential (intrinsic rate of increase), (b)environmental resistance, (c)carrying capacity 8. high 9. disease 10. (a)r selection, (b)K selection 11. zero population growth 12. (a)doubling time, (b)less 13. (a)replacement-level fertility, (b)lower, (c)higher 14. (a)total fertility rate, (b)higher

Definitions: 1. intraspecific competition 2. interspecific competition

Matching: 1. m 2. h 3. k 4. b 5. e 6. n 7. g 8. a 9. f 10. l

Table: 1. density-dependent 2. environmental event that affects size of population but is not influenced by changes in population density 3. density-dependent 4. dominant individuals obtain adequate supply of a limited resource at expense of others   **5.** density-dependent   **6.** as density increases, encounters between population members increase as does opportunity to spread disease   **7.** density-independent   **8.** random weather event that reduces vulnerable population   **9.** scramble competition   **10.** density-dependent

Multiple Choice: **1.** g   **2.** h   **3.** d   **4.** f,i   **5.** b   **6.** b   **7.** b   **8.** a,e   **9.** a   **10.** d   **11.** c   **12.** a   **13.** b   **14.** b,c   **15.** a-e   **16.** a,d   **17.** c

Chapter 53

Concept Questions: **1.** ecosystem   **2.** (a)autotrophs, (b)heterotrophs, (c)primary (or herbivores), (d)secondary (or carnivores), (e)omnivores, (f)saprophytes   **3.** detritivores (detritus feeders)   **4.** (a)fundamental, (b)realized   **5.** interspecific competition   **6.** (a)geographical isolation, (b)edge effect   **7.** climax community   **8.** (a)primary succession, (b)pioneer, (c)mosses, (d)shrubs, (e)trees   **9.** secondary succession   **10.** (a)organismic, (b)individualistic

Definitions: **1.** carnivore   **2.** omnivore   **3.** autotroph   **4.** heterotroph   **5.** herbivore

Matching: **1.** m   **2.** l   **3.** j   **4.** c   **5.** g   **6.** k   **7.** n   **8.** i   **9.** e   **10.** b

Table: **1.** secondary consumer   **2.** eat primary consumers   **3.** secondary consumer   **4.** eat primary consumers, producers, or secondary consumers   **5.** detritivore

Multiple Choice: **1.** e   **2.** e   **3.** a   **4.** b   **5.** a,d,e   **6.** a,c,e   **7.** a

Chapter 54

Concept Questions: **1.** (a)ecosystem, (b)ecosphere   **2.** (a)biogeochemical cycles, (b)carbon, nitrogen, phosphorus, water   **3.** (a)photosynthesis, (b)respiration   **4.** (a)nitrogen fixation, (b)nitrification, (c)assimilation, (d)ammonification, (e)denitrification   **5.** (a)rocks, (b)soil   **6.** (a)runoff, (b)groundwater   **7.** (a)radiant, (b)chemical, (c)heat   **8.** (a)primary producers (autotrophs), (b)primary consumers (herbivores), (c)secondary consumers (carnivores)   **9.** food web   **10.** (a)numbers, (b)biomass, (c)energy   **11.** (a)gross primary productivity, (b)net primary productivity   **12.** (a)distance between Earth and sun, (b)inclination   **13.** (a)high, (b)low, (c)Coriolis Effect   **14.** (a)currents, (b)gyres   **15.** climate   **16.** (a)equatorial, (b)poles   **17.** precipitation   **18.** deserts   **19.** microclimate

Definitions: **1.** biomass   **2.** atmosphere   **3.** hydrologic cycle   **4.** microclimate

Matching: **1.** m   **2.** i   **3.** e   **4.** g   **5.** k   **6.** l   **7.** d   **8.** f   **9.** c   **10.** j

Table: **1.** carbon   **2.** carbon dioxide   **3.** nitrogen   **4.** nitrogen   **5.** water   **6.** enters atmosphere by transpiration, evaporation; exits as precipitation

Multiple Choice: **1.** c   **2.** b,d,e   **3.** c,e   **4.** a-e   **5.** b   **6.** a,d   **7.** a,c   **8.** c   **9.** c   **10.** b   **11.** a,c   **12.** d   **13.** a   **14.** a,d,e

Chapter 55

Concept Questions: **1.** (a)endangered, (b)threatened   **2.** habitat destruction   **3.** (a)nature (the wild), (b)human controlled settings   **4.** (a)Amazon, (b)Congo River basin, (c)slash-and-burn   **5.** (a)commercial logging, (b)cattle rangeland   **6.** (a)CO_2, CH_4, O_3, N_2O, chlorofluorocarbons (CFCs) (b)CO_2, O_3, N_2O (c)CH_4, (d)CFCs   **7.** greenhouse effect   **8.** (a)2-5, (b)50   **9.** (a)chlorine, fluorine, carbon (b)chlorine   **10.** ultraviolet radiation

Definitions: **1.** deforestation   **2.** stratosphere

Matching: **1.** g   **2.** o   **3.** b   **4.** e   **5.** h   **6.** a   **7.** j   **8.** n   **9.** c   **10.** l

Table: **1.** deforestation   **2.** permanent removal of forests for agriculture, commercial logging, ranching, use for fuel   **3.** reduce current practices   **4.** global warming   **5.** atmosphere pollutants that trap solar heat   **6.** prevent build-up of gases, mitigate effects, adapt to the effects   **7.** chemical destruction of stratospheric ozone   **8.** reduce use of chlorofluorocarbons and similar chlorine containing compounds